Collins

White Rose
MATHS

White Rose Maths
AQA GCSE 9-1
Foundation
Student Book 1

Caroline Hamilton and Ian Davies

William Collins' dream of knowledge for all began with the publication of his first book in 1819. A self-educated mill worker, he not only enriched millions of lives, but also founded a flourishing publishing house. Today, staying true to this spirit, Collins books are packed with inspiration, innovation and practical expertise.
They place you at the centre of a world of possibility and give you exactly what you need to explore it.

Collins. Freedom to teach.

Published by Collins
An imprint of HarperCollins*Publishers*
The News Building
1 London Bridge Street
London
SE1 9GF

HarperCollins*Publishers*
Macken House
39/40 Mayor Street Upper
Dublin 1
D01 C9W8
Ireland

> **Browse the complete Collins catalogue at www.collins.co.uk**

© HarperCollins*Publishers* Limited 2024

10 9 8 7 6 5 4 3 2 1

ISBN: 978-0-00-866957-7

British Library Cataloguing-in-Publication Data
A catalogue record for this publication is available from the British Library.

Series editors: Caroline Hamilton and Ian Davies
Authors: Jennifer Clasper, Mary-Kate Connolly, Emily
    Fox and James Lansdale-Clegg
Publisher: Katie Sergeant
Product manager: Richard Toms
Development editor: Karl Warsi
Editorial: Amanda Dickson, Richard Toms and
    Deborah Dobson
Proofreading and answer checking: Eric Pradel,
    Deborah Dobson, Amanda Dickson and Anna Cox
Cover designer: Sarah Duxbury
Typesetter: Jouve India Private Limited
Production controller: Alhady Ali
Printed and bound in India

MIX
Paper | Supporting responsible forestry
FSC™ C007454

This book contains FSC™ certified paper and other controlled sources to ensure responsible forest management.

For more information visit: www.harpercollins.co.uk/green

Text acknowledgements
The publishers gratefully acknowledge the permission granted to reproduce the copyright material in this book. Every effort has been made to trace copyright holders and to obtain their permission for the use of copyright material.

Contents

Number   Algebra   Ratio, proportion and rates of change

Contents

Number   **Algebra**   **Ratio, proportion and rates of change**

Introduction

How to use this book

Welcome to the **Collins White Rose Maths AQA GCSE 9–1 Foundation tier** course.

There are two Student Books in the series:

- **Student Book 1** covers Number, Algebra, and Ratio, proportion and rates of change.
- **Student Book 2** covers Geometry and measures, Probability, and Statistics.

Sometimes you will need some knowledge of a different area of mathematics within the topic you are studying. For example, you may need to set up and solve an algebraic equation when solving a geometry problem. You will often be able to use your earlier knowledge and skills from Key Stage 3 to help you do this.

Here is a short guide to how to get the most out of this book. We hope you enjoy continuing your learning journey.

Caroline Hamilton and Ian Davies, series editors

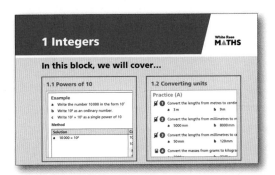

Block overviews Each block of related chapters starts with a visual introduction to the key concepts and learning you will encounter.

Are you ready? Before you start each part of a chapter, remind yourself of the maths you should already know with these questions. If you need more practice, refer to the *Collins White Rose Maths Key Stage 3* course.

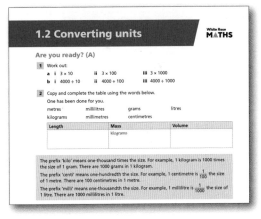

Explanatory text Key words and concepts are explained before moving on to worked examples.

Using your calculator Where appropriate, you are given advice on how to use the features of your calculator to find or check answers. Not all calculators work in the same way, so make sure you know how your model works.

Worked examples Learn how to approach different types of questions with worked examples that clearly walk you through the process of answering. Visual representations are provided to help when necessary.

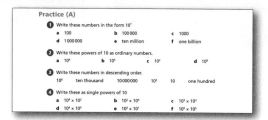

Practice Put what you have just learned into practice. Sometimes symbols are used in questions or whole sections to show when you should, or should not, use a calculator. If there is no symbol, the question or section can be approached in either way.

Many of the Practice sections conclude with a **What do you think?** exercise to encourage further exploration.

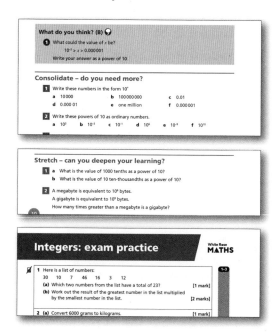

Consolidate Reinforce what you have learned in the chapter with additional practice questions.

Stretch Take your learning further and challenge yourself to apply it in new ways or different areas of maths.

Exam practice At the end of each block, you will find exam-style questions to practise your learning. These are organised into grade bands of 1–3 and 3–5. There is extra practice at the end of the three main parts of the book.

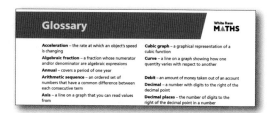

Glossary Look up the meanings of any key words or phrases you are not sure about.

Answers Check your work using the answers provided at the back of the book.

1 Integers

White Rose MATHS

In this block, we will cover...

1.1 Powers of 10

Example

a Write the number 10 000 in the form 10^n

b Write 10^6 as an ordinary number.

c Write $10^2 \times 10^3$ as a single power of 10

Method

Solution	Co
a 10 000 = 10^4	10
	10
	N
	r

1.2 Converting units

Practice (A)

1 Convert the lengths from metres to centim

    a 3 m   b 9 m

2 Convert the lengths from millimetres to m

    a 5000 mm   b 8000 mm

3 Convert the lengths from millimetres to ce

    a 50 mm   b 120 mm

4 Convert the masses from grams to kilogra

1.3 Place value and ordering

Consolidate – do you need more

1 What is the value of 3 in each number?

    a 1378   b 13 904   c 340 205   c

2 Write the numbers in figures.

    a Two thousand, four hundred and fiftee

    b Thirteen thousand, one hundred and s

    c Five hundred and thirty-four thousand

    d Five million, sixty-three thousand and e

1.4 Arithmetic

Stretch – can you deepen your le

1 A rule for generating a sequence is:

    Multiply the previous term by 13 and the

    Work out the fourth term of the sequence

2 The area of the rectangle is 56 cm².

    Work out the perimeter of the rectangle.

1.5 Using inverse operations

Example 1

Given that 50 365 + 2376 = 52 741, write down th

a 52 741 – 2376   b 52 741 – 50 365   c

Method

Solution	Commentary
a 52 741 – 2376 = 50 365	You can represe the parts and wh
	52 741
	50 365
b 52 741 – 50 365 = 2376	52 741

1.1 Powers of 10

Are you ready? (A)

1 Work out:

   a 10×10 **b** $10 \times 10 \times 10$ **c** $10 \times 10 \times 10 \times 10$

2 Work out the missing numbers.

   $10^2 = \boxed{} \times \boxed{} = \boxed{}$

Using your calculator

To write a power in your calculator, use the $\boxed{x^{\blacksquare}}$ button.

For example, to work out 2^5 type $\boxed{2}$ $\boxed{x^{\blacksquare}}$ $\boxed{5}$

Example

a Write the number 10 000 in the form 10^n

b Write 10^6 as an ordinary number.

c Write $10^2 \times 10^3$ as a single power of 10

Method

Solution	Commentary
a $10\,000 = 10^4$	$10\,000 = 10 \times 10 \times 10 \times 10$ $10 \times 10 \times 10 \times 10 = 10^4$ Notice that the power indicates how many times the 10 is used in the multiplication.
b $10^6 = 1\,000\,000$	$10^6 = 10 \times 10 \times 10 \times 10 \times 10 \times 10$ $\quad = 1\,000\,000$ Remember, when working out powers of 10, the power indicates the number of zeros.
c $10^2 \times 10^3 = 10^5$	$10^2 = 100$ $10^3 = 1000$ $100 \times 1000 = 100\,000$ $100\,000 = 10^5$

Practice (A)

1 Write these numbers in the form 10^n

    a 100   **b** 100 000   **c** 1000

    d 1 000 000   **e** ten million   **f** one billion

2 Write these powers of 10 as ordinary numbers.

    a 10^4 **b** 10^5 **c** 10^2 **d** 10^9

3 Write these numbers in descending order.

    10^6 ten thousand   10 000 000 10^3 10   one hundred

4 Write these as single powers of 10

    a $10^4 \times 10^2$ **b** $10^2 \times 10^4$ **c** $10^3 \times 10^3$

    d $10^4 \times 10^5$ **e** $10^3 \times 10^1$ **f** $10^5 \times 10^5$

What do you think? (A) 🌐

1 Is the statement **true** or **false**? $10^0 = 0$ | Use your calculator to check. |

Are you ready? (B)

1 Work out the following. Write your answers as decimals.

    a $1 \div 10$ **b** $1 \div 100$ **c** $1 \div 1000$

2 Write <, > or = to make these statements correct.

    a $0.01 \bigcirc 0.001$ **b** $0.01 \bigcirc \dfrac{1}{10}$ **c** $\dfrac{1}{1000} \bigcirc 0.001$

3 Write these powers of 10 as ordinary numbers. Use your calculator to check your answers.

    a 10^2 **b** 10^1 **c** 10^0

Example

a Write the number 0.01 in the form 10^n

b Write 10^{-6} as an ordinary number.

Method

Solution	Commentary
a $0.01 = 10^{-2}$	$0.01 = 1 \div 10 \div 10$ $\qquad = 10^{-2}$
b $10^{-6} = 0.000001$	$10^{-6} = 1 \div 10 \div 10 \div 10 \div 10 \div 10 \div 10$ $\qquad = \dfrac{1}{10^6}$ Notice there are the same number of zeros as the numeral in the power. $\qquad = 0.000001$

Practice (B)

1. Write these numbers in the form 10^n

   a 0.1
   b 0.001
   c 0.000001
   d 0.00000001

2. Write these powers of 10 as ordinary numbers.

   a 10^{-4}
   b 10^{-5}
   c 10^{-8}
   d 10^{-10}

3. Write these numbers in ascending order.

   10^{-7} one-tenth   0.00001 10^{-4} 0   one-hundredth

What do you think? (B)

1. What could the value of x be?

   $10^{-3} > x > 0.000001$

   Write your answer as a power of 10

Consolidate – do you need more?

1. Write these numbers in the form 10^n

   a 10000
   b 100000000
   c 0.01
   d 0.00001
   e one million
   f 0.000001

2. Write these powers of 10 as ordinary numbers.

   a 10^3
   b 10^{-3}
   c 10^{-1}
   d 10^6
   e 10^{-9}
   f 10^{10}

3. Write these as a single power of 10

   a $10^3 \times 10^2$
   b $10^2 \times 10^3$
   c $10^3 \times 10^1$
   d $10^4 \times 10^4$
   e $10^5 \times 10^{-2}$
   f $10^5 \times 10^0$

4. Write < or > to make these statements correct.

   a $10^2 \bigcirc 10^4$
   b $10^{-4} \bigcirc 10^{-2}$
   c $10^2 \bigcirc 10^{-4}$

5. Write these numbers in descending order.

   10^6 one hundred   0.01 10^{-6} 1   one-thousandth

Stretch – can you deepen your learning?

1. a What is the value of 1000 tenths as a power of 10?

   b What is the value of 10 ten-thousandths as a power of 10?

2. A megabyte is equivalent to 10^6 bytes.

   A gigabyte is equivalent to 10^9 bytes.

   How many times greater than a megabyte is a gigabyte?

1.2 Converting units

Are you ready? (A)

1 Work out:

a **i** 3 × 10   **ii** 3 × 100   **iii** 3 × 1000

b **i** 4000 ÷ 10   **ii** 4000 ÷ 100   **iii** 4000 ÷ 1000

2 Copy and complete the table using the words below.

One has been done for you.

metres   millilitres   grams   litres

kilograms   millimetres   centimetres

Length	Mass	Volume
	kilograms	

The prefix 'kilo' means one-thousand times the size. For example, 1 kilogram is 1000 times the size of 1 gram. There are 1000 grams in 1 kilogram.

The prefix 'centi' means one-hundredth the size. For example, 1 centimetre is $\frac{1}{100}$ the size of 1 metre. There are 100 centimetres in 1 metre.

The prefix 'milli' means one-thousandth the size. For example, 1 millilitre is $\frac{1}{1000}$ the size of 1 litre. There are 1000 millilitres in 1 litre.

Example

Convert:

a 6 metres to millimetres   **b** 7000 grams to kilograms   **c** 4.2 litres to millilitres.

Method

Solution	Commentary		
a 6 m = 6000 mm	× 100 1 m = 100 cm 6 m × 100 = 600 cm	× 10 1 cm = 10 mm 600 cm × 10 = 6000 mm	*Alternatively:* × 1000 1 m = 1000 mm 6 m × 1000 = 6000 mm
b 7000 g = 7 kg	÷ 1000 1000 g = 1 kg 7000 g ÷ 1000 = 7 kg		
c 4.2 l = 4200 ml	× 1000 1 l = 1000 ml 4.2 l × 1000 = 4200 ml		

Practice (A)

1 Convert the lengths from metres to centimetres.

   a   3 m   **b**   9 m   **c**   17 m

2 Convert the lengths from millimetres to metres.

   a   5000 mm   **b**   8000 mm   **c**   11 000 mm

3 Convert the lengths from millimetres to centimetres.

   a   50 mm   **b**   120 mm   **c**   500 mm

4 Convert the masses from grams to kilograms.

   a   3200 g   **b**   3240 g   **c**   43 500 g

5 Convert the volumes from litres to millilitres.

   a   1.2 l   **b**   2.43 l   **c**   12.43 l

What do you think? (A)

1 Amina thinks that 2600 metres is greater than 2.6 km.

Do you agree with Amina? Explain your answer.

Are you ready? (B)

1 Copy and complete the table by writing the imperial units of measurement.

One has been done for you.

miles   stones   gallons   inches   feet   pounds   pints

Length	Mass	Volume
		gallons

2 Write a sensible imperial unit for each measurement.

   a   The amount of fuel in a car fuel tank

   b   The distance from London to Paris

   c   The mass of a horse

The following facts are helpful for converting between imperial units:

1 foot = 12 inches   1 pound = 16 ounces

1 gallon = 8 pints   1 stone = 14 pounds

The following facts are helpful for converting between metric and imperial units:

1 inch ≈ 2.5 cm   5 miles ≈ 8 km   The symbol ≈ means 'approximately equal to'.

Example

a Convert 3 stones to pounds.

b Convert 96 pints to gallons.

c Use the fact 5 miles ≈ 8 kilometres to complete the conversion.

12.5 miles ≈ _____ kilometres.

Method

Solution	Commentary
a 3 stones = 42 pounds	× 14 1 stone = 14 pounds 3 stones × 14 = 42 pounds
b 96 pints = 12 gallons	÷ 8 8 pints = 1 gallon 96 pints ÷ 8 = 12 gallons
c 12.5 miles ≈ 20 kilometres	5 miles ≈ 8 kilometres so 10 miles ≈ 16 kilometres 5 miles ≈ 8 kilometres so 2.5 miles ≈ 4 kilometres 16 kilometres + 4 kilometres = 20 kilometres

Practice (B)

1 Convert the volumes from gallons to pints.

    a 7 gallons   **b** 17 gallons   **c** 47 gallons

2 Convert the volumes from pints to gallons.

    a 64 pints   **b** 144 pints   **c** 296 pints

3 Convert the masses from stones to pounds.

    a 6 stones   **b** 23 stones   **c** 31 stones

4 Convert the masses from ounces to pounds.

    a 96 ounces   **b** 224 ounces   **c** 528 ounces

5 Copy and complete the approximate conversions.

    a 8 inches ≈ _____ centimetres

    b 32 inches ≈ _____ centimetres

    c 246 inches ≈ _____ centimetres

6 Copy and complete the approximate conversions.

    a 20 miles ≈ _____ kilometres

    b 22.5 miles ≈ _____ kilometres

    c 27.5 miles ≈ _____ kilometres

Consolidate – do you need more?

1 Convert the lengths from centimetres to metres.

a 200 cm   **b** 800 cm   **c** 1200 cm

2 Convert the lengths from centimetres to millimetres.

a 6 cm   **b** 13 cm   **c** 26 cm

3 Convert the masses from kilograms to grams.

a 4 kg   **b** 5.4 kg   **c** 8.31 kg

4 Convert the volumes from millilitres to litres.

a 5000 ml   **b** 6500 ml   **c** 11 630 ml

5 Convert the masses from pounds to stones.

a 42 lbs   **b** 196 lbs   **c** 350 lbs

6 Convert the lengths from feet to inches.

a 9 ft   **b** 16 ft   **c** 31 ft

7 Copy and complete the approximate conversions.

a 5 centimetres ≈ _____ inches

b 7.5 centimetres ≈ _____ inches

c 25 centimetres ≈ _____ inches

8 Copy and complete the approximate conversions.

a 16 kilometres ≈ _____ miles

b 32 kilometres ≈ _____ miles

c 88 kilometres ≈ _____ miles

Stretch – can you deepen your learning?

1 Sven is 6 ft 2 in tall. Lida is 158 cm tall.

Who is taller? How much taller are they?

2 A bus driver travels 65 miles over the course of three days.

On day 1, they travel 40 km. On day 2, they travel 10 miles less than they did on day 1.

How far do they travel on day 3?

Are you ready? (A)

1 Write down the numbers represented in the place value grids.

a

Thousands			Ones		
H	T	O	H	T	O

b

Thousands			Ones		
H	T	O	H	T	O

c

Millions			Thousands			Ones		
H	T	O	H	T	O	H	T	O

2 Work out the missing numbers.

    a $26\,983 = 20\,000 + 6000 + \boxed{} + 80 + 3$

    b $103\,862 = \boxed{} + 3000 + 800 + 60 + 2$

    c $71\,415 = 5 + 400 + 70\,000 + 10 + \boxed{}$

3 Write the numbers in figures.

    a Eight thousand, five hundred and twenty-one

    b Ninety thousand, four hundred and six

    c Two hundred and fifteen thousand, three hundred and fifty

---

You can use a place value grid and counters to represent numbers.

The number of counters in each column tells you the value of that column. If there are no counters in a column, it represents zero. Zero in a number acts as a place holder.

You can also use a place value grid to help you write a number in words.

Thousands			Ones		
H	T	O	H	T	O

In digits: 407 361

In words: Four hundred and seven thousand, three hundred and sixty-one

Example

a What is the value of the 7 in the number 207 352?

b Write the number 12 603 054 in words.

Method

Solution	Commentary
a 7 is in the thousands column so this represents 7000 or 7 thousands.	Representing the number in a place value grid can help you identify the correct column. Reading the number aloud will also help.
b	Put the number into a place value grid.

a

Thousands			Ones		
H	T	O	H	T	O

7 is in the thousands column so this represents 7000 or 7 thousands.

b

Millions			Thousands			Ones		
H	T	O	H	T	O	H	T	O
		1	2	6	0	3	0	5	4

There are 12 millions, 603 thousands and 54 ones.

The number is twelve million, six hundred and three thousand, and fifty-four.

Practice (A)

1 Write the place value of 6 in each number (for example, hundreds).

    a 2461   **b** 16 304   **c** 289 650

    d 1 324 706   **e** 2 609 904   **f** 36 241 098

2 Write the numbers in figures.

    a Six thousand, three hundred and forty-two

    b Sixty thousand, three hundred and two

    c Three hundred and sixteen thousand, five hundred

    d Two million, six thousand and six

3 Write the numbers in words.

    a 3457   **b** 32 908   **c** 4 561 382   **d** 14 002 740

What do you think? (A)

1 Marta uses the digit cards to make a number.

| 3 | 0 | 3 | 6 |

She says,

"My number is an odd number.
The digit in the thousands column is the same as the digit in the ones column.
The digit in the hundreds column is greater than the digit in the tens column."

What number did Marta make? Write the number in words.

Are you ready? (B)

1 Write down the number shown in each place value grid.

Which number is greater in each pair?

a

Thousands			Ones		
H	T	O	H	T	O

Thousands			Ones		
H	T	O	H	T	O

b

Thousands			Ones		
H	T	O	H	T	O

Thousands			Ones		
H	T	O	H	T	O

2 Which number is smaller in each pair?

  a 204 or 240   **b** 1302 or 3001   **c** 12 302 or 897

3 Use the words **equal to**, **greater than** or **less than** to complete the statements.

  a 4563 is _____ 567

  b Three thousand, two hundred is _____ three thousand, one hundred and ninety-nine.

  c 13 054 is _____ thirteen thousand and fifty-four.

Example

a Use >, < or = to compare 3162 and 3181

b Write the following numbers in **ascending** order.

3417, 2521, 3186, 3452

Method

Solution	Commentary
a Highest place value	Write both numbers in a place value grid. Thousands is the highest place value. The numbers have the same number of thousands.

Thousands	Hundreds	Tens	Ones
3	1	6	2
3	1	8	1

Look at the next highest place value, which is hundreds. They have the same number of hundreds.

So look at the tens. **6** tens is less than **8** tens, so 3162 is less than 3181

3162 < 3181

The less than symbol is <

b Thousands	Hundreds	Tens	Ones
3	4	1	7
2	5	2	1
3	1	8	6
3	4	5	2

You could write the numbers in a place value grid to compare the place value of each digit before ordering.

Thousands	Hundreds	Tens	Ones
3	4	1	7
2	5	2	1
3	1	8	6
3	4	5	2

First check the thousands column as this has the highest place value.

They all have the same number of thousands apart from 2521, which makes this the smallest number.

Thousands	Hundreds	Tens	Ones
3	4	1	7
2	5	2	1
3	1	8	6
3	4	5	2

For the other three numbers, move to the hundreds column.

The number with the fewest hundreds is 3186 so that is the next smallest number.

Thousands	Hundreds	Tens	Ones
3	4	1	7
2	5	2	1
3	1	8	6
3	4	5	2

The other two numbers have the same number of hundreds so move on to comparing the tens.

3417 has fewer tens.

In ascending order: 2521, 3186, 3417, 3452

'Ascending' means increasing in size, so write the numbers from smallest to largest.

Practice (B)

1 Use <, > or = to compare the numbers.

   a 4501 ◯ 6874   **b** 397 ◯ 3970   **c** 11042 ◯ 9856

2 Here are some numbers in a place value grid.

TTh	Th	H	T	O
4	1	6	8	7
4	1	7	1	2
2	1	8	0	6
4	0	9	9	8

Write the numbers in: **a** ascending order   **b** descending order.

What do you notice?

3 **a** Write the numbers in **descending** order.

    2841   2481   1840   2840   2418

  b Write the numbers in **ascending** order.

    3670   30670   367   36007   376

4 The table shows the lengths, in miles, of some rivers.

Write the rivers in order of length, starting with the longest.

River	Length
Amazon	3976 miles
Nile	4130 miles
Mississippi	3902 miles
Yangtze	3917 miles

What do you think? (B)

1 Here are three cards:

    (3)   (1)   (6)

Using all three cards once only, how many different numbers can be made?
Write the numbers in ascending order.

Consolidate – do you need more?

1 What is the value of 3 in each number?

   a 1378   **b** 13904   **c** 340205   **d** 5234640   **e** 6606134   **f** 43215255

2 Write the numbers in figures.

  a Two thousand, four hundred and fifteen

  b Thirteen thousand, one hundred and seven

  c Five hundred and thirty-four thousand, six hundred

  d Five million, sixty-three thousand and eleven

3 Write each number in words.

a 2198   b 1 432 746   c 24 018   d 11 030 902

4 Here are four cards:

a What is the **smallest** number that could be made using all four cards?

b What is the **greatest** number that could be made using all four cards?

5 Here are some numbers in a place value grid:

HTh	TTh	Th	H	T	O
1	2	4	6	0	3
2	1	4	6	3	0
1	4	2	6	3	3
2	1	4	3	6	0
1	2	5	0	0	2

Write the numbers in **ascending** order.

6 a Write the numbers in **descending** order.

7980   6790   7960   7690   7691

b Write the numbers in **ascending** order.

9342   124 359   12 935   12 395   100 359

7 The table shows the prices of four cars.

Car	Price (£)
A	11 568
B	13 998
C	13 702
D	9970

Write the cars in order, starting with the least expensive.

Stretch – can you deepen your learning?

1 The price of house A is £$2\frac{1}{2}$ million.

The price of house B is £2 490 000

House C is less expensive than house A but more expensive than house B.

What could the price of house C be?

2 Write these numbers in order of size. Start with the greatest number.

−14   41   −41   −4   0   4

Are you ready? (A)

1 Copy and complete the addition grids.

a

+	3	6	1
9		15	
7			8
5	8		

b

+		8	
	8	10	
9			13
		15	11

2 Use the fact that 8 + 5 = 13 to work out:

a 80 + 50   **b** 5000 + 8000   **c** 1300 − 800

Example 1

Find the sum of 843 and 757   Remember that 'find the sum of' means 'add'.

Method

Solution	Commentary

```   H T O ```
```   8 4 3 ```
```   +   7 5 7 ```
```   1   6 0 0 ```
```   1   1   1 ```
	Use zeros as place holders.
	Add this to the other numbers in the tens column:
	1 + 4 + 5 = 10
	Add this to the other numbers in the hundreds column:
	1 + 8 + 7 = 16
	You need an extra column for thousands.

Example 2

Using the column method, work out:

a 47 − 23   **b** 470 − 123   **c** 4700 − 123

Method

	Solution	Commentary

a	``` 4 7 ```
	```− 2 3 ```
	``` 2 4 ```
		You do not need to exchange as 7 > 3

b	$4\ ^6\not{7}\ ^10$
	```− 1 2 3 ```
	``` 3 4 7 ```
		As 0 < 3, you need to exchange 1 ten for 10 ones.

c	$4\ ^6\not{7}\ ^{\not{9}}\not{8}\ ^10$
	```− 1 2 3 ```
	``` 4 5 7 7 ```
		Ensure digits are written in the correct place value column when values have a different number of digits. Here make two exchanges as there are currently no tens to exchange.

Practice (A)

1 Work out:

a 4231 + 3705   **b** 67 + 54   **c** 384 + 252

d 453 + 37   **e** 2451 + 269   **f** 685 + 817

2 Work out:

a 95 – 43   **b** 82 – 36   **c** 325 – 117

d 45851 – 30510   **e** 453 – 76   **f** 5014 – 2338

3 There are 17 marbles in a bag.

Jakub puts 19 marbles in the bag.

Samira puts 21 marbles in the bag.

How many marbles are in the bag now?

4 What is the greatest total you can get by adding three of the numbers together?

37   19   46   41   84   53

5 Class A raises £214 for charity.

Class B raises £46 less than Class A.

How much do they raise in total?

6 Part of an electricity bill is shown. 277 units of electricity were used between March 1st and April 1st.

March 1st	38974 units
April 1st	39251 units
May 1st	40012 units

How many units of electricity were used between March 1st and May 1st?

What do you think? (A)

1 The perimeter of the rectangle is 50 mm.

Work out the value of x.

17 mm

x mm

2 Work out the missing digits.

a

```
    2   4   □
  +   6   □   3
  ─────────────
    □   0   1
```

b

```
    □   4   2
  −   1   □   6
  ─────────────
    2   3   □
```

3 Work out the value of b.

$32 + b = 91 - 28$

Are you ready? (B)

1 Copy and complete the multiplication grid.

×	3	6	9
5			
7			
4			

2 Use the fact that 3 × 4 = 12 to work out:

a 30 × 4   **b** 3 × 40   **c** 30 × 40   **d** 4 × 3000

3 Work out:

a 36 ÷ 3   **b** 36 ÷ 4   **c** 36 ÷ 6

d 36 ÷ 2   **e** 36 ÷ 18   **f** 36 ÷ 12

4 Work out the remainders.

a 37 ÷ 3   **b** 38 ÷ 3   **c** 38 ÷ 4

d 39 ÷ 6   **e** 35 ÷ 3   **f** 34 ÷ 3

The **grid method** and the **column method** are two common ways of multiplying numbers together. Use whichever method you prefer.

Here is 46 × 73

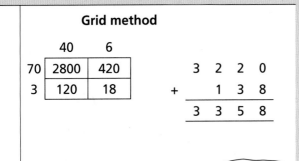

Grid method
40   6
70 | 2800 | 420
3 | 120 | 18

```
        3  2  2  0
   +       1  3  8
      _____
        3  3  5  8
```

Column method

```
           4  6
   ×       7  3
      _____
        1  3  8
   +  3  2  2  0
      _____
        3  3  5  8
```

Example 1

Calculate:

a 382 × 4   **b** 382 × 23

Method A

Solution									Commentary
a	300	80	2		1	2	0	0	You can use known facts to work out mutliplications of larger numbers. For example, $4 \times 3 = 12$ so $4 \times 300 = 1200$
4	1200	320	8			3	2	0	
				+				8	
382 × 4 = 1528					1	5	2	8	Then add the parts together.
b	300	80	2						When multiplying by a two-digit number, you need an extra row so that 23 can be partitioned into 20 and 3
20	6000	1600	40		7	6	4	0	
3	900	240	6	+	1	1	4	6	
382 × 23 = 8786					8	7	8	6	

Method B

Solution	Commentary
a 3 8 2 × 4 1 5 2 8 ₁ ₃ 382 × 4 = 1528	Remember to add on any exchanged amounts after you have worked out each multiplication fact.
b 3 8 2 × 2 3 1 1₂ 4 6 7₁ 6 4 0 + 8 7 8 6 382 × 23 = 8786	Don't forget to add 0 as a place holder in the second part of the multiplication as you are multiplying by 2 tens and not 2 ones.

Example 2

Calculate:

a 612 ÷ 4 **b** 273 ÷ 4 **c** 504 ÷ 14

Method

Solution	Commentary
a $$\begin{array}{r} 1\ 5\ 3 \\ \hline 4\,|\,6\ ^2 1\ ^1 2 \end{array}$$ $612 \div 4 = 153$	There is 1 group of 4 hundreds in 6 hundreds, and 2 hundreds remaining, which can be exchanged for 20 tens. There are 5 groups of 4 tens in 21 tens, with 1 ten remaining, which can be exchanged for 10 ones. There are 3 groups of 4 ones in 12 ones.
b $$\begin{array}{r} 6\ 8\ \ \text{r}1 \\ \hline 4\,|\,2\ 7\ ^3 3 \end{array}$$ or $$\begin{array}{r} 6\ 8\cdot 2\ 5 \\ \hline 4\,|\,2\ 7\ ^3 3\cdot ^1 0\ ^2 0 \end{array}$$ $273 \div 4 = 68$ remainder 1, or 68.25	There are ones remaining. This can be left as a remainder or the number could be extended to include decimal parts. The answer could also be written as a fraction, $68\frac{1}{4}$
c $$\begin{array}{r} 3\ 6 \\ \hline 14\,|\,5\ 0\ ^8 4 \end{array}$$ $504 \div 14 = 36$	When dividing by a two-digit number, it can be helpful to write out the multiples of the divisor: in this case, 14, 28, 42, 56, 70, 84, ...

Practice (B)

1 Estimate and then find the answer to:

   a 23 × 3   **b** 36 × 4   **c** 32 × 46

   d 246 × 23   **e** 1528 × 34   **f** 18 × 2016

2 Calculate:

   a 78 ÷ 3   **b** 822 ÷ 6   **c** 1024 ÷ 4   **d** 3141 ÷ 9

3 Calculate:

   a 312 ÷ 12   **b** 238 ÷ 14   **c** 294 ÷ 21   **d** 805 ÷ 35

4 Work out:

   a 634 ÷ 5   **b** 995 ÷ 7   **c** 1368 ÷ 13   **d** 584 ÷ 22

5 A pencil has a mass of 7 grams.

There are 234 pencils in a box.

The mass of the box alone is 250 grams.

What is the total mass of the box of pencils?

6 A packet contains 28 straws.

A box can hold 7 packets of straws.

There are 25 boxes on a crate.

How many straws are there altogether?

7 561 fans are travelling to a football match.

A coach can seat 34 fans.

How many coaches are needed to take all of the fans?

What do you think? (B)

1 Work out the area of each shape.

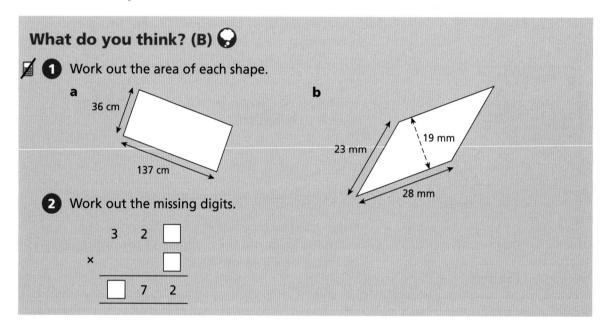

a
36 cm
137 cm

b
23 mm
19 mm
28 mm

2 Work out the missing digits.

```
      3   2   ☐
  ×           ☐
  ─────────────
      ☐   7   2
```

Are you ready? (C)

1 Without working them out, state which calculations will give an answer less than 0

$8 - 3$ $0 - 3$ $0 + 3$ $3 - 8$

2 Write down the numbers you would say in each case.

a Start at 6 and count down in ones to −6

b Start at 6 and count down in twos to −6

c Start at 6 and count down in threes to −6

3 Use the number line to help calculate:

a $3 - 5$

b $-3 + 5$

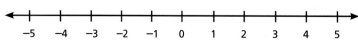

−5 −4 −3 −2 −1 0 1 2 3 4 5

Example 1

Work out: **a** 4 + −7   **b** 5 − −3

Method

Solution	Commentary
a (+1)(+1)(+1)(+1) (−1)(−1)(−1)(−1)(−1)(−1)(−1)	Start with four '+1' counters. Now add in seven '−1' counters.
(+1̷)(+1̷)(+1̷)(+1̷) (−1̷)(−1̷)(−1̷)(−1̷)(−1)(−1)(−1) 4 + −7 = −3	The counters in each zero pair cancel each other out. So you are left with −3
b (+1)(+1)(+1)(+1)(+1)	Start with five '+1' counters. You need to subtract three '−1' counters. However, you do not have any '−1' counters to subtract.
(+1)(+1)(+1)(+1)(+1)(+1)(+1)(+1) (−1)(−1)(−1)	To subtract three '−1' counters, add in three zero pairs. As they are zero pairs, they do not change the value of the original number.
(+1)(+1)(+1)(+1)(+1)(+1)(+1)(+1) (−1̷)(−1̷)(−1̷) 5 − −3 = 8	You can now subtract three '−1' counters. You are left with eight '+1' counters.

Example 2

Work out: **a** 4 × 3   **b** 4 × −3   **c** −4 × 3   **d** −4 × −3

Method

Solution	Commentary
a 4 × 3 = 12	4 × 3 is the same as '4 lots of 3' or 3 + 3 + 3 + 3 = 12 You can show this on a number line. −12 −11 −10 −9 −8 −7 −6 −5 −4 −3 −2 −1 0 1 2 3 4 5 6 7 8 9 10 11 12
b 4 × −3 = −12	4 × −3 is the same as '4 lots of −3' or (−3) + (−3) + (−3) + (−3) = −12 −12 −11 −10 −9 −8 −7 −6 −5 −4 −3 −2 −1 0 1 2 3 4 5 6 7 8 9 10 11 12 When one of the numbers you are multiplying is negative, the arrows change direction and go to the left on the number line.

c	$-4 \times 3 = -12$	What about -4×3? Having 'negative 4 groups of 3' does not necessarily make sense.

-4×3 is the same as 3×-4 or '3 groups of -4', which is $(-4) + (-4) + (-4) = -12$

> Multiplication is commutative.

$$-12 \; -11 \; -10 \; -9 \; -8 \; -7 \; -6 \; -5 \; -4 \; -3 \; -2 \; -1 \; 0 \; 1 \; 2 \; 3 \; 4 \; 5 \; 6 \; 7 \; 8 \; 9 \; 10 \; 11 \; 12$$

d	$-4 \times -3 = 12$	4×3

$$-12 \; -11 \; -10 \; -9 \; -8 \; -7 \; -6 \; -5 \; -4 \; -3 \; -2 \; -1 \; 0 \; 1 \; 2 \; 3 \; 4 \; 5 \; 6 \; 7 \; 8 \; 9 \; 10 \; 11 \; 12$$

> One number being negative means the arrows change direction.

4×-3

$$-12 \; -11 \; -10 \; -9 \; -8 \; -7 \; -6 \; -5 \; -4 \; -3 \; -2 \; -1 \; 0 \; 1 \; 2 \; 3 \; 4 \; 5 \; 6 \; 7 \; 8 \; 9 \; 10 \; 11 \; 12$$

> Both numbers being negative means the arrows need to change direction twice (or not change direction at all).

-4×-3

$$-12 \; -11 \; -10 \; -9 \; -8 \; -7 \; -6 \; -5 \; -4 \; -3 \; -2 \; -1 \; 0 \; 1 \; 2 \; 3 \; 4 \; 5 \; 6 \; 7 \; 8 \; 9 \; 10 \; 11 \; 12$$

So -4×-3 must equal positive 12

> In general, any negative number multiplied by any other negative number gives a positive answer.

Practice (C)

1 Work out:

    a $3 + -8$ **b** $6 - -5$ **c** $-2 + 5$

    d $-2 + -5$ **e** $36 + -18$ **f** $17 - -19$

2 Here are some number cards:

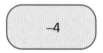 -4 6 -3 5

Use the cards to make the statements correct.

You can use each card only once.

$\square + \square = 2$ $\square - \square = 10$

3 Copy and complete the multiplication grid.

×	−4	−3	−2	−1	0	1	2	3	4
4					0				
3					0				
2					0				
1					0				
0	0	0	0	0	0	0	0	0	0
−1					0				
−2					0				
−3					0				
−4					0				

4 Work out:

a 5×-4

b -6×7

c -9×-5

d 23×-6

e -413×-4

f 32×-48

5 Work out:

a $12 \div -3$

b $-18 \div 6$

c $-39 \div -13$

d $-828 \div 23$

Consolidate – do you need more?

1 Calculate:

a $457 + 618$

b $1284 - 148$

c $803 - 164$

d $3267 + 136$

e $5405 - 567$

f $218 + 34 + 256$

2 Work out:

a $5 + -7$

b $4 - -2$

c $-1 + 9$

d $-3 + -8$

e $23 + -34$

f $107 - -142$

3 Work out:

a 31×4

b 28×6

c 26×43

d 152×32

e 2106×21

f 17×1425

4 Work out:

a $785 \div 5$

b $1416 \div 6$

c $2982 \div 14$

d $7904 \div 26$

e $448 \div 3$

f $1707 \div 12$

5 An adult ticket to the cinema costs £14

A child ticket costs £9

How much do 3 adult tickets and 4 child tickets cost altogether?

6 Work out:

a 7×-8

b $27 \div -3$

c -12×11

d -13×-21

e $-48 \div 4$

f $-156 \div -12$

7 The table shows how many cars and motorbikes were sold over 3 months.

	January	February	March
Number of cars	28	17	32
Number of motorbikes	9	23	16

How many more cars were sold than motorbikes over the 3 months?

8 The temperature in London is 3°C.

The temperature in Oslo is 8 degrees lower than in London.

The temperature in Edinburgh is 4 degrees higher than in Oslo.

What is the temperature in Edinburgh?

Stretch – can you deepen your learning?

1 A rule for generating a sequence is:

> Multiply the previous term by 13 and then subtract 10

Work out the fourth term of the sequence if the first term is 17

2 The area of the rectangle is 56 cm².

Work out the perimeter of the rectangle.

14 cm

3 Mario has these number cards.

    –3   5   3   –7

What is the greatest product Mario could make using two of the cards?

White Rose
M▲THS

Are you ready?

1 Which of the statements are **false**?

| 57 + 28 = 85 | 85 − 28 = 57 | 28 = 85 − 57 |

| 57 − 28 = 85 | 85 = 28 + 57 | 57 = 85 + 28 |

2 Copy and complete the multiplication and division facts shown by the array.

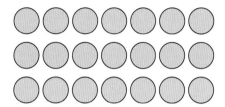

3 × ☐ = 21

7 × 3 = ☐

21 ÷ 7 = ☐

☐ ÷ 3 = 7

3 Work out the missing numbers in this multiplication grid.

×	2		
3		24	
9			45
	8		

Addition and subtraction are **inverse operations**, and so are multiplication and division.

Inverse operations can be a useful tool for working out related calculations and checking accuracy.

Example 1

Given that 50 365 + 2376 = 52 741, write down the value of:

a 52 741 − 2376 **b** 52 741 − 50 365 **c** 503 650 + 23 760

Method

Solution	Commentary
a 52 741 − 2376 = 50 365	You can represent the calculation as a bar model to see the parts and whole. 52 741 / 50 365 | 2376 52 741 / 50 365 | ~~2376~~
b 52 741 − 50 365 = 2376	52 741 / ~~50 365~~ | 2376
c 503 650 + 23 760 = 527 410	Both numbers being added are 10 times the size of the original numbers. So the sum will also be 10 times the size. 52 741 × 10 = 527 410

Example 2

Given that 213 × 84 = 17 892, write down the value of:

a 17 892 ÷ 213   **b** $\dfrac{17\,892}{84}$ **c** 213 × 840

Method

Solution	Commentary
a 17 892 ÷ 213 = 84	Division is the inverse operation of multiplication. If $a \times b = c$ then $c \div b = a$ and $c \div a = b$
b $\dfrac{17\,892}{84} = 213$	A fraction is a division so $\dfrac{17\,892}{84} = 17\,892 \div 84$
c 213 × 840 = 178 920	840 is 10 times the size of 84, therefore the product of 213 and 840 will be 10 times the size of the product of 213 and 84 17 892 × 10 = 178 920

Practice

1 Write down all the additions and subtractions you can see in the bar model.

One is done for you: 406 – 139 = 267

406	
139	267

2 Use the facts from question 1 to work out:

a 1390 + 2670   **b** 40 600 – 13 900

3 Given that 3218 + 1954 = 5172, write down the value of:

a 5172 – 3218   **b** 1954 + 3218   **c** 5172 – 1954

d 321 800 + 195 400   **e** 51 720 – 19 540

4 Use the fact that 4 × 6 = 24 to work out:

a 40 × 6   **b** 60 × 40   **c** 240 ÷ 4   **d** 2400 ÷ 600

5 Given the fact that 140 × 67 = 9380, write down the value of:

a 9380 ÷ 140   **b** $\dfrac{9380}{67}$ **c** 1400 × 67

d 67 × 14   **e** 938 ÷ 14

6 Use inverse operations to check whether each calculation below is correct or not.

a 2845 + 6319 = 9163   **b** 87 421 – 9563 = 77 858

c 245 × 34 = 8330   **d** 34 368 ÷ 23 = 1432

What do you think?

1 Flo says, "b subtract c is equal to a."

Do you agree with Flo? Explain your answer.

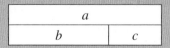

a	
b	c

Consolidate – do you need more?

1 Use the fact 2871 + 6085 = 8956 to write two subtraction facts.

2 Given that 3310 − 847 = 2463, write down the value of:

a 847 + 2463

b 3310 − 2463

c 33 100 − 24 630

d 84 700 + 246 300

3 Use the fact that 8 × 4 = 32 to work out:

a 80 × 4

b 400 × 800

c 3200 ÷ 4

d 3200 ÷ 400

4 Given the fact that 9576 ÷ 63 = 152, write down the value of:

a 63 × 152

b 152 × 63

c $\dfrac{9576}{152}$

d 63 × 1520

e 95 760 ÷ 63

5 Given the bar model, which of the statements are **true**?

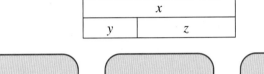

	x	
y		z

$z + y = x$ $z - y = x$ $z = y - x$

$x = y + z$ $x + y = z$ $z = x - y$

Stretch – can you deepen your learning?

1 Work out the input.

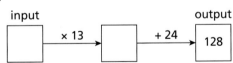

input → ☐ → × 13 → ☐ → + 24 → output 128

2 Junaid thinks of a number.

He divides his number by 9 and then subtracts 13

His answer is 18

What was Junaid's original number?

3 Tiff buys a computer that costs £140

She also buys three games that each cost the same amount.

She pays £260 in total.

How much does each game cost?

Integers: exam practice

---

1 Here is a list of numbers:

  30   10   7   46   16   3   12

  (a) Which two numbers from the list have a total of 23?   [1 mark]

  (b) Work out the result of the greatest number in the list
  multiplied by the smallest number in the list.   [2 marks]

[1–3]

---

2 (a) Convert 6000 grams to kilograms.   [1 mark]

  (b) Convert 8 metres to centimetres.   [1 mark]

---

3 Write the numbers in order of size, starting with the smallest.

  15, 6, –1, 9, 19, –3   [1 mark]

---

4 (a) Write the number four thousand and seventy-one in figures.   [1 mark]

  (b) Write down the value of the digit 8 in the number 86 320   [1 mark]

---

5 Work out 357 × 14   [3 marks]

[3–5]

---

6 A group of 6 friends rent a holiday home for 7 days.

  The cost is £1680

  What is the cost per person **per day**?   [3 marks]

---

7 Given that   5232 ÷ 8 = 654

  work out the value of:

  (a) 654 × 8   [1 mark]

  (b) 52 320 ÷ 8   [1 mark]

  (c) 65 400 × 80   [1 mark]

---

8 At midnight, the temperature was –3°C.

  By 8 am the next morning, the temperature had increased by 4°C.

  (a) Work out the temperature at 8 am the next morning.   [1 mark]

  At midday, the temperature was 9°C.

  (b) Work out the difference between the temperature at midday
  and the temperature at midnight.   [2 marks]

2 Fractions and decimals

In this block, we will cover...

2.1 Equivalence

Example 2

Convert $2\frac{1}{3}$ to an improper fraction.

R
n

Method A

Solution	Commentary
	All of the bar models are
	Two whole bars and one-t
	Seven of the parts are sha
$2\frac{1}{3} = \frac{7}{3}$	or $\frac{7}{3}$

2.2 Ordering

Practice (A)

1. Write <, > or = to compare the fractions.

   a $\frac{1}{3} \bigcirc \frac{1}{8}$ b $\frac{5}{7} \bigcirc \frac{5}{9}$

   e $\frac{3}{4} \bigcirc \frac{5}{8}$ f $\frac{2}{3} \bigcirc \frac{7}{12}$

   i $\frac{7}{8} \bigcirc \frac{5}{6}$

2. Write the fractions in order of size. Start w

   $\frac{2}{3}$ $\frac{5}{9}$ $\frac{11}{18}$ $\frac{8}{9}$

2.3 Fraction arithmetic

Consolidate – do you need more

1. a Work out the additions. Write the ans

   i $\frac{2}{3} + \frac{7}{12}$ ii $\frac{3}{4} + \frac{7}{12}$

   b Work out the subtractions. Write the

   i $\frac{5}{4} - \frac{5}{8}$ ii $\frac{6}{5} - \frac{7}{10}$

2. Work out these additions and subtraction

   a $\frac{1}{2} + \frac{3}{7}$ b $\frac{3}{4} - \frac{1}{3}$

2.4 Decimal arithmetic

Stretch – can you deepen your le

1. A 5p coin is 1.7 mm thick.

   ‡ 1.7 mm

   Benji builds a tower of 5p coins.

   The total value of the tower is £1.35

   What is the height of the tower?

2. a A rule for generating a sequence is:

2.1 Equivalence

Are you ready? (A)

1 What fraction of each diagram is shaded?

a 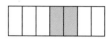   b   c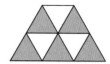

2 What fraction is each red arrow pointing to?

a

b

c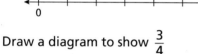

3 Draw a diagram to show $\frac{3}{4}$

Equivalent fractions are equal in value. A fraction is in its simplest form when the numerator and denominator only have 1 as a common factor, for example $\frac{3}{7}$

The **reciprocal** of a number is its multiplicative inverse. This means that the product of a number and its reciprocal is always 1. For example, the reciprocal of $\frac{1}{6}$ is 6 because $\frac{1}{6} \times 6 = 1$

Example 1

Write the missing numbers in these equivalent fractions.

a $\frac{1}{2} = \frac{\square}{8}$ b $\frac{3}{5} = \frac{9}{\square}$ c $\frac{2}{8} = \frac{3}{\square}$

Method A

Solution	Commentary
a $\begin{array}{c} \frac{1}{2} \\ \vdash\!\!-\!\!-\!\!-\!\!-\!\!-\!\!-\!\!-\!\!-\!\!-\!\dashv \\ 0 \qquad\qquad \frac{4}{8} \qquad\qquad 1 \end{array}$ $\frac{1}{2}$ is equivalent to $\frac{4}{8}$ so the missing number is 4	Draw a number line and split it into two equal parts on top and eight equal parts on the bottom. Mark $\frac{1}{2}$ on the number line. Look at the corresponding fraction on the bottom.

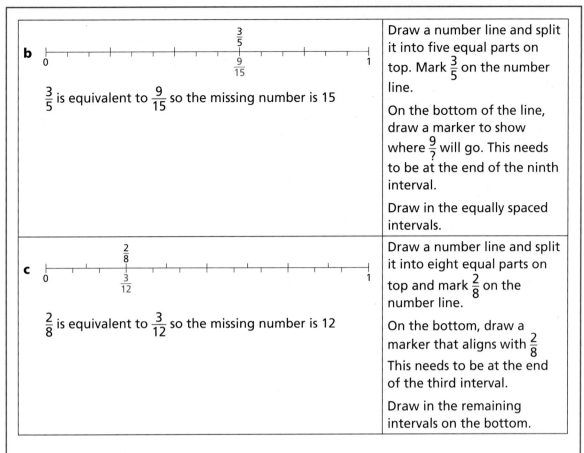

b

$\frac{3}{5}$ is equivalent to $\frac{9}{15}$ so the missing number is 15

Draw a number line and split it into five equal parts on top. Mark $\frac{3}{5}$ on the number line.

On the bottom of the line, draw a marker to show where $\frac{9}{?}$ will go. This needs to be at the end of the ninth interval.

Draw in the equally spaced intervals.

c

$\frac{2}{8}$ is equivalent to $\frac{3}{12}$ so the missing number is 12

Draw a number line and split it into eight equal parts on top and mark $\frac{2}{8}$ on the number line.

On the bottom, draw a marker that aligns with $\frac{2}{8}$ This needs to be at the end of the third interval.

Draw in the remaining intervals on the bottom.

Method B

Solution	Commentary
$\times 4$ $\frac{1}{2} = \frac{4}{8}$ $\times 4$	The denominator has been multiplied by 4 so multiply the numerator by 4 $\frac{1}{2}$ is equivalent to $\frac{4}{8}$
$\times 3$ $\frac{3}{5} = \frac{9}{15}$ $\times 3$	The numerator has been multiplied by 3 so the denominator also needs to be multiplied by 3 $\frac{3}{5}$ is equivalent to $\frac{9}{15}$
$\times 4 \; \frac{2}{8} = \frac{3}{12} \; \times 4$	2 isn't a factor of 3, so instead consider the relationship between the numerator and the denominator. The denominator of the first fraction is four times the numerator, so the same must be true in the second fraction. $\frac{2}{8}$ is equivalent to $\frac{3}{12}$

Example 2

Convert $2\frac{1}{3}$ to an improper fraction.

Remember, in an improper fraction the numerator is bigger than the denominator.

Method A

Solution	Commentary
$2\frac{1}{3} = \frac{7}{3}$	All of the bar models are split into thirds. Two whole bars and one-third are shaded to represent $2\frac{1}{3}$ Seven of the parts are shaded, so this is equivalent to seven-thirds or $\frac{7}{3}$

Method B

Solution	Commentary
$2\frac{1}{3} = \frac{7}{3}$	$2\frac{1}{3}$ = 2 wholes + 1 third 2 wholes = 6 thirds 6 thirds + 1 third = 7 thirds $2\frac{1}{3} = \frac{7}{3}$

Example 3

Write $\frac{13}{5}$ as a mixed number.

Remember, a mixed number is made up of a whole number and a fraction.

Method A

Solution	Commentary
$\frac{13}{5} = 2\frac{3}{5}$	Five-fifths make a whole one so $\frac{13}{5}$ makes two whole bars and three-fifths of a bar. You can write this as $2\frac{3}{5}$

Method B

Solution	Commentary
$\frac{13}{5} = 2\frac{3}{5}$	13 ÷ 5 = 2 r3 So there are two wholes and three-fifths left over, which can be written as the mixed number $2\frac{3}{5}$

Practice (A)

1. Write the missing number in each equivalent fraction.

   a $\frac{1}{3} = \frac{\square}{15}$

   b $\frac{1}{6} = \frac{3}{\square}$

   c $\frac{3}{5} = \frac{\square}{15}$

   d $\frac{2}{3} = \frac{14}{\square}$

   e $\frac{12}{18} = \frac{\square}{3}$

   f $\frac{11}{55} = \frac{1}{\square}$

2. Write the missing number in each equivalent fraction.

   a $\frac{2}{6} = \frac{5}{\square}$

   b $\frac{6}{18} = \frac{\square}{21}$

   c $\frac{\square}{9} = \frac{4}{12}$

3. Simplify:

   a $\frac{4}{8}$

   b $\frac{6}{18}$

   c $\frac{18}{24}$

4. Write the reciprocal of each fraction.

   a $\frac{1}{6}$

   b $\frac{1}{3}$

   c $\frac{1}{18}$

   d $\frac{4}{5}$

   e $\frac{2}{3}$

   f $\frac{5}{6}$

5. Convert each mixed number to an improper fraction.

   a $2\frac{3}{4}$

   b $2\frac{5}{6}$

   c $3\frac{4}{7}$

6. Convert each improper fraction to a mixed number.

   a $\frac{11}{3}$

   b $\frac{9}{4}$

   c $\frac{19}{5}$

Are you ready? (B)

1. What fraction of each hundred square is shaded?

   a
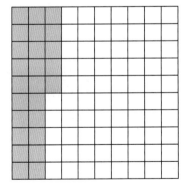

   b
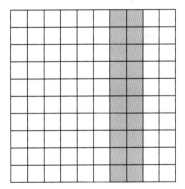

   c
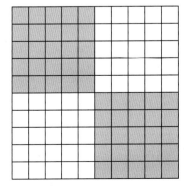

   d
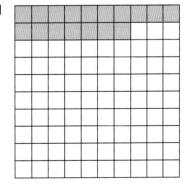

2 What fraction of the thousand square is shaded?

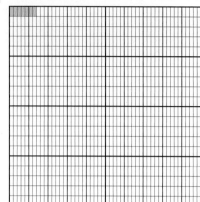

3 Work out the divisions. Write your answers as decimals.

    a $1 \div 2$ **b** $3 \div 4$ **c** $3 \div 5$ **d** $1 \div 8$

Using your calculator

Fractions can be converted to decimals using a calculator.

A fraction is a division so $\frac{3}{8}$ can be interpreted as $3 \div 8$

Using a calculator, $3 \div 8 = 0.375$ so $\frac{3}{8} = 0.375$

Example 1

Write each decimal as a fraction.   **a**   0.8   **b**   0.75   **c**   0.375

Method

Solution	Commentary	
a $0.8 = \frac{4}{5}$	$0.8 = \frac{8}{10} = \frac{4}{5}$	0.8 is eight-tenths, which can be written as the fraction $\frac{8}{10}$ This can be simplified to $\frac{4}{5}$
b $0.75 = \frac{3}{4}$	$0.75 = \frac{75}{100} = \frac{3}{4}$	0.75 is 75 hundredths, which can be written as $\frac{75}{100}$ This can be simplified to $\frac{3}{4}$
c $0.375 = \frac{3}{8}$	$0.375 = \frac{375}{1000} = \frac{3}{8}$	0.375 is 375 thousandths, which can be written as $\frac{375}{1000}$ This can be simplified to $\frac{3}{8}$

Example 2

Write each fraction as a decimal. **a** $\frac{1}{4}$ **b** $\frac{1}{8}$

Method A

Solution	Commentary
a $\frac{1}{4} = 0.25$	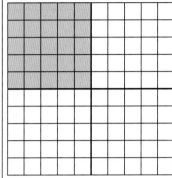 Split a hundred square into four equal parts. One-quarter of the hundred square is shaded. You can divide the hundred in several different ways as long as you shade 25 small squares.
b $\frac{1}{8} = 0.125$	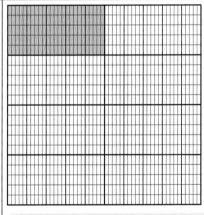 A hundred square cannot be equally split into eight equal parts so instead you can use a thousand square. Divide the thousand square into eight equal parts. Shade any one of these eight parts or any 125 small sections.

Method B

Solution	Commentary	
a $\frac{1}{4} = 0.25$	$\times 25$ $\frac{1}{4} = \frac{25}{100} = 0.25$ $\times 25$	Find an equivalent fraction with a denominator of 100. One-quarter is equivalent to 25 hundredths, which can be written as 0.25

b $\dfrac{1}{8} = 0.125$	$\times 125$ $\dfrac{1}{8} = \dfrac{125}{1000} = 0.125$ $\times 125$	If you cannot find an equivalent fraction with a denominator of 100, you could find one with a denominator of 1000. One-eighth is equivalent to 125 thousandths, which can be written as 0.125

Practice (B)

1 Write each fraction as a decimal.

a $\dfrac{1}{10}$ **b** $\dfrac{3}{10}$ **c** $\dfrac{9}{10}$ **d** $\dfrac{1}{100}$ **e** $\dfrac{87}{100}$

f $\dfrac{21}{100}$ **g** $\dfrac{1}{1000}$ **h** $\dfrac{179}{1000}$ **i** $\dfrac{991}{1000}$

2 Write each fraction as a decimal.

a $\dfrac{1}{2}$ **b** $\dfrac{2}{5}$ **c** $\dfrac{3}{4}$ **d** $\dfrac{7}{20}$

e $\dfrac{4}{25}$ **f** $\dfrac{49}{50}$ **g** $\dfrac{51}{200}$ **h** $\dfrac{5}{8}$

3 Write each decimal as a fraction.

a 0.1   **b** 0.3   **c** 0.01   **d** 0.03   **e** 0.71

f 0.23   **g** 0.001   **h** 0.009   **i** 0.107

4 Write each decimal as a fraction in its simplest form.

a 0.8   **b** 0.5   **c** 0.25   **d** 0.24   **e** 0.95   **f** 0.125

5 Match each fraction to the equivalent decimal.

$\dfrac{3}{5}$	$\dfrac{3}{4}$	$\dfrac{3}{10}$	$\dfrac{3}{8}$

0.3	0.6	0.75	0.375

Consolidate – do you need more?

1 Write the missing numbers in the fractions.

a $\dfrac{1}{4} = \dfrac{\square}{16}$ **b** $\dfrac{1}{8} = \dfrac{3}{\square}$ **c** $\dfrac{4}{5} = \dfrac{\square}{25}$

d $\dfrac{2}{5} = \dfrac{14}{\square}$ **e** $\dfrac{\square}{8} = \dfrac{3}{12}$ **f** $\dfrac{\square}{9} = \dfrac{5}{15}$

2 a What fraction of the shape below is shaded? Give your answer in its simplest form.

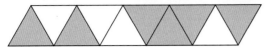

b What fraction of the shape is unshaded? Give your answer in its simplest form.

3 Match each decimal to the equivalent fraction.

0.5	0.55	0.005	0.05

$\frac{1}{20}$	$\frac{11}{20}$	$\frac{1}{2}$	$\frac{1}{200}$

4 Write each fraction as a decimal.

a $\frac{9}{10}$ **b** $\frac{1}{4}$ **c** $\frac{3}{6}$ **d** $\frac{21}{100}$ **e** $\frac{13}{20}$ **f** $\frac{431}{500}$

5 Write each decimal as a fraction.

a 0.7   **b** 0.3   **c** 0.07   **d** 0.59   **e** 0.871   **f** 0.013

6 Write each decimal as a fraction in its simplest form.

a 0.6   **b** 0.2   **c** 0.75   **d** 0.04   **e** 0.55   **f** 0.525

7 Convert each mixed number to an improper fraction.

a $3\frac{2}{3}$ **b** $2\frac{5}{8}$ **c** $11\frac{1}{6}$

8 Convert each improper fraction to a mixed number.

a $\frac{11}{2}$ **b** $\frac{11}{5}$ **c** $\frac{21}{6}$

9 Out of 100 people surveyed, 65 say they can speak a second language.

Write the proportion of people who cannot speak a second language as a fraction. Give your answer in its simplest form.

Stretch – can you deepen your learning?

1 There are 80 cars in a car park.

There are 7 red cars.

There are 17 black cars.

There are 8 blue cars.

The rest of the cars are silver.

What fraction of the cars are silver? Give your answer in its simplest form.

2 Write the reciprocal of 0.6

3 Write the reciprocal of $\frac{4}{5}$ as a mixed number.

Are you ready? (A)

1 a Write <, > or = to compare the fractions.

    i

$$\frac{1}{4} \bigcirc \frac{1}{5}$$

    ii

$$\frac{2}{5} \bigcirc \frac{2}{3}$$

  b Copy and complete the sentence using the word **greater** or **smaller**.

When the numerators are equal, the greater the denominator, the _____ the fraction.

2 a Write <, > or = to compare the fractions.

    i

$$\frac{2}{9} \bigcirc \frac{7}{9}$$

    ii

$$\frac{5}{8} \bigcirc \frac{3}{8}$$

  b Copy and complete the sentence using the word **greater** or **smaller**.

When the denominators are equal, the greater the numerator, the _____ the fraction.

3 Write <, > or = to compare the fractions.

    a

$$\frac{2}{3} \bigcirc \frac{5}{9}$$

    b

$$\frac{2}{3} \bigcirc \frac{3}{4}$$

Example 1

Write the fractions in order of size, starting with the smallest.

$$\frac{2}{3} \qquad \frac{1}{2} \qquad \frac{3}{4} \qquad \frac{5}{12}$$

Method A

Solution	Commentary
$\frac{2}{3}$ $\frac{1}{2}$ $\frac{3}{4}$ $\frac{5}{12}$ $\times 4$ $\times 6$ $\times 3$ $\frac{8}{12}$ $\frac{6}{12}$ $\frac{9}{12}$ $\frac{5}{12}$	To compare the fractions, find a common denominator of all the fractions. The lowest common multiple (LCM) of 3, 2, 4 and 12 is 12
$\frac{5}{12}$ $\frac{1}{2}$ $\frac{2}{3}$ $\frac{3}{4}$	Now compare the numerators to order the fractions.

Method B

Solution	Commentary
$\frac{5}{12}$ is the only fraction smaller than $\frac{1}{2}$ as the numerator is less than half of the denominator.	When ordering fractions, start by comparing them to a simple, known value such as one-half.
$\frac{2}{3}$ and $\frac{3}{4}$ are both greater than $\frac{1}{2}$ $\frac{3}{4} = \frac{9}{12}$ or $\frac{6}{8}$ $\frac{2}{3} = \frac{8}{12}$ or $\frac{6}{9}$	Then compare $\frac{2}{3}$ and $\frac{3}{4}$ by finding a common numerator or denominator, or by comparing how far from 1 whole they are.
$\frac{5}{12}$ $\frac{1}{2}$ $\frac{2}{3}$ $\frac{3}{4}$	

Example 2

Write the numbers in descending order.

'Descending' means getting smaller. Note that 'ascending' would mean getter larger.

$$2\frac{1}{3} \qquad \frac{7}{2} \qquad 2\frac{7}{12} \qquad \frac{17}{6}$$

Method A

Solution	Commentary
$\frac{7}{2} = 3\frac{1}{2} \qquad \frac{17}{6} = 2\frac{5}{6}$	Start by converting all the improper fractions to mixed numbers. $3\frac{1}{2}$ is the only number greater than 3 so it must be the greatest.
$2\frac{1}{3} = 2\frac{4}{12} \qquad 2\frac{5}{6} = 2\frac{10}{12} \qquad 2\frac{7}{12}$ $2\frac{10}{12} > 2\frac{7}{12} > 2\frac{4}{12}$	Then find a common denominator. 12 is the LCM of 3, 6 and 12
$\frac{7}{2} \qquad \frac{17}{6} \qquad 2\frac{7}{12} \qquad 2\frac{1}{3}$	Remember to write the numbers in their original form.

Method B

Solution	Commentary
$2\frac{7}{12} = \frac{31}{12} \qquad 2\frac{1}{3} = \frac{7}{3}$	Start by converting all of the mixed numbers to improper fractions.
$\frac{7}{3} = \frac{28}{12} \qquad \frac{7}{2} = \frac{42}{12} \qquad \frac{17}{6} = \frac{34}{12} \qquad \frac{31}{12}$ $\frac{42}{12} > \frac{34}{12} > \frac{31}{12} > \frac{28}{12}$	Then find a common denominator. 12 is the LCM of 2, 3, 6 and 12
$\frac{7}{2} \qquad \frac{17}{6} \qquad 2\frac{7}{12} \qquad 2\frac{1}{3}$	Remember to write the numbers in their original form.

Practice (A)

1 Write <, > or = to compare the fractions.

a $\frac{1}{3} \bigcirc \frac{1}{8}$ b $\frac{5}{7} \bigcirc \frac{5}{9}$ c $\frac{1}{6} \bigcirc \frac{5}{6}$ d $\frac{3}{8} \bigcirc \frac{3}{5}$

e $\frac{3}{4} \bigcirc \frac{5}{8}$ f $\frac{2}{3} \bigcirc \frac{7}{12}$ g $\frac{2}{9} \bigcirc \frac{4}{13}$ h $\frac{4}{5} \bigcirc \frac{3}{4}$

i $\frac{7}{8} \bigcirc \frac{5}{6}$

2 Write the fractions in order of size. Start with the smallest.

$$\frac{2}{3} \qquad \frac{5}{9} \qquad \frac{11}{18} \qquad \frac{8}{9}$$

3 Write the fractions in order of size. Start with the greatest.

$$\frac{4}{5} \qquad \frac{3}{8} \qquad \frac{1}{2} \qquad \frac{3}{16}$$

4 Write the mixed numbers in ascending order.

$$4\frac{5}{6} \qquad 3\frac{7}{11} \qquad 4\frac{3}{4} \qquad 2\frac{2}{9}$$

5 Write the improper fractions in descending order.

$$\frac{7}{2} \qquad \frac{13}{6} \qquad \frac{11}{3} \qquad \frac{17}{12}$$

6 Write the numbers in ascending order.

$$4\frac{11}{16} \qquad \frac{31}{8} \qquad \frac{17}{4} \qquad 4\frac{3}{8}$$

7 Three friends completed the same maths test.

Amina answered $\frac{3}{4}$ of the questions correctly.

Bev answered 27 out of 32 questions correctly.

Samira answered $\frac{11}{16}$ of the questions correctly.

Write the friends in order, starting with the person who answered the most questions correctly.

What do you think? (A) 💡

1 The value of Ed's house increases by $\frac{2}{3}$

The value of Benji's house increases by $\frac{3}{4}$

Benji says, "My house has increased by more money than Ed's house."

Do you agree with Benji? Explain your answer.

2 Which of these two fractions is closer to 3?

$$\frac{17}{6} \qquad \frac{10}{3}$$

Are you ready? (B)

1 Identify the decimal in each list.

a   4, −4, 0.4

b $75\frac{1}{2}$, 126.3, 875

2 Write <, > or = to compare the decimals.

a   0.7 ◯ 0.3

b   0.43 ◯ 0.46

c   0.265 ◯ 0.301

Example

The table shows the distances achieved by four competitors in a throwing competition.

Rhys	Zach	Abdullah	Sven
9.27 m	10.2 m	9.6 m	9.181 m

Write the competitors in order, starting with the person who threw the furthest distance.

Method

Solution						Commentary

Tens	Ones	tenths	hundredths	thousandths
	9	2	7	0
1	0	2	0	0
	9	6	0	0
	9	1	8	1

Starting with the greatest place value column, there is only one number with a digit in the tens column so 10.2 m is the furthest distance.

6 is greater than 2 and 1, so 9.6 m is the next greatest distance.

2 > 1 so 9.27 m > 9.181 m

Zach, Abdullah, Rhys, Sven

Commentary

It can help to write the numbers you are comparing in a place value grid. Take care to write the digits in the correct place value columns.

You can add zeros as place holders to help compare.

When numbers have the same digit in a column, look to the next place value column.

Practice (B)

1 Write <, > or = to compare the decimals.

   a 0.5 ◯ 0.9   **b** 0.5 ◯ 0.09   **c** 0.64 ◯ 0.67

   d 3.24 ◯ 3.22   **e** 0.978 ◯ 1.2   **f** 1.068 ◯ 1.203

2 Write the numbers in order of size, starting with the smallest.

   0.4   0.9   0.08   0.005   1.1

3 Write the numbers in descending order.

   6.3   6.61   6.03   6.67   6.76

4 The table shows the heights of four students.

Huda	64.3 inches
Rob	64.19 inches
Chloe	64.34 inches
Filipo	66.01 inches

Write the students in order of their height, starting with the shortest.

5 Five athletes ran a 100-metre race.

The table shows the time it took for them to finish.

Marta	11.8 seconds
Amina	10.95 seconds
Beca	11.08 seconds
Emily	11.86 seconds
Samira	11.68 seconds

Write the athletes in order, starting with the person who finished first.

6 Write these numbers in ascending order.

0.09 $\frac{1}{4}$ 0.8 $\frac{19}{20}$ 0.84

What do you think? (B)

1 Ed is ordering these three numbers, starting with the greatest.

2.1   2.09   2.13

Ed says, "13 is greater than 9, which is greater than 1, so the answer is 2.13, 2.09, 2.1"

Do you agree with Ed? Explain your answer.

Consolidate – do you need more?

1 Write <, > or = to compare the numbers.

a $\frac{3}{4} \bigcirc \frac{3}{11}$ **b** $\frac{5}{7} \bigcirc \frac{6}{7}$ **c** $0.03 \bigcirc 0.07$

d $\frac{2}{3} \bigcirc \frac{8}{15}$ **e** $0.2 \bigcirc 0.08$ **f** $4.53 \bigcirc 4.39$

g $\frac{5}{7} \bigcirc \frac{2}{3}$ **h** $3.015 \bigcirc 3.021$

2 Write the fractions in order of size, starting with the smallest.

$\frac{3}{4}$ $\frac{19}{24}$ $\frac{7}{12}$ $\frac{5}{6}$

3 Write the fractions in order of size, starting with the greatest.

$\frac{2}{5}$ $\frac{5}{7}$ $\frac{2}{3}$ $\frac{5}{9}$

4 Write the numbers in ascending order.

2.1   2.07   2.34   2.39   1.989

5 Write the numbers in descending order.

$5\frac{1}{4}$ $\frac{10}{3}$ $5\frac{7}{12}$ $\frac{35}{6}$

6 Three boys see how far they can run.

Sven runs 3.6 km.

Abdullah runs $3\frac{3}{4}$ km.

Jackson runs $\frac{25}{8}$ km.

Write the boys in order, starting with the person who runs the furthest.

7 The table shows the masses of four dogs.

Rufus	13.4 kg
Patch	13.15 kg
Millie	12.9 kg
Barney	13.08 kg

Write the dogs in order, starting with the heaviest.

8 Write these numbers in ascending order.

0.64 $\frac{1}{2}$ 0.6 $\frac{7}{10}$ 0.13

Stretch – can you deepen your learning?

1 Zach completes five long jumps in a national competition. The table shows the distances of each jump.

Jump 1	Jump 2	Jump 3	Jump 4	Jump 5
7.4 m	6.98 m	7.5 m	7.19 m	7.6 m

Zach's final score is the mean of his best three jumps.

What is his final score?

2 Here are the heights of five plants:

48.6 cm   46.86 cm   46.8 cm   48.06 cm   40.68 cm

Work out the median height of the plants.

3 Here are four numbers:

$4\frac{3}{4}$ 3.95 $3\frac{3}{10}$ 4.2

a   Write the numbers in ascending order.

b   Is the greatest number or the smallest number closer to 4?

2.3 Fraction arithmetic

Are you ready? (A)

1 Copy and complete the calculation for each representation.

a $\frac{2}{7} + \frac{3}{7} = \frac{\square}{7}$

b $\frac{4}{5} - \frac{3}{5} = \frac{\square}{5}$

2 Work out:

a $\frac{3}{5} + \frac{1}{5}$ **b** $\frac{2}{9} + \frac{5}{9}$ **c** $\frac{4}{11} + \frac{4}{11}$

d $\frac{6}{7} - \frac{1}{7}$ **e** $\frac{5}{8} - \frac{2}{8}$ **f** $\frac{12}{13} - \frac{2}{13}$

3 Convert the mixed numbers to improper fractions.

a $2\frac{3}{5}$ **b** $4\frac{5}{6}$ **c** $6\frac{3}{8}$

4 Work out the additions. Write your answers as mixed numbers.

a $\frac{3}{5} + \frac{4}{5}$ **b** $\frac{2}{3} + \frac{2}{3}$ **c** $\frac{7}{8} + \frac{10}{8}$

5 Work out the lowest common multiple (LCM) of each set of numbers.

a 3, 12   **b** 4, 5   **c** 3, 4, 8

It is important when adding and subtracting fractions that they have a common denominator. You can use your knowledge of equivalent fractions and lowest common multiples to add and subtract fractions.

Example 1

Work out $\frac{2}{5} + \frac{3}{10}$

Method

Solution	Commentary
$\frac{2}{5}$ $\frac{4}{10}$ $\frac{3}{10}$ $\frac{2}{5} + \frac{3}{10} = \frac{4}{10} + \frac{3}{10} = \frac{7}{10}$	To add two fractions, the denominators need to be the same. Start with a bar model split into five equal parts. Shade two parts. Then divide each part in half to get 10 equal parts. Shade three more parts. The bar models show that $\frac{2}{5}$ is equivalent to $\frac{4}{10}$ When the fractions have a common denominator, you can then add them together. Remember to add the numerators only – not the denominators as well.

Example 2

Work out $\dfrac{5}{6} - \dfrac{3}{4}$

> The fractions must have the same denominator before you can subtract.

Method

Solution	Commentary
$\dfrac{5}{6} - \dfrac{3}{4} = \dfrac{?}{12} - \dfrac{?}{12}$	Find the lowest common multiple of 6 and 4 to find the lowest common denominator.
	You could do this by writing out some multiples of 6 and 4
	6, (12,) 18, 24…
	4, 8, (12,) 16 …
	This means that the common denominator is 12
Multiply the numerator by the same number as the denominator. $$\overset{\times 2 \quad \times 3}{\dfrac{5}{6} - \dfrac{3}{4}} = \dfrac{10}{12} - \dfrac{9}{12} = \dfrac{1}{12}$$ $\underset{\times 2 \quad \times 3}{}$ Think about what you multiply each denominator by to get 12	Now change both fractions into equivalent fractions with a denominator of 12
	When the fractions have the same denominator, you can subtract.
	Only subtract the numerators; the denominator is still 12

Example 3 📝

Work out $2\frac{4}{5} + 1\frac{7}{10}$

Think about the most appropriate method for each question rather than always using the same method. Choose the method that is more efficient.

Solution	Commentary
Method A $2 + 1 = 3$	Add the whole number parts first.
$\frac{4}{5} + \frac{7}{10} = \frac{8}{10} + \frac{7}{10} = \frac{15}{10}$	Then add the fractions.
$\frac{15}{10} = \frac{10}{10} + \frac{5}{10} = 1\frac{5}{10} = 1\frac{1}{2}$	Convert your answer to a mixed number and simplify if necessary.
$3 + 1\frac{1}{2} = 4\frac{1}{2}$ So $2\frac{4}{5} + 1\frac{7}{10} = 4\frac{1}{2}$	Add the two parts of the answer together to get the final answer.
Method B $2\frac{4}{5} = \frac{14}{5} \qquad 1\frac{7}{10} = \frac{17}{10}$	Convert the mixed numbers to improper fractions.
$\frac{14}{5} + \frac{17}{10} = \frac{28}{10} + \frac{17}{10} = \frac{45}{10}$	Add the fractions by finding a common denominator.
$\frac{45}{10} = 4\frac{5}{10} = 4\frac{1}{2}$	Convert the improper fraction to a mixed number and simplify.

Practice (A) 📝

1 Work out these additions and subtractions. Write the answers in their simplest form.

   a $\frac{1}{3} + \frac{5}{9}$ **b** $\frac{1}{4} + \frac{5}{12}$ **c** $\frac{5}{6} - \frac{3}{12}$

   d $\frac{2}{7} + \frac{1}{21}$ **e** $\frac{4}{5} - \frac{3}{10}$ **f** $\frac{17}{18} - \frac{7}{9}$

2 **a** Work out the additions. Write the answers as mixed numbers.

     i $\frac{2}{3} + \frac{5}{9}$ **ii** $\frac{3}{4} + \frac{9}{16}$ **iii** $\frac{11}{15} + \frac{1}{3}$

   b Work out the subtractions. Write the answers in their simplest form.

     i $\frac{3}{2} - \frac{5}{8}$ **ii** $\frac{8}{5} - \frac{14}{15}$ **iii** $\frac{15}{9} - \frac{4}{3}$

3 Work out these additions and subtractions. Write the answers in their simplest form.

   a $\frac{1}{3} + \frac{1}{4}$ **b** $\frac{2}{5} + \frac{1}{4}$ **c** $\frac{1}{2} - \frac{1}{3}$

   d $\frac{2}{7} + \frac{2}{5}$ **e** $\frac{7}{8} - \frac{2}{3}$ **f** $\frac{8}{9} - \frac{1}{6}$

4 Work out the additions. Write your answers as mixed numbers.

   a $\frac{2}{3} + \frac{3}{4}$ **b** $\frac{5}{8} + \frac{3}{5}$ **c** $\frac{7}{9} + \frac{1}{4}$

5 Work out these additions and subtractions. Write the answers in their simplest form.

a $2\frac{2}{3} + 3\frac{1}{9}$ **b** $5\frac{7}{8} - 2\frac{1}{4}$ **c** $4\frac{1}{2} - 1\frac{5}{6}$

d $6\frac{7}{10} + 1\frac{2}{5}$ **e** $5\frac{2}{9} + 3\frac{1}{4}$ **f** $2\frac{2}{3} - 1\frac{4}{5}$

6 On Monday, Benji fed his dog $\frac{3}{4}$ of a can of food.

On Tuesday, he fed his dog $\frac{1}{2}$ of a can of food.

On Wednesday, he fed his dog $\frac{5}{8}$ of a can of food.

How many cans of food did Benji feed his dog altogether?

7 A race is $4\frac{1}{4}$ km long.

Zach has run $2\frac{4}{5}$ km of the race.

How much further does he still have to run?

What do you think?

1 Work out the perimeter of the hexagon.

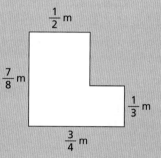

2 Here is a square and an equilateral triangle.

Work out the difference in the perimeters of the two shapes.

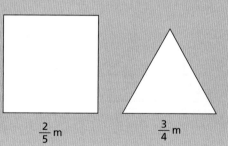

Are you ready? (B)

1 Work out: **a** 8×4 **b** 3×9 **c** 5×6 **d** 6×7 **e** 9×8

2 Work out the missing numbers.

a $\frac{1}{4} + \frac{1}{4} + \frac{1}{4} = \square \times \frac{1}{4}$ **b** $\frac{1}{8} + \frac{1}{8} + \frac{1}{8} = 3 \times \frac{\square}{\square}$

c $\frac{2}{3} + \frac{2}{3} + \frac{2}{3} + \frac{2}{3} = \square \times \frac{\square}{3}$ **d** $5 \times \frac{1}{6} = \frac{1}{\square} + \frac{1}{\square} + \frac{1}{\square} + \frac{1}{\square} + \frac{1}{\square} + \frac{1}{\square}$

e $4 \times \frac{3}{4} = \frac{\square}{\square} + \frac{\square}{\square} + \frac{\square}{\square} + \frac{\square}{\square}$

3   What fraction of each grid is shaded? Write the fractions in their simplest form.

a

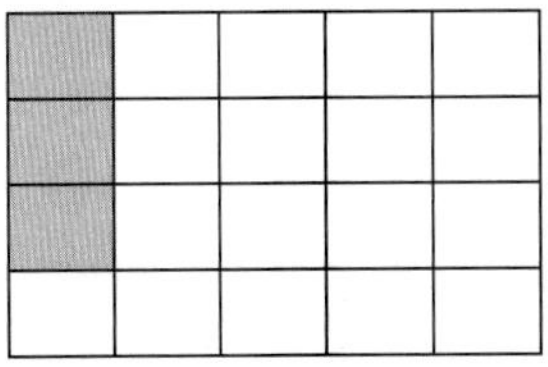

b

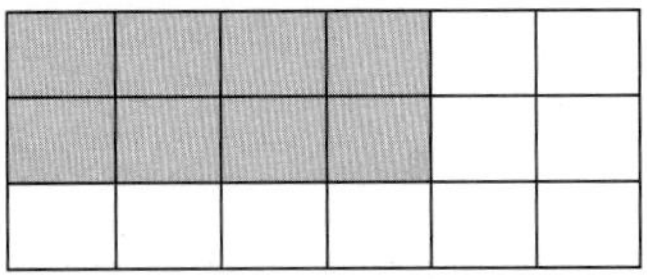

You can think of multiplication as repeated addition.

You know that $3 \times 5 = 5 + 5 + 5$ so, similarly, $3 \times \dfrac{2}{5} = \dfrac{2}{5} + \dfrac{2}{5} + \dfrac{2}{5}$

Example 1

a   Draw a diagram to represent $\dfrac{2}{9} \times 4$ **b**   Use your diagram to calculate $\dfrac{2}{9} \times 4$

Method

Solution	Commentary
a 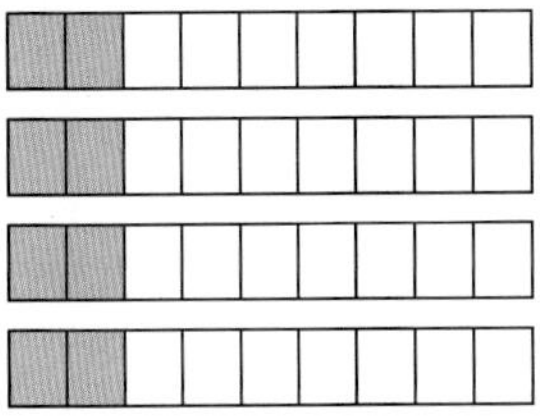	Each bar model is split into nine equal parts and two of these parts are shaded. So $\dfrac{2}{9}$ of each bar model is shaded. There are four bar models, so there are four lots of $\dfrac{2}{9}$ The diagram represents $4 \times \dfrac{2}{9}$ or $\dfrac{2}{9} \times 4$
b $\dfrac{8}{9}$ 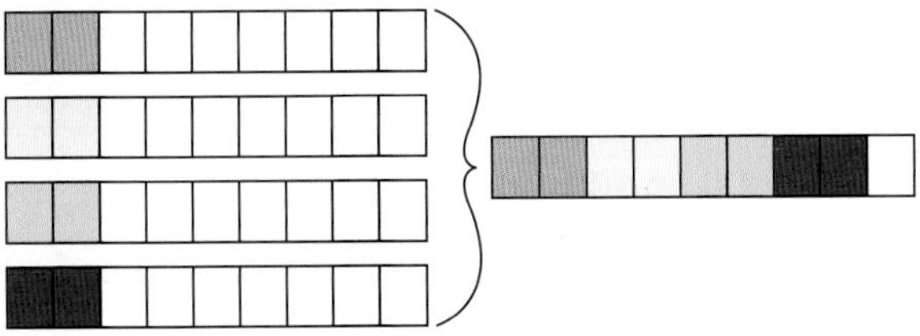	Each of the $\dfrac{2}{9}$ are shown here in different colours. You can put them all on the same bar model to show that there are $\dfrac{8}{9}$ shaded in total. Do not be tempted to think that there are now 36 equal parts altogether and that the answer is $\dfrac{8}{36}$. This is not correct because each *whole* is split into ninths, so you are working with ninths.

Example 2 📝

Calculate $\frac{2}{3} \times \frac{3}{5}$

Method A

Solution	Commentary
1 / 1	Start with a square.
$\frac{2}{3}$ / 1	Split the square into 3 equal sections horizontally so that each row represents $\frac{1}{3}$. Draw an arrow of length $\frac{2}{3}$
$\frac{2}{3}$ / $\frac{3}{5}$	Split the square into five equal sections vertically so that each column represents $\frac{1}{5}$. Draw an arrow of length $\frac{3}{5}$
$\frac{2}{3}$ / $\frac{3}{5}$ $\frac{2}{3} \times \frac{3}{5} = \frac{6}{15}$	To work out $\frac{2}{3}$ multiplied by $\frac{3}{5}$, look at the part of the representation where $\frac{2}{3}$ and $\frac{3}{5}$ overlap. This is the shaded part of the diagram. Six out of 15 equal parts are shaded.
$\frac{6}{15} = \frac{2}{5}$	Using prior knowledge of fractions, simplify your answer. 3 is a common factor of 6 and 15, so divide both the numerator and the denominator by 3 $\quad \begin{array}{c} \div 3 \\ \frac{6}{15} = \frac{2}{5} \\ \div 3 \end{array}$

Method B

Solution	Commentary
$\frac{2}{3} \times \frac{3}{5} = \frac{6}{15}$ $= \frac{2}{5}$	To multiply two fractions, multiply the numerators and also multiply the denominators. $2 \times 3 = 6$ so the numerator of the product is 6 $3 \times 5 = 15$ so the denominator of the product is 15 $\frac{6}{15}$ can be simplified by dividing the numerator and $\quad \begin{array}{c} \div 3 \\ \frac{6}{15} = \frac{2}{5} \\ \div 3 \end{array}$ denominator by the highest common factor, 3

Example 3

Work out $3\frac{2}{5} \times \frac{21}{4}$

Give your answer as a mixed number.

Method

Solution	Commentary
$3\frac{2}{5} \times \frac{21}{4} = \frac{17}{5} \times \frac{21}{4}$	Convert $3\frac{2}{5}$ into an improper fraction and rewrite the calculation.
$\frac{17}{5} \times \frac{21}{4} = \frac{357}{20}$	Find the product of the fractions.
$\frac{357}{20} = 17\frac{17}{20}$	Convert the answer into a mixed number.

Practice (B)

1 Work out the multiplications. Write your answers in their simplest form.

a $4 \times \frac{1}{7}$ b $4 \times \frac{1}{6}$ c $\frac{1}{9} \times 5$

d $2 \times \frac{3}{7}$ e $3 \times \frac{2}{9}$ f $\frac{2}{10} \times 4$

2 Work out the multiplications. Write your answers as mixed numbers.

a $4 \times \frac{1}{3}$ b $8 \times \frac{1}{3}$ c $\frac{1}{4} \times 9$

d $2 \times \frac{3}{5}$ e $4 \times \frac{3}{8}$ f $\frac{3}{4} \times 6$

3 Work out the multiplications.

a $\frac{1}{2} \times \frac{1}{3}$ b $\frac{1}{4} \times \frac{1}{3}$ c $\frac{1}{5} \times \frac{1}{5}$ d $\frac{1}{7} \times \frac{1}{4}$

4 Work out the multiplications. Write your answers in their simplest form.

a $\frac{2}{7} \times \frac{3}{5}$ b $\frac{5}{9} \times \frac{5}{8}$ c $\frac{3}{4} \times \frac{2}{5}$ d $\frac{3}{4} \times \frac{2}{3}$

5 Work out the multiplications. Write your answers as mixed numbers.

a $\frac{4}{3} \times \frac{5}{4}$ b $\frac{3}{2} \times \frac{7}{4}$ c $2\frac{1}{3} \times \frac{7}{4}$

d $\frac{7}{6} \times 2\frac{2}{5}$ e $2\frac{1}{4} \times 3\frac{1}{5}$ f $4\frac{2}{3} \times 6\frac{5}{6}$

6 It takes $\frac{3}{5}$ of a tin of paint to paint a fence panel.

How many tins of paint would it take to paint four fence panels?

7 A lawn measures $4\frac{3}{4}$ m by $6\frac{1}{2}$ m.

What is the area of the lawn?

Are you ready? (C) ▨

1 Write each improper fraction as a mixed number.

    a $\dfrac{7}{5}$ **b** $\dfrac{11}{4}$ **c** $\dfrac{30}{9}$

2 Write each mixed number as an improper fraction.

    a $2\dfrac{1}{3}$ **b** $6\dfrac{3}{4}$ **c** $3\dfrac{3}{7}$

3 Work out:

    a $\dfrac{1}{5} \times 7$ **b** $\dfrac{3}{4} \times 5$ **c** $9 \times \dfrac{1}{2}$ **d** $6 \times \dfrac{2}{5}$

    e $\dfrac{1}{4} \times \dfrac{1}{6}$ **f** $\dfrac{3}{4} \times \dfrac{5}{6}$ **g** $2\dfrac{1}{3} \times \dfrac{7}{4}$ **h** $\dfrac{11}{9} \times 3\dfrac{1}{6}$

4 Write the reciprocal of each number.

    a 6   **b** $\dfrac{1}{4}$ **c** 11   **d** $\dfrac{2}{5}$ **e** $\dfrac{3}{7}$ **f** 21

The next example shows how to divide by a fraction.

Example 1 ▨

Work out: **a** $6 \div \dfrac{1}{4}$ **b** $6 \div \dfrac{3}{4}$

Method

Solution	Commentary
a	There are four quarters in one whole as $\dfrac{4}{4} = 1$
	In six wholes, there are 6 lots of 1 whole. This means that there are 6 lots of 4 quarters.
$6 \div \dfrac{1}{4} = 6 \times 4 = 24$	$6 \times 4 = 24$ so $6 \div \dfrac{1}{4} = 24$

b $6 \div \frac{3}{4}$

	You want to work out how many lots of three-quarters there are in 6 wholes.

$6 \div \frac{3}{4}$

$6 \div \frac{3}{4} = 6 \times \frac{4}{3}$

$\qquad = (6 \times 4) \div 3 = 8$

You already know that there are 24 lots of one-quarter in 6 wholes.

You can group these quarters into threes to make three-quarters.

$24 \div 3 = 8$
so $6 \div \frac{3}{4} = 8$

Here are some important facts about reciprocals.

The reciprocal is the result of dividing 1 by a given number.	$\frac{1}{9}$ is the reciprocal of 9, and 9 is the reciprocal of $\frac{1}{9}$
The product of a number and its reciprocal is always 1	$9 \times \frac{1}{9} = 1$
Dividing by a fraction is the same as multiplying by its reciprocal.	$4 \div \frac{1}{3} = 4 \times 3 = 12$

Example 2

a Work out the missing number. $\qquad \frac{1}{6} \times \boxed{} = 1$

b Write down the reciprocal of $\frac{1}{6}$

Method

Solution	Commentary
a $\frac{1}{6} \times \boxed{6} = 1$	There are six-sixths in 1 whole.
b The reciprocal of $\frac{1}{6}$ is 6	The reciprocal of a number is its multiplicative inverse. This means that the product of a number and its reciprocal is always 1

Example 3

Calculate $\frac{3}{4} \div \frac{2}{7}$

Method

Solution	Commentary
$\frac{3}{4} \div \frac{2}{7} = \frac{3}{4} \times \frac{7}{2}$	To divide by a fraction, you multiply by its reciprocal.
$\frac{3 \times 7}{4 \times 2} = \frac{21}{8}$	You can rewrite the calculation as a multiplication.
$\frac{21}{8} = 2\frac{5}{8}$	Write $\frac{21}{8}$ as a mixed number.

Example 4

Work out $3\frac{2}{5} \div \frac{21}{4}$

Method A

Solution	Commentary
$3\frac{2}{5} \div \frac{21}{4} = \frac{17}{5} \div \frac{21}{4}$	Convert $3\frac{2}{5}$ into an improper fraction and rewrite the calculation.
$\frac{17}{5} \div \frac{21}{4} = \frac{17}{5} \times \frac{4}{21}$ $\frac{17}{5} \times \frac{4}{21} = \frac{68}{105}$	To divide by a fraction, you multiply by its reciprocal.
So $3\frac{2}{5} \div \frac{21}{4} = \frac{68}{105}$	

Method B

Solution	Commentary
$3\frac{2}{5} \div \frac{21}{4} = 3.4 \div 5.25$	You can rewrite $3\frac{2}{5}$ and $\frac{21}{4}$ as decimals.
$3.4 \div 5.25 = \frac{3.4}{5.25} = \frac{340}{525} = \frac{68}{105}$ Multiply the numerator and denominator by 100 so that the denominator is an integer.	This calculation can then be written as a single fraction, and simplified to work out the answer.

Practice (C)

1 **a** How many quarters are there in:

    i 1 whole   **ii** 3 wholes   **iii** 12 wholes?

1			
$\frac{1}{4}$	$\frac{1}{4}$	$\frac{1}{4}$	$\frac{1}{4}$

    b Work out: **i** $1 \div \frac{1}{4}$ **ii** $3 \div \frac{1}{4}$ **iii** $12 \div \frac{1}{4}$

2 **a** Write a multiplication equivalent to each division.

    i $5 \div \frac{1}{3}$ **ii** $9 \div \frac{1}{6}$ **iii** $13 \div \frac{1}{5}$

    b Work out: **i** $5 \div \frac{1}{3}$ **ii** $9 \div \frac{1}{6}$ **iii** $13 \div \frac{1}{5}$

3 Work out:

a $4 \div \frac{2}{3}$ **b** $6 \div \frac{3}{5}$ **c** $4 \div \frac{2}{9}$

d $5 \div \frac{5}{6}$ **e** $12 \div \frac{3}{8}$ **f** $4 \div \frac{2}{7}$

4 Work out the calculations. Give each answer as an improper fraction.

a $7 \div \frac{2}{3}$ **b** $5 \div \frac{3}{5}$ **c** $3 \div \frac{2}{9}$

d $8 \div \frac{5}{6}$ **e** $15 \div \frac{7}{8}$ **f** $7 \div \frac{3}{7}$

5 Work out the calculations. Give each answer as an improper fraction in its simplest form.

a $\frac{2}{3} \div \frac{5}{8}$ **b** $\frac{3}{5} \div \frac{1}{3}$ **c** $\frac{4}{5} \div \frac{3}{4}$

d $\frac{9}{10} \div \frac{5}{6}$ **e** $\frac{2}{7} \div \frac{4}{9}$ **f** $\frac{3}{8} \div \frac{11}{12}$

6 Work out the calculations. Give each answer as a mixed number in its simplest form.

a $\frac{2}{3} \div \frac{5}{8}$ **b** $\frac{4}{5} \div \frac{1}{4}$ **c** $\frac{5}{7} \div \frac{1}{6}$

d $\frac{7}{12} \div \frac{2}{9}$ **e** $\frac{7}{8} \div \frac{4}{9}$ **f** $\frac{8}{11} \div \frac{2}{7}$

7 Work out:

a $\frac{7}{3} \div \frac{1}{4}$ **b** $2\frac{3}{5} \div \frac{1}{3}$ **c** $4\frac{2}{7} \div \frac{3}{4}$

d $3\frac{2}{3} \div \frac{5}{9}$ **e** $\frac{8}{5} \div \frac{7}{4}$ **f** $3\frac{5}{8} \div \frac{3}{5}$

8 Work out:

a $4\frac{2}{5} \div 2\frac{3}{4}$ **b** $5\frac{2}{3} \div 2\frac{4}{9}$ **c** $1\frac{5}{7} \div 1\frac{1}{5}$

d $6\frac{3}{4} \div 3\frac{5}{6}$ **e** $8\frac{3}{8} \div 2\frac{1}{2}$ **f** $6\frac{8}{9} \div 4\frac{4}{5}$

Consolidate – do you need more? 📝

1 **a** Work out the additions. Write the answers as mixed numbers.

     i $\frac{2}{3} + \frac{7}{12}$ **ii** $\frac{3}{4} + \frac{7}{12}$ **iii** $\frac{13}{20} + \frac{4}{5}$

   b Work out the subtractions. Write the answers in their simplest form.

     i $\frac{5}{4} - \frac{5}{8}$ **ii** $\frac{6}{5} - \frac{7}{10}$ **iii** $\frac{19}{12} - \frac{4}{3}$

2 Work out these additions and subtractions.

   a $\frac{1}{2} + \frac{3}{7}$ **b** $\frac{3}{4} - \frac{1}{3}$ **c** $\frac{8}{9} - \frac{3}{4}$ **d** $\frac{5}{6} + \frac{1}{4}$

3 Work out these additions and subtractions. Write the answers in their simplest form.

a $4\frac{1}{3} + 2\frac{4}{9}$ **b** $6\frac{5}{8} - 3\frac{1}{4}$ **c** $4\frac{2}{3} - 1\frac{3}{4}$ **d** $6\frac{5}{6} + 3\frac{3}{5}$

4 Work out the multiplications. Write your answers as mixed numbers.

a $9 \times \frac{1}{4}$ **b** $\frac{1}{5} \times 13$ **c** $4 \times \frac{2}{5}$ **d** $\frac{3}{4} \times 7$

5 Work out the multiplications.

a $\frac{2}{3} \times \frac{4}{5}$ **b** $\frac{1}{4} \times \frac{5}{6}$ **c** $\frac{2}{7} \times \frac{3}{5}$ **d** $\frac{7}{9} \times \frac{7}{10}$

6 Work out the multiplications. Write your answers as mixed numbers.

a $\frac{3}{2} \times \frac{5}{4}$ **b** $\frac{5}{3} \times \frac{4}{3}$ **c** $3\frac{1}{4} \times \frac{7}{2}$

d $\frac{11}{6} \times 1\frac{4}{5}$ **e** $2\frac{3}{5} \times 3\frac{2}{3}$ **f** $4\frac{3}{4} \times 3\frac{7}{8}$

7 Work out the calculations.

a $6 \div \frac{3}{4}$ **b** $4 \div \frac{2}{5}$ **c** $5 \div \frac{5}{8}$ **d** $9 \div \frac{3}{7}$

8 Work out the calculations. Give each answer as a mixed number in its simplest form.

a $\frac{3}{4} \div \frac{1}{3}$ **b** $\frac{5}{6} \div \frac{1}{2}$ **c** $\frac{7}{9} \div \frac{1}{5}$ **d** $\frac{9}{10} \div \frac{3}{4}$

9 Work out:

a $\frac{7}{4} \div \frac{1}{3}$ **b** $3\frac{1}{4} \div \frac{1}{2}$ **c** $5\frac{3}{8} \div \frac{2}{3}$ **d** $4\frac{5}{6} \div \frac{4}{5}$

e $\frac{13}{5} \div \frac{4}{3}$ **f** $2\frac{2}{7} \div \frac{2}{5}$ **g** $4\frac{7}{8} \div 2\frac{1}{3}$ **h** $8\frac{4}{5} \div 2\frac{3}{4}$

Stretch – can you deepen your learning? 🖊

1 The next term in a sequence is found by multiplying the previous term by $\frac{3}{5}$

The first term in the sequence is $\frac{3}{4}$

Work out the third term in the sequence.

2 Here is a function machine:

Input $\longrightarrow$ $\boxed{\times \frac{4}{5}}$ $\longrightarrow$ $\boxed{\div \frac{5}{6}}$ $\longrightarrow$ Output

a Work out the output if the input is 3

b Work out the input if the output is $\frac{1}{4}$

3 The wall of a classroom is to be painted.

The wall has an area of $16.625\,\text{m}^2$.

A tin of paint costs £5 and covers $2\,\text{m}^2$.

How much will it cost to paint the wall?

2.4 Decimal arithmetic

Are you ready? (A) 📝

1   Work out:
   a 324 + 263   **b** 147 + 325   **c** 368 + 271
   d 459 + 184   **e** 2508 + 3693   **f** 4179 + 227

2   Work out:
   a 563 – 212   **b** 437 – 219   **c** 608 – 271
   d 332 – 184   **e** 6513 – 1074   **f** 1142 – 875

3   Write two addition facts and two subtraction facts shown by the bar model.

8.9	
6.4	2.5

4   Use the fact that 7 + 6 = 13 to work out:
   a 70 + 60   **b** 7000 + 6000   **c** 1300 – 700   **d** 130 – 70

5   Work out:
   a 7 ÷ 10   **b** 5 ÷ 100   **c** 80 ÷ 100   **d** 130 ÷ 1000

Example 1 📝

Show how to work out:

a 7.2 + 0.93   **b** 7.2 + 18   **c** 0.93 + 0.68

Method

Solution	Commentary
a 7.2 + 0.93 <table><tr><td></td><td>O</td><td>•</td><td>t</td><td>h</td></tr><tr><td></td><td>7</td><td>•</td><td>2</td><td></td></tr><tr><td>+</td><td>0</td><td>•</td><td>9</td><td>3</td></tr><tr><td></td><td>8</td><td>•</td><td>1</td><td>3</td></tr></table> 1	The column method is efficient here, but be careful to line up the digits with the same place value. You could add a zero to 7.2 to help you as 7.2 and 7.20 are equal in value.
b 7.2 + 18 <table><tr><td></td><td>T</td><td>O</td><td>•</td><td>t</td><td>h</td></tr><tr><td></td><td></td><td>7</td><td>•</td><td>2</td><td></td></tr><tr><td>+</td><td>1</td><td>8</td><td>•</td><td></td><td></td></tr><tr><td></td><td>2</td><td>5</td><td>•</td><td>2</td><td></td></tr></table> 1	Again the column method works well, but you could work this out mentally if you know that 7 + 18 = 25

c 0.93 + 0.68

	O	•	t	h
	0	•	9	3
+	0	•	6	8
	1	•	6	1
			1	1

It is again sensible to use the column method here.

Example 2

Work out: **a** 6.7 − 3.2   **b** 0.67 − 0.38

Method

Solution	Commentary
a 6.7 − 3.2 <table><tr><td></td><td>O</td><td>•</td><td>t</td></tr><tr><td></td><td>6</td><td>•</td><td>7</td></tr><tr><td>−</td><td>3</td><td>•</td><td>2</td></tr><tr><td></td><td>3</td><td>•</td><td>5</td></tr></table> So 6.7 − 3.2 = 3.5	You subtract decimals in the same way as you subtract integers.
b 0.67 − 0.38 <table><tr><td></td><td>O</td><td>•</td><td>t</td><td>h</td></tr><tr><td></td><td>0</td><td>•</td><td>⁵6̸</td><td>¹7</td></tr><tr><td>−</td><td>0</td><td>•</td><td>3</td><td>8</td></tr><tr><td></td><td>0</td><td>•</td><td>2</td><td>9</td></tr></table> So 0.67 − 0.38 = 0.29	The relationship between adjacent columns is the same as for integers, so you can use the same process. Make an exchange between the tenths and the hundredths columns. 6 tenths and 7 hundredths is partitioned into 5 tenths and 17 hundredths.

Example 3

Using the column method, work out:

a 47 − 23   **b** 47 − 2.3   **c** 47 − 0.23

Method

Solution	Commentary

a   4 7
  − 2 3
      2 4
	You do not need to exchange as 7 > 3

b   4 ^{6}7•^{1}0
  −   2•3
     4 4•7
	Write 47 as 47.0 and line up the decimal points.
	As 0 < 3, exchange 1 one for 10 tenths.

c   4 ^{6}7•^{1}0̸ ^{9}^{1}0
  −   0•2 3
     4 6•7 7
	Write 47 as 47.00 and line up the decimal points. Then make two exchanges.

Practice (A)

1 Work out:

  a   53.82 + 31.41   **b**   3.24 + 2.57   **c**   4.54 + 3.68

  d   5.803 + 3.447   **e**   3.08 + 11.93   **f**   0.341 + 0.607 + 0.458

2 Work out:

  a   9.5 − 4.3   **b**   45.85 − 30.54   **c**   4.53 − 1.36

  d   8.37 − 5.34   **e**   7.03 − 2.84   **f**   0.301 − 0.153

3 Work out:

  a   3.4 + 8.63   **b**   8.73 − 5.9   **c**   20.04 + 5.962

  d   301.87 + 58.6   **e**   34.6 − 1.82   **f**   10.74 − 3.953

4 Work out the perimeter of the triangle.

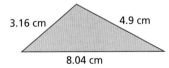

3.16 cm   4.9 cm

8.04 cm

5 The table shows the price list at Sam's grocery.

  a   Rob buys two apples, a bag of grapes and a bag of carrots.

      Work out the total cost.

  b   Rob pays with a £10 note.

      Work out how much change he should get.

Sam's grocery	
Apple	55p
Pineapple	£1.30
Bag of grapes	95p
Bag of carrots	£1.05
Potatoes	£2.20

6 Building A is 20.1 m tall.

Building B is 4.96 m taller than building A.

Building C is 184 cm shorter than building B.

Work out the heights of buildings B and C.

7 The perimeter of the quadrilateral is 30.64 cm.

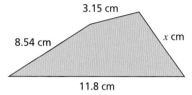

3.15 cm

8.54 cm

x cm

11.8 cm

Work out the value of x.

What do you think? (A)

1 Work out the missing digits.

a
```
    3 • 8 □
  +
    5 • □ 6
  ─────────
    □ • 0 1
```

b
```
    □ 6 • 3
  -
    2 □ • 8
  ─────────
    2 5 • □
```

2 Work out the value of b.

$3.2 + b = 9.1 - 2.8$

Are you ready? (B)

1 a Write the number that is 10 times greater than:

    i 21   **ii** 2.1   **iii** 0.21   **iv** 2.01

b Write the number that is one-tenth the size of:

    i 21   **ii** 2.1   **iii** 0.21   **iv** 2.01

2 Copy and complete the multiplication grid.

×	4	8	6
2			
7			
3			

3 Use the fact $6 \times 3 = 18$ to work out:

a 60×3 **b** 60×30 **c** 3×6000

d $180 \div 6$ **e** $180 \div 60$ **f** $1800 \div 30$

4 Work out:

    a **i** $24 \div 4$ **ii** $240 \div 40$ **iii** $2400 \div 400$

    b **i** $63 \div 21$ **ii** $630 \div 210$ **iii** $6300 \div 2100$

What do you notice about your answers? Discuss with a partner.

Example 1

Estimate the answers to these calculations and then work out the exact answers.

a 28×3 **b** 2.8×3 **c** 36×24 **d** 0.36×2.4

Method

Solution	Commentary
a 28×3 $\approx 30 \times 3 = 90$ $28 \times 3 = 84$	Estimate by rounding the numbers to 1 significant figure. $28 \approx 30$ 3 is already to 1 significant figure. You know $3 \times 3 = 9$ so 30×3 will be 10 times greater. To work out the exact answer, use the column method as that is the most efficient.
b 2.8×3 $\approx 3 \times 3 = 9$ $2.8 \times 3 = 8.4$	$2.8 \approx 3$ Now use the fact that $28 \times 3 = 84$ 2.8×3 will be 10 times smaller, which is 8.4 This is close to the estimate of 9 To work out the exact answer, you could use the column method with decimals.
c 36×24 $\approx 40 \times 20 = 800$ $36 \times 24 = 864$	Estimate: $36 \approx 40$ $24 \approx 20$ To work out the exact answer, use the column method as that is the most efficient.
d 0.36×2.4 $\approx 0.4 \times 2 = 0.8$ 0.36×2.4 $= 36 \times 24 \div (100 \times 10)$ $= 864 \div 1000$ $= 0.864$	Estimate: $0.36 \approx 0.4$ $2.4 \approx 2$ Now use the answer to 36×24 to work out the exact answer. $\div (100 \times 10)$ is the same as $\div 1000$ 0.864 is very close to the estimate of 0.8. That suggests the answer is probably correct.

Example 2

Work out: **a** $61.2 \div 4$ **b** $6.12 \div 4$

Method

Solution	Commentary
a $\begin{array}{r} 1\ \ 5\ .\ \ 3 \\ 4\,\overline{)\,6\ \ ^21\ .\ \ ^12} \end{array}$	You can use the short division method to divide decimals, as you would with integers. Just remember to include the decimal point.
	There is 1 group of 4 tens in 6 tens and 2 tens remaining, which can be exchanged for 20 ones.
	There are 5 groups of 4 ones in 21 ones with 1 one remaining, which can be exchanged for 10 tenths.
$61.2 \div 4 = 15.3$	There are 3 groups of 4 tenths in 12 tenths.
b $\begin{array}{r} 1\ .\ \ 5\ \ 3 \\ 4\,\overline{)\,6\ .\ \ ^21\ \ ^12} \end{array}$	The same process can be followed for $6.12 \div 4$. Just remember to write the decimal point in the correct place.
$6.12 \div 4 = 1.53$	

Example 3

Work out: **a** $272 \div 0.8$ **b** $27.2 \div 0.08$ **c** $27.2 \div 0.8$

Method

Solution	Commentary
a $272 \div 0.8$ $\times 10 \downarrow \quad \downarrow \times 10$ $2720 \div 8$	When dividing either an integer or a decimal by a decimal, it can help to multiply both the dividend and the divisor by the same power of 10 so that they are both integers. Note that the answer will be the same.
	Multiply both 272 and 0.8 by 10
$272 \div 0.8 = 340$	$\begin{array}{r} 0\ \ 3\ \ 4\ \ 0 \\ 8\,\overline{)\,2\ \ ^27\ ^32\ \ 0} \end{array}$
b $27.2 \div 0.08$ $\times 100 \downarrow \quad \downarrow \times 100$ $2720 \div 8$	Multiply both 27.2 and 0.08 by 100
	$\begin{array}{r} 0\ \ 3\ \ 4\ \ 0 \\ 8\,\overline{)\,2\ \ ^27\ ^32\ \ 0} \end{array}$
$27.2 \div 0.08 = 340$	
c $27.2 \div 0.8$ $\times 10 \downarrow \quad \downarrow \times 10$ $272 \div 8$	Multiply both 27.2 and 0.8 by 10
	$\begin{array}{r} 0\ \ 3\ \ 4 \\ 8\,\overline{)\,2\ \ ^27\ ^32} \end{array}$
$27.2 \div 0.8 = 34$	

Practice (B)

1 Use the fact 24 × 8 = 192 to work out:

    a 2.4 × 8   **b** 24 × 0.8   **c** 2.4 × 0.08   **d** 0.24 × 0.8

2 **a** Work out 916 × 17

    b Use your answer to part **a** to work out 9.16 × 1.7

3 Estimate and then find the answer to:

    a 2.3 × 3   **b** 0.36 × 4   **c** 3.2 × 46

    d 24.6 × 23   **e** 1.2 × 3.4   **f** 0.18 × 21.6

4 Use the fact 28 ÷ 4 = 7 to work out:

    a 2.8 ÷ 4   **b** 0.28 ÷ 4   **c** 2.8 ÷ 7   **d** 2.8 ÷ 0.4

5 **a** Work out 984 ÷ 8

    b Use your answer to part **a** to work out:

        i 984 ÷ 0.8   **ii** 98.4 ÷ 0.8   **iii** 9.84 ÷ 0.8

6 Work out:

    a 378 ÷ 0.3   **b** 924 ÷ 0.04   **c** 19.6 ÷ 0.08

    d 41.58 ÷ 0.09   **e** 53.2 ÷ 1.4   **f** 333.5 ÷ 0.23

7 A paperclip has a mass of 0.9 grams.

There are 234 paperclips in a box.

The mass of the box when empty is 320 grams.

What is the total mass of the box of paperclips?

8 The price of a football is £16.95

What is the total cost of 23 footballs?

What do you think? (B)

1 Calculate the area of each shape.

a

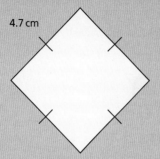

4.7 cm

b

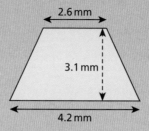

2.6 mm

3.1 mm

4.2 mm

2 A roll of wallpaper is 64.4 m long.

Flo is putting up wallpaper in a room.

The room is 2.3 m tall.

2.3 m

How many strips of wallpaper can Flo get from the roll?

Consolidate – do you need more?

1 Work out:

   a $4.57 + 6.18$ **b** $12.84 - 1.48$ **c** $80.3 - 1.64$

   d $26.7 + 13.06$ **e** $4.05 - 1.67$ **f** $0.034 + 2.18 + 2.506$

2 Estimate and then work out the answer to:

   a 6.1×4 **b** 4.7×6 **c** 2.6×4.3

   d 1.52×0.3 **e** 21.06×2.1 **f** 3.7×14.25

3 Work out:

   a $7.65 \div 5$ **b** $1.422 \div 6$ **c** $29.96 \div 14$ **d** $7.956 \div 26$

4 Work out:

   a $2248 \div 0.4$ **b** $129.6 \div 0.06$ **c** $24.32 \div 0.08$

   d $35.91 \div 0.7$ **e** $185.9 \div 1.3$ **f** $53.56 \div 0.26$

5 An adult ticket to the cinema costs £12.45

   A child ticket costs £7.95

   How much do 3 adult tickets and 4 child tickets cost altogether?

6 Tiff gets paid £8.94 per hour.

   On a Saturday she gets an extra £2.36 per hour.

   Tiff works 24 hours during the week and 6 hours on Saturday.

   How much does Tiff get paid altogether?

7 Jakub buys three books that cost £3.65 each.

   He pays with a £20 note.

   How much change should Jakub get?

Stretch – can you deepen your learning?

1 A 5p coin is 1.7 mm thick.

 ↕ 1.7 mm

Benji builds a tower of 5p coins.

The total value of the tower is £1.35

What is the height of the tower?

2 a A rule for generating a sequence is:

> Multiply the previous term by 0.13 and then add 10

Work out the third term in the sequence if the first term is 40

b A rule for generating a sequence is:

> Divide the previous term by 0.5 and then subtract 5

Work out the fourth term in the sequence if the first term is 65

3 The diagram shows a square inside a rectangle.

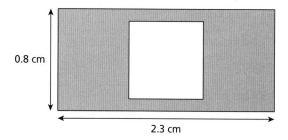

0.8 cm

2.3 cm

The perimeter of the square is 2.4 cm.

What is the area of the shaded part?

Fractions and decimals: exam practice

1 **(a)** Write $\frac{1}{2}$ as a decimal. [1 mark]

(b) Write $\frac{3}{10}$ as a percentage. [1 mark]

(c) Write 0.67 as a fraction. [1 mark]

2 Write 45% as a fraction.

Give your answer in its simplest form. [2 marks]

3 $\frac{1}{5}$ is equivalent to 0.2

What decimal number is $\frac{4}{5}$ equivalent to? [1 mark]

4 Shade 0.75 of a copy of this shape.

[1 mark]

5 Work out $3 \times \frac{1}{7}$ [1 mark]

6 A pizza is cut into 5 equal slices.

3 people eat one slice each.

What fraction of the pizza is left? [2 marks]

7 Work out:

(a) $\frac{3}{7} + \frac{4}{5}$

Write your answer as a mixed number. [3 marks]

(b) $\frac{1}{6} \div \frac{3}{4}$

Write your answer in its simplest form. [2 marks]

8 Write the numbers in order of size. Start with the smallest number.

(a) $\frac{1}{2}, \frac{2}{3}, \frac{2}{5}, \frac{3}{4}$ [2 marks]

(b) 0.82, $\frac{8}{10}, \frac{3}{5}$, 70%, $\frac{1}{2}$ [2 marks]

9 Work out the area of the rectangle.

3.4 cm

12.7 cm

[3 marks]

3 Moving on with number

In this block, we will cover...

3.1 Powers and roots

Example 1

a Work out $\sqrt{81}$

b Solve $x^2 = 81$

Method

Solution	Commentary
a 9	The square root symbol mea
	Since $9 \times 9 = 81$, the square
b $x = \pm 9$	A number, x, multiplied by i
	From part **a**, $9 \times 9 = 81$
	However, $9 \times 9 = 81$ since

3.2 Order of operations

Practice 🖩

1. Work out:

   a $15 + 3 - 7$ **b** $15 - 7 + 3$

   d $15 + 7 - 3$ **e** $15 - (7 + 3)$

2. Work out:

   a $54 \div 9 \times 6$ **b** $54 \times 6 \div 9$

   d $54 \div 6 \times 9$ **e** $9 \times 54 \div 6$

3. Work out:

   a $21 - 6 \times 3$ **b** $(21 - 6) \times 3$

3.3 Standard form

Consolidate – do you need more

1. Write these numbers in standard form.

   a 30 000   **b** 14 000

   d 23 000 000   **e** 2 590 000

2. Write these numbers in the form $A \times 10^n$

   a 0.2   **b** 0.06

   d 0.0041   **e** 0.000 000 89

3. Write these numbers in ordinary form.

Are you ready?

1 **a** Work out the area of each large square.

i

3 cm
3 cm

ii
5 cm
5 cm

b Copy and complete the calculations. **i** $3^2 = 3 \times 3 =$ ☐ **ii** $5^2 = 5 \times 5 =$ ☐

2 **a** Work out the volume of each large cube.

i
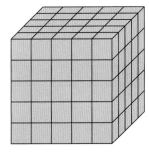

ii

b Copy and complete the calculations. **i** $3^3 = 3 \times 3 \times 3 =$ ☐ **ii** $5^3 = 5 \times 5 \times 5 =$ ☐

3 Calculate: **a** $5 \times 5 \times 5 \times 5$ **b** $3 \times 3 \times 3 \times 3 \times 3$ **c** $7 \times 7 \times 7$

A square root of a number is a value that, when multiplied by itself, gives the number.

Example: $6 \times 6 = 36$, so a square root of 36 is 6

Note that $(-6) \times (-6) = 36$ too, so -6 is also a square root of 36

The square root symbol is $\sqrt{\ }$, and it always means the positive square root.

Example: $\sqrt{36} = 6$ (because $6 \times 6 = 36$)

You will also look at powers and roots that are greater than 2

Just like the inverse of squaring is finding the square root, the inverse of higher powers also involves finding a root. The inverse of cubing is finding the cube root ($\sqrt[3]{\ }$); the inverse of raising a number to the power of 4 is finding the fourth root ($\sqrt[4]{\ }$), and so on for higher powers.

For example, $\sqrt[3]{27} = 3$ because $3 \times 3 \times 3 = 27$, or $3^3 = 27$

And $\sqrt[5]{1024} = 4$ because $4 \times 4 \times 4 \times 4 \times 4 = 1024$, or $4^5 = 1024$

Example 1

a Work out $\sqrt{81}$

b Solve $x^2 = 81$

Method

Solution	Commentary
a 9	The square root symbol means the positive square root of the number. Since $9 \times 9 = 81$, the square root of 81 is 9
b $x = \pm 9$	A number, x, multiplied by itself is equal to 81
	From part **a**, $9 \times 9 = 81$
	However, $-9 \times -9 = 81$ since the product of two negative numbers is positive. So x could be positive or negative 9

Example 2

Calculate:

a 5^4 **b** $\sqrt[4]{625}$

Method

Solution	Commentary
a 625	To calculate 5 to the power of 4, find the product of four 5s.
	$5 \times 5 \times 5 \times 5$
	$= 25 \times 5 \times 5$
	$= 125 \times 5$
	$= 625$ so $5^4 = 625$
b 5	In part **a**, 5 raised to the power 4 is equal to 625. The 4th root is the inverse of raising to the power 4, so the 4th root of 625 is equal to 5

Practice

1 Match each box in the first row to its equivalent in the second row.

2^4		4^3		4^2		3^4		2^3

$2 \times 2 \times 2$		$4 \times 4 \times 4$		$3 \times 3 \times 3 \times 3$		$2 \times 2 \times 2 \times 2$		4×4

2 Work out the value of:

a the square of 6   **b** 8^2 **c** 12 squared.

3 Work out the value of:

a 10 cubed   **b** 2^3 **c** 3 cubed.

4 Work out the value of:

a 4^2 **b** 4^3 **c** 4^4 **d** 4^6

5 Work out the value of:

a $3^2 + 2^3$ **b** $4^3 - 2^2$ **c** $2^5 + 7^2$

6 Identify all the square numbers in this list.

4 10 25 100 121 99 16 9

7 Write down the value of:

a $\sqrt{9}$ **b** the square root of 36 **c** $\sqrt{100}$

d the cube root of 8 **e** $\sqrt[3]{64}$ **f** $\sqrt[3]{125}$

8 Calculate the following. Write down the full calculator display.

a 26^3 **b** $\sqrt{1369}$ **c** $\sqrt[3]{4913}$

d $\sqrt{240}$ **e** $\sqrt[3]{316}$ **f** 6.4^2

g 10.71^3 **h** $\sqrt{82.4}$ **i** 3.2^5

What do you think? 💡

1 Junaid says, "When you cube a number, the answer is always greater."

Do you agree with Junaid? Explain your reason.

Consolidate – do you need more?

1 Work out the value of:

a the square of 4 **b** 7^2 **c** 11 squared

d the square of 1 **e** 9^2 **f** 13 squared.

2 Work out the value of:

a the cube of 1 **b** 5^3 **c** 4 cubed

d the cube of 2 **e** 10^3 **f** 9 cubed.

3 Work out the value of:

a 6^2 **b** 6^3 **c** 6^4 **d** 6^5

4 Work out the value of:

a $4^2 + 3^3$ **b** $5^3 - 6^2$ **c** $3^4 + 8^2$

5 Identify all the cube numbers in this list.

4 100 6 8 64 1 16 9 27

6 Write down the value of:

   a $\sqrt{144}$ **b** the square root of 49   **c** $\sqrt{1}$

   d the cube root of 216   **e** $\sqrt[3]{343}$ **f** $\sqrt[3]{1000}$

7 Calculate the following.

   a 31^3 **b** $\sqrt{729}$ **c** $\sqrt[3]{4096}$

Stretch – can you deepen your learning?

1 Flo is thinking of two integers.

She squares one of the numbers and cubes the other.

The sum of the numbers is 100

Write down the two numbers Flo is thinking of.

2 Jackson is thinking of a square number and a cube number.

The difference between the numbers is 25

Write down the two numbers Jackson is thinking of.

3 Write these in descending order.

2^3 $\sqrt{16}$ $\sqrt[3]{125}$ 3^2

4 The sum of three square numbers is 194

Write down the three numbers.

3.2 Order of operations

Are you ready?

1 Work out:

    a **i** 8 + 4 − 2   **ii** 8 − 2 + 4   **b** **i** −6 + 4 − 2   **ii** 4 − 2 − 6

    What do you notice about your answers to parts **a** and **b**? Discuss with a partner.

2 Work out:

    a **i** 8 × 4 ÷ 2   **ii** 8 ÷ 2 × 4   **b** **i** 30 ÷ 5 × 4   **ii** 30 × 4 ÷ 5

    What do you notice about your answers to parts **a** and **b**? Discuss with a partner.

3 **a** The counters show the calculation 3 + 4 × 2

       Copy and complete the calculation.   3 + 4 × 2 = ☐

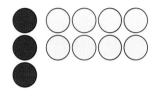

    b The counters show the calculation (3 + 4) × 2

       Copy and complete the calculation.   (3 + 4) × 2 = ☐

    What is the same and what is different about parts **a**
and **b**? Discuss with a partner.

4 Draw counters to represent each calculation.

    a 6 + 2 × 3   **b** (6 + 2) × 3

There are several rules of **priority** in calculations you need to know.

Addition and *subtraction* have equal priority. If a calculation has only additions and subtractions, you just work from left to right: $\underline{8 + 4} - 7 - 2 + 5$ $= \underline{12 - 7} - 2 + 5$ $= \underline{5 - 2} + 5$ $= 3 + 5$ $= 8$	*Multiplication* and *division* have equal priority too. If a calculation has only multiplications and divisions, you just work from left to right: $\underline{8 \times 4} \div 2 \times 5$ $= \underline{32 \div 2} \times 5$ $= 16 \times 5$ $= 80$

Multiplication and division take priority over *addition and subtraction.*

In 3 + 4 × 5, do the multiplication first:

3 + 4 × 5

= 3 + 20

= 23

Powers and *roots* take priority over *multiplication and division*.

$3 \times \underline{5^2} = 3 \times \mathbf{25} = 75$

If you want the calculations in a different order, put brackets around the part you want to do first.

$(3 \times 5)^2$ means do the multiplication first.

$\underline{(3 \times 5)}^2 = \mathbf{15}^2 = 225$

This is different from 3×5^2

$(3 + 4) \times 5$ means do the addition first.

$\underline{(3 + 4)} \times 5 = \mathbf{7} \times 5 = 35$

This is different from $3 + 4 \times 5$

The order can be summarised in this pyramid.

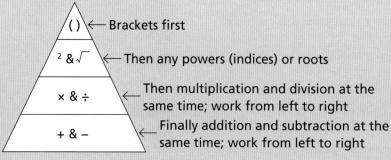

( ) ← Brackets first

^{2} & $\sqrt{}$ ← Then any powers (indices) or roots

× & ÷ ← Then multiplication and division at the same time; work from left to right

+ & − ← Finally addition and subtraction at the same time; work from left to right

This is called the **order of operations**.

Example

Work out:

a $6 \div 2 + 4$	**b** $6 \div (2 + 4)$	**c** $6^2 \div 2 + 4$
d $6 \div 2^2 + 4$	**e** $6 \div (2^2 + 4)$	**f** $(6 \div 2)^2 + 4$

It can be helpful to underline the parts you are working out first as you do them.

Method

Solution	Commentary
a $\underline{6 \div 2} + 4$ $= \underline{3 + 4}$ $= 7$	Division takes priority over addition.
b $6 \div \underline{(2 + 4)}$ $= \underline{6 \div 6}$ $= 1$	Calculations in brackets take priority, so do the addition first.
c $\underline{6^2} \div 2 + 4$ $= \underline{36 \div 2} + 4$ $= \underline{18 + 4}$ $= 22$	The squaring takes priority. The division is next. Finally the addition.

d	$6 \div \underline{2^2} + 4$	The squaring takes priority.
	$= 6 \div \underline{4} + 4$	The division is next.
	$= \underline{1.5 + 4}$	Finally the addition.
	$= 5.5$	
e	$6 \div (\underline{2^2} + 4)$	Calculations in brackets take priority and, in the brackets, squaring takes priority.
	$= 6 \div \underline{(4 + 4)}$	Next do the addition in the brackets.
	$= \underline{6 \div 8}$	Finally do the division.
	$= 0.75$	
f	$(\underline{6 \div 2})^2 + 4$	Calculations in brackets take priority, so do the division first.
	$= \underline{3^2} + 4$	Next, squaring takes priority over addition.
	$= \underline{9 + 4}$	Finally the addition.
	$= 13$	

Practice

1 Work out:

    a $15 + 3 - 7$ **b** $15 - 7 + 3$ **c** $3 + 15 - 7$

    d $15 + 7 - 3$ **e** $15 - (7 + 3)$ **f** $15 - (3 + 7)$

2 Work out:

    a $54 \div 9 \times 6$ **b** $54 \times 6 \div 9$ **c** $6 \times 54 \div 9$

    d $54 \div 6 \times 9$ **e** $9 \times 54 \div 6$ **f** $54 \div (9 \times 6)$

3 Work out:

    a $21 - 6 \times 3$ **b** $(21 - 6) \times 3$ **c** $21 \times 6 - 3$

    d $21 \times (6 - 3)$ **e** $21 - 6 \div 3$ **f** $(21 - 6) \div 3$

4 It costs £25 to rent a bike plus £3 for every hour it is rented for.

Which of the calculations shows the cost of renting a bike for 4 hours?

    $(25 + 3) \times 4$ $25 \times 4 + 3 \times 4$ $25 + 3 \times 4$ $25 \times 4 + 3$

5 Work out:

    a $5 + 4^2 \times 3$ **b** $3 + 4^2 \times 5$ **c** $(3 + 4^2) \times 5$

    d $(4^2 + 5) \div 3$ **e** $3 \times (4^2 + 5)$ **f** $(3 \times 4)^2 + 5$

6 Rhys works out $2 \times 6 + 4 \times 5$

He gets the answer 80

    a Explain what Rhys has done wrong.

    b Work out the correct answer to $2 \times 6 + 4 \times 5$

7 Work out:

a $16 + 9 \times \sqrt{4}$ **b** $(\sqrt{16} + 9) \times 4$ **c** $\sqrt{16} + 9 \times 4$

d $\sqrt{(16 + 9)} \times 4$ **e** $\sqrt{9} + (16 \times 4)$ **f** $16 + \sqrt{(9 \times 4)}$

8 Insert brackets, if needed, to make each calculation correct.

a $6 + 4 \times 2 + 3 = 23$ **b** $6 + 4 \times 2 + 3 = 17$

c $6 + 4 \times 2 + 3 = 50$ **d** $6 + 4 \times 2 + 3 = 26$

What do you think? 💭

1 Bev is working out $9 - 3 + 4$

She uses BIDMAS to remember the order of operations.

She says, "In BIDMAS, addition comes before subtraction so $9 - 3 + 4 = 9 - 7 = 2$"

a Explain why Bev is incorrect.

b Work out the correct answer to $9 - 3 + 4$

2 a Insert brackets to make a total of 23

$5 + 3^2 \times 4 - 2$

b Insert brackets to find different totals.

How many different totals can you make? Compare your answers with a partner.

Consolidate – do you need more?

1 Work out:

a $12 - 6 + 4$ **b** $12 + 4 - 6$ **c** $4 + 12 - 6$

d $4 + (12 - 6)$ **e** $12 \times 6 \div 4$ **f** $12 \div 4 \times 6$

g $12 \times 4 \div 6$ **h** $12 \div 6 \times 4$ **i** $6 \times 4 \div 12$

2 Work out:

a $15 - 3 \times 4$ **b** $15 \times 4 - 3$ **c** $(15 - 3) \times 4$

d $15 - 4 \times 3$ **e** $(15 - 4) \times 3$ **f** $15 \times (4 - 3)$

3 Work out:

a $2 \times 6^2 - 5$ **b** $6^2 - 5 \times 2$ **c** $5 \times (6^2 + 2)$

d $5 \times (6^2 \div 2)$ **e** $2 \times (6^2 - 5)$ **f** $(6^2 - 5) \times 2$

4 Work out:

a $25 + 11 \times \sqrt{4}$ **b** $(\sqrt{25} + 11) \times 4$ **c** $\sqrt{25} + 11 \times 4$

d $\sqrt{(25 + 11)} \times 4$ **e** $11 + (\sqrt{25} - \sqrt{4})$ **f** $11 + \sqrt{(25 \times 4)}$

5 A machine makes 24 toys every hour.

Three of the toys that are made every hour are red. The rest are blue.

Which calculation shows how many blue toys are made in 6 hours?

$24 - (3 \times 6)$ $(24 - 3) \times 6$ $24 - 3 \times 6$

6 Work out:

a $(5 + 3) \times 4 + 6$ **b** $5 + 3 \times 4 + 6$ **c** $5 + 3 \times (4 + 6)$ **d** $(5 + 3) \times (4 + 6)$

7 Insert brackets, if needed, to make each calculation correct.

a $8 + 12 \div 4 - 2 = 14$ **b** $8 + 12 \div 4 - 2 = 3$

c $8 + 12 \div 4 - 2 = 9$ **d** $8 + 12 \div 4 - 2 = 10$

8 A car salesperson is paid a weekly wage of £950 plus a bonus of £50 for every car they sell.

Last week, the car salesperson sold 7 cars.

Their pay is found by working out $950 + 50 \times 7$

Marta thinks that their weekly pay will be £7000

Explain why Marta is incorrect.

Stretch – can you deepen your learning?

1 Use the digits 1, 2, 3 and 4, together with $+, -, \times, \div, \sqrt{}$ and brackets, to make numbers in as many ways as you can.

For example, to make the number 1: $1 = (2 + 3) - 4 \times 1$

Can you make the numbers 2 to 20 using all four digits?

Are you ready? (A)

1 Work out: **a** 10^2 **b** 7^2 **c** 2^5 **d** 5^3

2 Write $10 \times 10 \times 10 \times 10$ using powers.

3 Copy and complete the table.

Using powers	10^4		10^2	10^1			10^{-2}	
Ordinary number		1000			1	0.1		0.001	0.0001

Using your calculator

A calculator can be used to change numbers from ordinary to standard form.

First you will need to set your calculator to SCI mode.

Press **SHIFT** then **SET UP** then press **7** for SCI mode. Choose the number of significant figures (recommend 5 or above).

Type the number you want to convert (e.g. 300000) into your calculator then press **=**

Your calculator will display

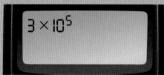

You can set your calculator back to normal mode when finished.

Sometimes maths and science deal with very large or very small numbers. For example, the average distance from the Earth to the Sun is about 150 000 000 000 metres.

It is quite easy to make a mistake when reading or comparing numbers like these. There is a way to write these numbers that avoids these issues. This is normally known as **standard form** or **standard index form**.

A number that is in standard form is written in the form $A \times 10^n$ where A is a number that is at least 1 and less than 10, and n is an integer.

For example, the large number $150\,000\,000\,000 = 1.5 \times 10^{11}$

And the small number $0.000\,000\,235 = 2.35 \times 10^{-7}$

The following examples show how to write numbers like this in standard form.

Example 1

Write $300\,000$ in the form $A \times 10^n$

Method

Solution	Commentary
$300\,000$ is equal to $3 \times 100\,000$	Rewrite $300\,000$ as a number that is at least 1 and less than 10, multiplied by a power of 10
$100\,000 = 10 \times 10 \times 10 \times 10 \times 10 = 10^5$	Write the power of 10 in index form.
$300\,000 = 3 \times 10^5$	Write the answer in the form $A \times 10^n$ $A = 3$ and $n = 5$

Example 2

Write $761\,000\,000$ in the form $A \times 10^n$

Method

Solution	Commentary
$761\,000\,000 = 7.61 \times 100\,000\,000$	All the non-zero digits need to form the number between 1 and 10 In this case A is 7.61
$100\,000\,000 = 10 \times 10 \times 10 \times 10 \times 10 \times 10 \times 10 \times 10$ $\qquad = 10^8$	Write the power of 10 in index form.
$761\,000\,000 = 7.61 \times 10^8$	There are eight digits after the digit 7, so multiply 7.61 by 10^8

Example 3

Write $0.000\,208$ in standard form.

Method

Solution	Commentary
$2.08 \times \dfrac{1}{10} = 0.208$	A will be the number 2.08 as the decimal point will be placed after the first non-zero digit.
$2.08 \times \dfrac{1}{100} = 0.0208$	
$2.08 \times \dfrac{1}{1000} = 0.00208$	
$2.08 \times \dfrac{1}{10\,000} = 0.000\,208$	Remember, $\dfrac{1}{10\,000}$ can be written as 10^{-4}
2.08×10^{-4}	So the answer is 2.08×10^{-4}

Practice (A)

1 Write these numbers in the form $A \times 10^n$
- **a** 60 000
- **b** 900 000
- **c** 4000
- **d** 500 000 000
- **e** 2 000 000 000
- **f** 7 000 000

2 Write these numbers in standard form.
- **a** 630 000
- **b** 19 000
- **c** 3500
- **d** 210 000 000
- **e** 1 570 000
- **f** 729 000

3 Write these numbers in standard form.
- **a** 0.3
- **b** 0.03
- **c** 0.000 07
- **d** 0.0004
- **e** 0.000 000 09
- **f** 0.000 000 6

4 Write these numbers in the form $A \times 10^n$
- **a** 0.32
- **b** 0.032
- **c** 0.000 64
- **d** 0.0214
- **e** 0.000 908
- **f** 0.002 067

5 Lida is writing the number 2 340 000 in standard form.

She says, "The answer is 23.4×10^5"

Do you agree with Lida? Explain your answer.

6 The distance from Earth to the Sun is approximately 150 000 000 km.
Write the distance in standard form.

7 The volume of a neuron cell is approximately 0.000 000 007 mm^3.
Write the volume of the neuron cell in the form $A \times 10^n$

Are you ready? (B)

1 Write these powers of 10 as ordinary numbers.
- **a** 10^3
- **b** 10^5
- **c** 10^0

2 Write these powers of 10 as ordinary numbers.
- **a** 10^{-3}
- **b** 10^{-6}
- **c** 10^{-10}

3 Work out:
- **a** 3.2×1000
- **b** $0.4 \times 10 000$
- **c** $1.05 \times 100 000$
- **d** 4×0.001
- **e** 4.6×0.01
- **f** 3.02×0.00001

Example 1

Write these numbers in ordinary form. **a** 9×10^5 **b** 5.47×10^4

Method

Solution	Commentary
a $9 \times 10^5 = 900\,000$	$10^5 = 10 \times 10 \times 10 \times 10 \times 10 = 100\,000$
b $5.47 \times 10^4 = 5.47 \times 10\,000$ $= 54\,700$	A common mistake is to write four zeros in the answer to a problem like this. This is incorrect.

Example 2

Write 8.06×10^{-4} as a decimal.

Method

Solution	Commentary
$10^{-4} = 0.0001$ So $8.06 \times 10^{-4} = 8.06 \times 0.0001$ $= 0.000806$	The power is -4 so the number will be less than 1

Practice (B)

1. Write these numbers in ordinary form.
   - **a** 2×10^4
   - **b** 6×10^6
   - **c** 3×10^3
   - **d** 4×10^7
   - **e** 9×10^9
   - **f** 5×10^1

2. Write these numbers in ordinary form.
   - **a** 2.4×10^4
   - **b** 5.1×10^5
   - **c** 3.3×10^2
   - **d** 7.41×10^6
   - **e** 1.08×10^3
   - **f** 5.25×10^7

3. Write these as decimals.
   - **a** 2×10^{-4}
   - **b** 6×10^{-6}
   - **c** 9×10^{-3}
   - **d** 3×10^{-5}
   - **e** 8×10^{-2}
   - **f** 1×10^{-10}

4. Write these numbers in ordinary form.
   - **a** 2.6×10^{-4}
   - **b** 4.5×10^{-3}
   - **c** 1.9×10^{-6}
   - **d** 3.24×10^{-5}
   - **e** 8.01×10^{-2}
   - **f** 3.056×10^{-8}

5. Jakub is writing the number 7.4×10^2 in ordinary form.

   He says, "The answer is 7.400"

   Do you agree with Jakub? Explain your answer.

6. The circumference, in metres, of the planet Mars is approximately 2.13×10^7

   Write the circumference of Mars in metres as an ordinary number.

Are you ready? (C)

1 Which number is greater in each pair? Compare your method with a partner.

    a 3×10^4 or 9×10^3 **b** 2×10^{-6} or 4×10^{-2}

2 Work out: **a** $6 + 3 \times 1000$ **b** $(6 + 3) \times 1000$

3 Write each of these as a single power of 10

    a $10^3 \times 10^2$ **b** $10^2 \times 10^4$ **c** $10^3 \times 10^3$

    d $10^6 \div 10^5$ **e** $10^6 \div 10^3$ **f** $10^3 \div 10^4$

Example 1

Work out each calculation, giving your answers in standard form.

a $9 \times 10^3 + 4 \times 10^3$ **b** $4 \times 10^3 + 8 \times 10^5$

Method A

Solution	Commentary
a $9000 + 4000 = 13\,000$	Change each number from standard form to an ordinary number.
$13\,000 = 13 \times 10^3 = 1.3 \times 10^4$	Convert your answer back to standard form.
b $4000 + 800\,000 = 804\,000$	
$804\,000 = 8.04 \times 10^5$	

Method B

Solution	Commentary
a $(9 + 4) \times 10^3 = 13 \times 10^3$	The powers are the same, so re-order the calculation and add the numbers.
$13 \times 10^3 = 1.3 \times 10^4$	Make sure that your final answer is in standard form.
b $4 \times 10^3 = 0.04 \times 10^5$	Adjust one of the numbers so that the powers are the same.
$0.04 \times 10^5 + 8 \times 10^5 = (0.04 + 8) \times 10^5$ $= 8.04 \times 10^5$	Then use the same method as in part **a**.

Example 2

Work out: **a** $(3.1 \times 10^2) \times (2 \times 10^2)$ **b** $\dfrac{8 \times 10^5}{2 \times 10^6}$

Method

Solution	Commentary
a $(3.1 \times 10^2) \times (2 \times 10^2)$ $= (3.1 \times 2) \times (10^2 \times 10^2)$	Multiplication is **commutative**, so the calculation can be rearranged to make it easier.
$3.1 \times 2 = 6.2$ $10^2 \times 10^2 = 10^4$	
So $(3.1 \times 10^2) \times (2 \times 10^2) = 6.2 \times 10^4$	Use your knowledge of **index** laws: when multiplying powers of 10 you add the powers.

b $\dfrac{8 \times 10^5}{2 \times 10^6}$

$8 \div 2 = 4$ $\qquad$ $10^5 \div 10^6 = 10^{-1}$

So $\dfrac{8 \times 10^5}{2 \times 10^6} = 4 \times 10^{-1}$

Use your knowledge of index laws: when dividing powers of 10, you subtract the powers.

Practice (C)

1 Work out these calculations. Give your answers in standard form.

a $4 \times 10^3 + 3 \times 10^3$ $\qquad$ **b** $2 \times 10^5 + 7 \times 10^5$ $\qquad$ **c** $2.1 \times 10^2 + 5.6 \times 10^2$

d $7 \times 10^3 - 3 \times 10^3$ $\qquad$ **e** $9 \times 10^5 - 8 \times 10^5$ $\qquad$ **f** $4.5 \times 10^4 - 3.2 \times 10^4$

2 Work out these calculations. Give your answers in standard form.

a $4 \times 10^4 + 3 \times 10^3$ $\qquad$ **b** $3 \times 10^5 + 5 \times 10^3$ $\qquad$ **c** $3.2 \times 10^2 + 6.1 \times 10^3$

d $7 \times 10^3 - 3 \times 10^2$ $\qquad$ **e** $8 \times 10^5 - 6 \times 10^3$ $\qquad$ **f** $4.8 \times 10^6 - 2 \times 10^5$

3 Work out these calculations. Give your answers in standard form.

a $(2 \times 10^3) \times (3 \times 10^3)$ $\qquad$ **b** $(4 \times 10^3) \times (3 \times 10^3)$ $\qquad$ **c** $(4 \times 10^2) \times (3 \times 10^4)$

d $(4.2 \times 10^3) \times (2 \times 10^5)$ $\qquad$ **e** $(6 \times 10^5) \div (3 \times 10^3)$ $\qquad$ **f** $(9.9 \times 10^4) \div (3 \times 10^2)$

4 Write your answers to question 3 as ordinary numbers.

What do you think? 💡

1 Work out the missing numbers.

a $3 \times 10^3 + \boxed{} \times 10^3 = 8 \times 10^3$

b $(3.4 \times 10^3) \times (2 \times 10^{\boxed{}}) = 6.8 \times 10^5$

c $(9 \times 10^{\boxed{}}) \div (3 \times 10^4) = 3 \times 10^{-2}$

Consolidate – do you need more?

1 Write these numbers in standard form.

a $30\,000$ $\qquad$ **b** $14\,000$ $\qquad$ **c** 4200

d $23\,000\,000$ $\qquad$ **e** $2\,590\,000$ $\qquad$ **f** $104\,000$

2 Write these numbers in the form $A \times 10^n$

a 0.2 $\qquad$ **b** 0.06 $\qquad$ **c** $0.000\,05$

d 0.0041 $\qquad$ **e** $0.000\,000\,89$ $\qquad$ **f** $0.000\,601$

3 Write these numbers in ordinary form.

a 4×10^4 $\qquad$ **b** 5×10^7 $\qquad$ **c** 3.1×10^3

d 6.24×10^5 $\qquad$ **e** 3.05×10^4 $\qquad$ **f** 2.22×10^6

4 Write these as decimals.

  a 4×10^{-3}
  b 7×10^{-5}
  c 8×10^{-1}

  d 3.1×10^{-4}
  e 1.27×10^{-2}
  f 3.08×10^{-6}

5 Work out these calculations. Give your answers in standard form.

  a $5 \times 10^3 + 2 \times 10^3$
  b $2 \times 10^6 + 1 \times 10^6$
  c $3.2 \times 10^4 + 6.1 \times 10^4$

  d $9 \times 10^4 - 3 \times 10^4$
  e $8 \times 10^5 - 4 \times 10^5$
  f $5.8 \times 10^3 - 3.1 \times 10^3$

6 Work out these calculations. Give your answers in standard form.

  a $5 \times 10^5 + 2 \times 10^2$
  b $2 \times 10^5 + 5 \times 10^2$
  c $4.3 \times 10^2 + 3.4 \times 10^3$

  d $9 \times 10^4 - 3 \times 10^3$
  e $6 \times 10^4 - 2 \times 10^2$
  f $3.6 \times 10^3 - 5 \times 10^2$

7 Work out these calculations. Give your answers in standard form.

  a $(4 \times 10^4) \times (2 \times 10^4)$
  b $(3 \times 10^3) \times (5 \times 10^3)$
  c $(5 \times 10^1) \times (3 \times 10^5)$

  d $(3.4 \times 10^4) \times (2 \times 10^2)$
  e $(8 \times 10^2) \div (4 \times 10^2)$
  f $(8.6 \times 10^5) \div (2 \times 10^2)$

Stretch – can you deepen your learning?

1 Work out $\sqrt[3]{7.29 \times 10^5}$

2 The table shows the diameter, in metres, of some planets. The diameters are given to an accuracy of 3 significant figures.

  a Order the planets in size, starting with the planet with the smallest diameter.

Planet	Diameter (metres)
Earth	1.28×10^7
Mercury	4.88×10^6
Uranus	5.11×10^7
Saturn	1.21×10^8

  b Write the diameter of Earth, in metres, as an ordinary number.

  c Write the diameter of Earth, in kilometres, as an ordinary number.

  d Calculate the difference in diameter between Saturn and Uranus. Give your answer in metres in standard form.

3 The formula to work out time (t) is $t = \dfrac{\text{distance}}{\text{speed}}$

The speed of light is approximately 3×10^8 metres per second.

The distance from Earth to the Sun is approximately 1.496×10^{11} metres.

Calculate the approximate time it takes for light to travel from the Sun to Earth.

Write your answer in seconds in standard form.

Moving on with number: exam practice

1 (a) Work out the value of 4^2 **[1 mark]**
(b) Work out the value of $\sqrt{36}$ **[1 mark]**
(c) Write as a power of 2
$2 \times 2 \times 2 \times 2 \times 2$ **[1 mark]**

2 Work out $3 \times 5 + 4$ **[1 mark]**

3 What is the square of 7? **[1 mark]**

4 Sophie says $2^3 = 6$
Explain why Sophie is wrong. **[1 mark]**

5 Work out the value of $2^3 + 9^2$ **[2 marks]**

6 Chloe says $30 - 5 \times 3$ is 75
Ron says $30 - 5 \times 3$ is 15
(a) Who is right? Give a reason for your answer. **[2 marks]**
(b) Work out the value of $2 \times 3^2 + (10 - 3)$ **[2 marks]**
(c) Use brackets to make the statement correct.
$9 + 2 \times 6 - 3 = 33$ **[1 mark]**

7 Write in standard form:
(a) 45 700 **[1 mark]**
(b) 0.000 03 **[1 mark]**
(c) 7 million **[1 mark]**

8 A square has an area of 100 cm².
What is the side length of the square? **[1 mark]**

9 Write as ordinary numbers.
(a) 4.1×10^5 **[1 mark]**
(b) 6.04×10^{-6} **[1 mark]**

10 Work out:
$(5 \times 10^4) \times (6 \times 10^7)$
Give your answer in standard form. **[2 marks]**

1–3

3–5

4 Factors, multiples and primes

White Rose
M▲THS

In this block, we will cover...

4.1 HCF and LCM

Example 2

Work out the lowest common multiple of 15 and

Method

Solution	Co
Multiples of 15:	Th
15, 30, 45, 60, 75, 90, 105, 120, 135, 150 ...	the
Multiples of 6:	
6, 12, 18, 24, 30, 36, 42, 48, 54, 60 ...	
The lowest common multiple is 30	Th

4.2 Prime factorisation

Practice (B)

1. The prime factors of 24 and 36 are shown
   Venn diagram.

   a   What is the HCF of 24 and 36?

   b   What is the LCM of 24 and 36?

2. The prime factors of two numbers, A and A
   in the Venn diagram.

   a   What is the value of A?

Are you ready?

1 List all the factors of each number.

   a 18   **b** 25   **c** 48   **d** 84

2 List the first five multiples of each number.

   a 4   **b** 8   **c** 12   **d** 23

3 Choose two numbers from the box that have:

   a a common factor of 6

   b a common multiple of 32

16	12		30
	128	3	4
14	10	64	

Here are some important key words relating to factors and multiples.

Key words	Examples
A **factor** is a positive integer that divides exactly into another positive integer.	4 is a factor of 12 because $12 \div 4 = 3$ with no remainder.
A **common factor** is a factor that is shared by two or more numbers.	2 is a factor of both 6 and 8, so 2 is a common factor of 6 and 8
The **highest common factor (HCF)** of two or more numbers is their greatest shared factor.	8 is the HCF of 16 and 24 because it is the largest number that will go exactly into both: $16 \div 8 = 2$ $24 \div 8 = 3$
A **multiple** is the result of multiplying a number by a positive integer.	28 is a multiple of 7 because $7 \times 4 = 28$
A **common multiple** is a shared multiple of two or more numbers.	100 is a multiple of both 5 and 10, so 100 is a common multiple of 5 and 10
The **lowest common multiple (LCM)** is the lowest shared multiple of two or more numbers.	63 is the LCM of 7 and 9 because it is the lowest number that is in both of their times tables.

Example 1

Work out the highest common factor of 12 and 16

Method

Solution	Commentary
4	The factors of 12 are 1, 2, 3, 4, 6 and 12
	The factors of 16 are 1, 2, 4, 8 and 16
	The common factors of 12 and 16 are 1, 2 and 4
	The highest common factor is therefore 4

Example 2

Work out the lowest common multiple of 15 and 6

Method

Solution	Commentary
Multiples of 15: 15, 30, 45, 60, 75, 90, 105, 120, 135, 150 … Multiples of 6: 6, 12, 18, 24, 30, 36, 42, 48, 54, 60 …	The common multiples of 15 and 6 shown in the lists are 30 and 60
The lowest common multiple is 30	The lowest common multiple is therefore 30 You need only write out multiples of each number until you find one in both lists.

Practice

1 **a** List the factors of 24 and 42

   b What are the common factors of 24 and 42?

   c What is the highest common factor of 24 and 42?

2 Work out the highest common factor for each set of numbers.

   a 18 and 36   **b** 24 and 36   **c** 48 and 72

   d 80 and 120   **e** 56 and 98   **f** 36, 54 and 72

3 **a** List the first 10 multiples of 3 and 4

   b What is the lowest common multiple of 3 and 4?

4 **a** List the first 10 multiples of 9 and 6

   b What is the lowest common multiple of 9 and 6?

5 Work out the lowest common multiple for each set of numbers.

   a 3 and 5   **b** 8 and 12   **c** 15 and 20

   d 18 and 24   **e** 36 and 48   **f** 10, 15 and 20

6 A green light flashes every 6 seconds.

   A red light flashes every 8 seconds.

   Both lights flash at the same time.

   After how many seconds will both lights flash at the same time again?

What do you think? 🗨

1 A number 4 bus leaves the station every 12 minutes.

A number 6 bus leaves the station every 21 minutes.

A number 4 bus and number 6 bus leave the station at 2:05 pm.

At what time will both a number 4 bus and a number 6 bus next leave the station at the same time?

2 Flo says, "To find the lowest common multiple of two numbers, you just have to multiply them together."

Do you agree with Flo? Explain your answer.

Consolidate – do you need more?

1 Work out the highest common factor for each set of numbers.

a 14 and 28   **b** 30 and 45   **c** 63 and 84

d 50 and 125   **e** 27 and 81   **f** 90, 120 and 150

2 Work out the lowest common multiple for each set of numbers.

a 4 and 6   **b** 9 and 12   **c** 25 and 35

d 14 and 21   **e** 16 and 24   **f** 30, 60 and 45

3 Rob plays tennis every 3 days.

Junaid plays tennis every 8 days.

They both play tennis on Monday.

After how many days will they both play tennis on the same day?

4 Pens are sold in packs of 13

Pencils are sold in packs of 7

Mario buys the same number of pens and pencils.

What is the smallest number of each pack Mario can buy?

5 A train leaves Halifax station to travel to Leeds every 24 minutes.

A train leaves Halifax station to travel to Manchester every 40 minutes.

a A train to Leeds and a train to Manchester leave at the same time.

After how many minutes will a train to Leeds and a train to Manchester next leave at the same time?

b A train to Leeds and a train to Manchester leave Halifax at 10:50 am.

At what time will a train to Leeds and a train to Manchester next leave at the same time?

6 Ali is wrapping presents. He has some blue ribbon and some green ribbon.

The blue ribbon is 18m long.

The green ribbon is 24m long.

Ali wants to cut the ribbons into equal lengths.

What is the longest length he can cut the ribbons into without wasting any?

Stretch – can you deepen your learning?

1 A red light flashes every 18 minutes.

A blue light flashes every 27 minutes.

A yellow light flashes every 12 minutes.

All three lights flash together at 4:00 pm.

a At what time will the red and blue light next flash at the same time?

b At what time will the red and yellow light next flash together?

c At what time will all three lights next flash at the same time?

2 The areas of two rectangles are shown. The diagrams are not accurately drawn.

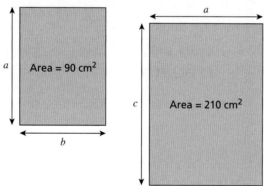

a, b and c are all integers.

What are the possible values of a, b and c?

4.2 Prime factorisation

Are you ready? (A)

1 Which of these numbers are prime? 1 2 3 4 5 6 7 8 9 10 11 12 13 14

2 Write the factors of each number.

   a 12   **b** 18   **c** 32   **d** 84

3 Work out:

   a 2 × 2 × 3   **b** 3 × 3 × 5   **c** 2 × 3 × 7   **d** 7 × 7

4 Write using powers:

   a 2 × 2 × 2   **b** 5 × 5   **c** 3 × 3 × 3 × 3

Any number can be written as a product of its factors, for example 8 = 2 × 4

However, as 4 is not prime, 2 × 4 is not a product of **prime factors**.

In fact, any number can be written as a product of its prime factors.

When writing a number as the product of its prime factors, all the numbers in the product are prime, for example   8 = 2 × 2 × 2   or   84 = 2 × 2 × 3 × 7

Writing a number in this way is often called the **prime factor decomposition** of the number.

A factor tree is useful when finding prime factors of numbers.

Example 1

Write 100 as the product of its prime factors.

Method A

Solution	Commentary
100 2 50 2 25 5 5	To start a prime factor tree, first identify a pair of factors. 100 = 2 × 50 2 is prime, so circle this and that branch ends there. 50 is not prime, so continue. 50 = 2 × 25 Again, 2 is prime so circle this and end that branch. 25 is not prime, so continue. 25 = 5 × 5 Now that all branches are ended with primes, you can write 100 as the product of its prime factors.
100 = 2 × 2 × 5 × 5	Write the prime factors in ascending order, so 100 = 2 × 2 × 5 × 5 This is the prime factor decomposition of 100

Method B

Solution	Commentary
100 4 25 (2) (2) (5) (5) $100 = 2 \times 2 \times 5 \times 5$	You could have started with a different pair of factors and you would still end up with the same final answer. Try it. It is always a good idea to check your answer. Work out $2 \times 2 \times 5 \times 5$ to make sure it gives 100

Example 2

Write 120 as a product of its prime factors. Give your answer in index form.

Method

Solution	Commentary
120 (3) 40 (2) 20 (2) 10 (2) (5)	Start by finding the prime factors of 120
$120 = 2 \times 2 \times 2 \times 3 \times 5$ $120 = 2^3 \times 3 \times 5$	Because multiple factors are the same, in this case three 2s, write these in index form as 2^3

Practice (A)

1 A factor tree has been drawn for each number.

Write each number as the product of its prime factors.

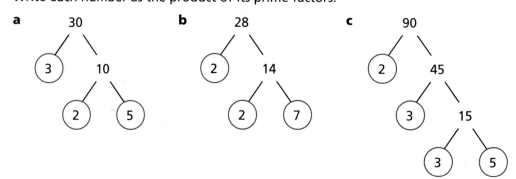

a   30   **b**   28   **c**   90

2 **a** Copy and complete the factor tree.

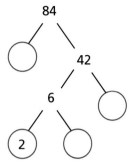

b Write 84 as a product of its prime factors.

3 Write each of the numbers as the product of its prime factors.

    a 48   **b** 18   **c** 25

    d 90   **e** 68   **f** 160

4 Write each of the numbers as the product of its prime factors.
Give your answers in index form.

    a 48   **b** 50   **c** 49

    d 96   **e** 140   **f** 248

5 The numbers have been written as products of their prime factors.
Work out the numbers.

    a $2 \times 2 \times 3$ **b** $3 \times 3 \times 5$ **c** $2 \times 3 \times 7$

    d $3 \times 3 \times 11$ **e** $2 \times 3^2 \times 7$ **f** $2^3 \times 5^2$

6 Zach has written 180 as the product of its prime factors.

    a Explain the mistake Zach has made.

    b Write 180 as the product of its prime factors.

$180 = 2 \times 2 \times 5 \times 9$

What do you think? (A) 🌑

1 Using the fact that $24 = 2^3 \times 3$, write each of the numbers as the product of its prime factors in index form.

    a 48   **b** 240   **c** 480   **d** 120

Are you ready? (B)

1 Write each number as a product of its prime factors.   **a** 12   **b** 36

2 **a** What is the HCF of 12 and 36?   **b** What is the LCM of 12 and 36?

3 Copy the Venn diagram and sort the numbers into it.

6   11   12   7   22   15

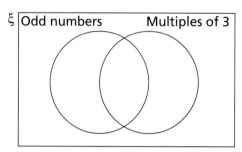

You can use the prime factors of a number, along with a Venn diagram, to calculate the HCF and LCM of two or more numbers.

Example

a Work out the highest common factor of 72 and 60

b Work out the lowest common multiple of 72 and 60

Method

Solution	Commentary
a $72 = \underline{2} \times \underline{2} \times 2 \times \underline{3} \times 3$ $60 = \underline{2} \times \underline{2} \times \underline{3} \times 5$ ξ 72 ⬭ 60 with 2, 3 (left), 2, 2, 3 (centre), 5 (right) HCF = 2 × 2 × 3 = 12	First, you need to write each number as the product of its prime factors. Looking at the prime factor decomposition of each number, 72 and 60 have three prime factors in common: two 2s and a 3 These go in the intersection of the Venn diagram. The highest common factor is the product of these three numbers. Check that you have correctly organised your prime factors by multiplying. The circle for 72 contains three 2s and two 3s 2 × 2 × 2 × 3 × 3 = 72 The circle for 60 contains two 2s, a 3 and a 5 2 × 2 × 3 × 5 = 60
b ξ 72 ⬭ 60 with 2, 3 (left), 2, 2, 3 (centre), 5 (right) LCM = 2 × 3 × 2 × 2 × 3 × 5 = 360	To calculate the lowest common multiple, find the product of all the numbers in the Venn diagram.

Practice (B)

1 The prime factors of 24 and 36 are shown in the Venn diagram.

   a What is the HCF of 24 and 36?

   b What is the LCM of 24 and 36?

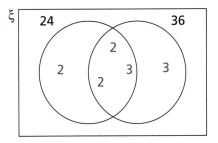

2 The prime factors of two numbers, A and B, are shown in the Venn diagram.

   a What is the value of A?

   b What is the value of B?

   c What is the HCF of A and B?

   d What is the LCM of A and B?

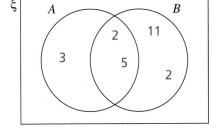

3 **a** Write 18 as a product of its prime factors.

   b Write 60 as a product of its prime factors.

   c Write the prime factors of 18 and 60 in a copy of the Venn diagram.

   d What is the HCF of 18 and 60?

   e What is the LCM of 18 and 60?

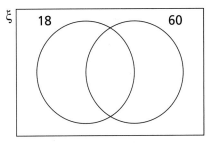

4 For each pair of numbers:

   i write each number as the product of its prime factors

   ii work out the HCF and LCM of the numbers using a Venn diagram.

     a 16 and 48   **b** 24 and 60   **c** 45 and 120

     d 40 and 75   **e** 110 and 115   **f** 96 and 132

What do you think? (B) 💭

1 Faith is thinking of two numbers, x and y.

She writes x as a product of its prime factors.

$x = 2^3 \times 3$

The HCF of x and y is 12

The LCM of x and y is 168

Write y as a product of its prime factors.

Consolidate – do you need more?

1 **a** Copy and complete the factor trees.

i 54

27

3

3

ii 126

2

3   21

3

b Write 54 and 126 as a product of their prime factors.

2 Write each of the numbers as the product of its prime factors.

a 24	**b** 36	**c** 72
d 80	**e** 112	**f** 320

3 Write each of the numbers as the product of its prime factors. Give your answers in index form.

a 24	**b** 60	**c** 84
d 125	**e** 250	**f** 360

4 The numbers have been written as products of their prime factors.

Work out the numbers.

a $3 \times 3 \times 2$	**b** $3 \times 5 \times 7$	**c** $2 \times 2 \times 11$
d $2 \times 3 \times 3 \times 5$	**e** $3 \times 5^2 \times 11$	**f** $3^3 \times 5^2$

5 The prime factors of two numbers, X and Y, are shown in the Venn diagram.

a What is the value of X?

b What is the value of Y?

c What is the HCF of X and Y?

d What is the LCM of X and Y?

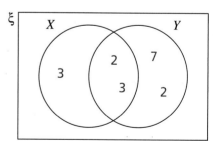

6 **a** Write 16 as a product of its prime factors.

b Write 80 as a product of its prime factors.

c Write the prime factors of 16 and 80 in a copy of the Venn diagram.

d What is the HCF of 16 and 80?

e What is the LCM of 16 and 80?

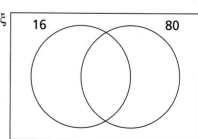

7 For each pair of numbers:

   i   write each number as the product of its prime factors

   ii work out the HCF and LCM of the numbers using a Venn diagram.

      a 21 and 56   **b** 32 and 40   **c** 55 and 100

      d 60 and 85   **e** 120 and 125   **f** 86 and 146

Stretch – can you deepen your learning?

1 $3960 = 2^3 \times 3^2 \times 5 \times 11$

Decide whether each number is a factor of 3960

Explain your answer for each.

   a 6   **b** 15   **c** 65   **d** 45   **e** 140

2 $a = 2 \times 2 \times 3 \times 3 \times 3 \times 5 \times 11 \times 13$

$b = 2^2 \times 3^2 \times 5 \times 11 \times 13$

How many times greater than b is a? Explain your answer.

3 You are given that $40 = 2^3 \times 5$

Write each of these numbers as the product of its prime factors in index form.

   a 20   **b** 80   **c** 200   **d** 400

Factors, multiples and primes: exam practice

1 Here is a list of numbers:

8   15   23   27   32   33

From the numbers in the list, write down:

(a) a number that is prime   [1 mark]

(b) a multiple of 8   [1 mark]

(c) a factor of 45   [1 mark]

2 Write down the first five prime numbers.   [1 mark]

3 Is 1 a prime number?

Explain your answer.   [1 mark]

4 List all the factors of 30   [2 marks]

5 (a) What is the lowest common multiple of 10 and 12?   [1 mark]

(b) What is the highest common factor of 10 and 12?   [1 mark]

6 Show that 3845 **is not** a prime number.   [1 mark]

7 A number is written as a product of its prime factors as $2 \times 5^2 \times 7$

Work out the number.   [1 mark]

8 A bell rings every 8 seconds.

A different bell rings every 6 seconds.

They both start ringing at the same time.

How many times in the first 60 seconds will the bells ring at the same time?   [3 marks]

9 (a) Write 120 as a product of its prime factors.   [2 marks]

(b) Work out the highest common factor of 120 and 35   [2 marks]

1–3

3–5

5 Percentages

In this block, we will cover...

5.1 Fraction, decimal and percentage equivalence

Example 1

Write the decimals as percentages.

a 0.15   **b** 0.2   **c** 1.2

Method

Solution	Commentary
a 0.15 = 15%	To convert a decimal to a pe
	0.15 × 100% = 15%
b 0.2 = 20%	It is important to remember the zero and decimal point,

5.2 Fractions of amounts

Practice (A)

1. Work out:
   a $\frac{1}{4}$ of 12   **b** $\frac{1}{3}$ of 21

   d $\frac{1}{5}$ of 60   **e** $\frac{1}{7}$ of 84

2. Work out:
   a $\frac{3}{4}$ of 12   **b** $\frac{3}{7}$ of 21

   d $\frac{5}{8}$ of 64   **e** $\frac{11}{12}$ of 72

3. Calculate:

5.3 Percentages of amounts

Consolidate – do you need more

1. Work out:
   a 50% of 56   **b** 25% of 132

2. Work out:
   a 75% of 36   **b** 70% of 400

   d 40% of 540   **e** $66\frac{2}{3}$% of 360

3. There are 24 000 fans at a rugby match. 40% of the fans support the away team.

5.4 Simple interest and compound interest

Stretch – can you deepen your le

1. A bank offers two different savings accou

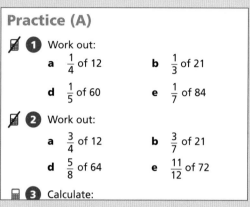

Super Saver
Simple interest
3.5% per annum

Gold
Compoun
3% pe

Ali wants to invest £12 000 in one of the a

5.5 Reverse percentages

Example 2

Beca's rent increases by 30%. After the increase,

Work out the cost of her rent before the increas

Method

Solution	C
100% + 30% = 130%	W
130%	re

100%   30%

Are you ready?

1 Copy and complete the equivalent fractions.

   a $\frac{1}{2} = \frac{\square}{100}$ **b** $\frac{1}{5} = \frac{\square}{100}$ **c** $\frac{4}{5} = \frac{\square}{100}$ **d** $\frac{21}{25} = \frac{\square}{100}$

2 Convert each fraction to a decimal.

   a $\frac{1}{10}$ **b** $\frac{7}{10}$ **c** $\frac{1}{100}$ **d** $\frac{19}{100}$

3 Write each decimal as a fraction in its simplest form.

   a 0.9   **b** 0.2   **c** 0.65   **d** 0.18

4 Convert each fraction to a decimal.

   a $\frac{1}{8}$ **b** $\frac{5}{8}$ **c** $\frac{7}{40}$ **d** $\frac{21}{24}$

5 Convert each fraction to a decimal.

   a $2\frac{3}{10}$ **b** $3\frac{23}{100}$ **c** $5\frac{1}{4}$ **d** $10\frac{3}{5}$

Using your calculator 🔲

You can convert between fractions and decimals on a calculator using the **S⇔D** button.

Simply type a fraction or a decimal into your calculator then press **S⇔D** to convert it.

Parts of a whole can be described using fractions, decimals or percentages. You can convert between these three forms since they are equivalent, meaning equal in value.

For example, to convert from a percentage to a fraction, remember that 'per cent' means 'out of a hundred'. You can therefore find an equivalent fraction with a denominator of 100.

20% $\frac{20}{100}$

$\frac{?}{100}$

Percentage ⟶ Fraction

Don't forget to simplify...

$\frac{20}{100} = \frac{2}{10} = \frac{1}{5}$

So 20% is equivalent to $\frac{1}{5}$

You can also convert from a decimal to a percentage.

Example 1

Write the decimals as percentages.

a 0.15   **b** 0.2   **c** 1.2

Method

Solution	Commentary
a $0.15 = 15\%$	To convert a decimal to a percentage, multiply the decimal by 100% $0.15 \times 100\% = 15\%$
b $0.2 = 20\%$	It is important to remember to multiply by 100%, and not just remove the zero and decimal point, as that would lead to an incorrect answer of 2% $0.2 \times 100\% = 20\%$
c $1.2 = 120\%$	1 whole is equal to 100% so you know that 1.2 as a percentage will be greater than 100% $1.2 \times 100\% = 120\%$

Example 2

Write these fractions as percentages.

a $\dfrac{3}{10}$ **b** $\dfrac{4}{25}$ **c** $\dfrac{2}{7}$

Method

Solution	Commentary
a $\dfrac{3}{10} = \dfrac{30}{100} = 30\%$ $\overset{\times 10}{\dfrac{3}{10} = \dfrac{30}{100}} = 30\%$ $\times 10$	Remember, 'per cent' means 'out of a hundred'. You can change $\dfrac{3}{10}$ to a fraction out of 100 by multiplying the numerator and the denominator by 10 $\dfrac{30}{100}$ means '30 out of 100', so it is the same as 30%. Alternatively, you could have written $\dfrac{3}{10}$ as a decimal, 0.3, and multiplied the decimal by 100% $0.3 \times 100\% = 30\%$
b $\dfrac{4}{25} = \dfrac{16}{100} = 16\%$ $\overset{\times 4}{\dfrac{4}{25} = \dfrac{16}{100}} = 16\%$ $\times 4$	Multiply the numerator and the denominator by 4, because $100 \div 25 = 4$ Alternatively, you could have converted $\dfrac{4}{25}$ to a decimal by working out $4 \div 25 = 0.16$, and multiplied the decimal by 100% $0.16 \times 100\% = 16\%$
c $\dfrac{2}{7} = 2 \div 7 = 0.285\,714\ldots$ $0.285\,714\ldots \times 100\%$ $= 28.5714\ldots\%$ $= 28.6\%$ (3 s.f.)	7 is not a factor of 100, so you could convert $\dfrac{2}{7}$ to a decimal first. You can then multiply the decimal by 100% and round the percentage to a sensible degree of accuracy.

Example 3

Write each percentage as a fraction and as a decimal.

a 91%   **b** 70%   **c** 75%

Method

Solution	Commentary
a $91\% = \dfrac{91}{100} = 0.91$	You can write any percentage as a fraction with a denominator of 100 $91\% = \dfrac{91}{100}$ $\dfrac{91}{100} = 0.91$ Check you have converted the fraction to a decimal correctly by using a calculator and dividing the numerator by the denominator. $91 \div 100 = 0.91$
b $70\% = \dfrac{7}{10} = 0.7$	$70\% = \dfrac{70}{100}$ $\dfrac{7}{10} = 0.7$ $\dfrac{70}{100}$ can be simplified to $\dfrac{7}{10}$ by dividing the numerator and denominator by 10
c $75\% = 0.75 = \dfrac{3}{4}$	$75\% = \dfrac{75}{100}$ $\dfrac{3}{4} = 0.75$ $\dfrac{75}{100}$ can be simplified to $\dfrac{3}{4}$ by dividing the numerator and denominator by 25

Practice

1 Write each decimal as a percentage.

   a 0.37   **b** 0.12   **c** 0.02   **d** 0.2   **e** 0.5   **f** 0.9

2 Write each fraction as a percentage.

   a $\dfrac{33}{100}$ **b** $\dfrac{9}{100}$ **c** $\dfrac{4}{10}$ **d** $\dfrac{1}{5}$ **e** $\dfrac{27}{50}$ **f** $\dfrac{3}{4}$

3 Convert each fraction to a percentage.

   a $\dfrac{1}{2}$ **b** $\dfrac{1}{4}$ **c** $\dfrac{1}{8}$ **d** $\dfrac{3}{8}$ **e** $\dfrac{5}{8}$ **f** $\dfrac{7}{8}$

4 Convert each fraction or decimal to a percentage.

   a 0.23   **b** 0.236   **c** $\dfrac{236}{1000}$ **d** 0.502   **e** $\dfrac{135}{1000}$ **f** $\dfrac{402}{1000}$

5 **a** Write each fraction as a decimal.

      i $\dfrac{12}{75}$ **ii** $\dfrac{33}{60}$ **iii** $\dfrac{102}{120}$

   b Write each answer as a percentage.

6 Write each percentage as a decimal.

   a 23%   **b** 98%   **c** 90%

   d 4%   **e** 9%   **f** 10%

7 Write each percentage as a fraction in its simplest form.

    a 21%   **b** 10%   **c** 50%   **d** 75%   **e** 80%   **f** 70%

8 The table shows the percentage of each make of car sold by a garage.

    a What percentage of cars sold were Ford?

    b Write your answer to part **a** as a decimal.

    c Write your answer to part **a** as a fraction in its simplest form.

Make of car	Percentage of overall sales
Jaguar	40%
BMW	8%
Ford	
Audi	27%

9 Write:

    a 10 as a percentage of 50   **b** 13 as a percentage of 20

    c 70 as a percentage of 280

What do you think? 💭

1 Huda thinks that 20% is equal to $\frac{1}{20}$

Show that Huda is incorrect.

2 Benji and Jackson completed a test.

Benji got 55% of the questions correct.

Jackson got $\frac{3}{5}$ of the questions correct.

Who did better in the test? Explain your answer.

3 $\frac{1}{4}$ of the counters in a bag are blue.

30% of the counters are red.

The rest of the counters are white.

    a What percentage of the counters are white?

    b Write the percentage of white counters as a fraction in its simplest form.

Consolidate – do you need more?

1 Write each decimal as a percentage.

    a 0.21   **b** 0.86   **c** 0.06   **d** 0.6   **e** 0.1   **f** 0.01

2 Write each fraction as a percentage.

    a $\frac{67}{100}$ **b** $\frac{7}{100}$ **c** $\frac{3}{10}$ **d** $\frac{1}{4}$ **e** $\frac{21}{25}$ **f** $\frac{4}{5}$

3 Write each fraction as a decimal and a percentage.

    a $\frac{68}{80}$ **b** $\frac{132}{240}$ **c** $\frac{13}{40}$

4 Write each percentage as a decimal.

    a 41%   **b** 99%   **c** 80%   **d** 8%   **e** 2%   **f** 30%

5 Write each percentage as a fraction in its simplest form.

    a 37%   **b** 30%   **c** 60%   **d** 25%   **e** 7%   **f** 5%

6 Copy and complete the table of equivalent fractions, decimals and percentages.

Fraction	Decimal	Percentage
$\frac{4}{5}$		
	0.73	
		45%

7 Write <, > or = to compare the fractions, decimals and percentages.

    a $\frac{1}{5}$ $\bigcirc$ 5%   **b** 0.9 $\bigcirc$ $\frac{9}{100}$ **c** 0.7 $\bigcirc$ 70%   **d** 40% $\bigcirc$ $\frac{3}{4}$

8 Write:

    a 9 as a percentage of 10   **b** 12 as a percentage of 25

    c 120 as a percentage of 320

9 Beca, Huda and Marta go for a meal.

Beca pays $\frac{2}{5}$ of the bill. Huda pays 35% of the bill.

What fraction of the bill does Marta have to pay? Write your answer in its simplest form.

10 11 children in a class wear glasses. There are 20 children in the class.

What percentage of children **do not** wear glasses?

Stretch – can you deepen your learning?

1 Write each fraction as a decimal and a percentage.

    a $\frac{1}{3}$ **b** $\frac{2}{3}$

2 Convert each percentage to a fraction. Write your answers in their simplest forms.

    a 12.5%   **b** 22.2%   **c** 99.9%

3 Copy and complete the table of equivalent fractions, decimals and percentages.

Fraction	Decimal	Percentage
	1.13	
$1\frac{1}{4}$		
		150%

4 Write a percentage to complete the comparison.

$$1.3 > \boxed{}\% > \frac{5}{4}$$

Is there more than one possible answer? Compare your answer with a partner.

Are you ready? (A)

1 What fraction of each bar is shaded?

a

b

c

d

2 Work out:

a 40 ÷ 5   b 32 ÷ 8   c 32 ÷ 4   d 84 ÷ 7

3 Work out:

a 8 × 3   b 7 × 5   c 9 × 4   d 6 × 7

4 Work out:

a 40 ÷ 5 × 3   b 49 ÷ 7 × 6   c 64 ÷ 8 × 7

Using your calculator 📱

You can work out fractions of amounts easily on a calculator by inputting the fraction using the fraction button ▣, then multiplying by the amount. For example, to work out $\frac{3}{4}$ of 20, type

---

Example

Work out:

a $\frac{1}{5}$ of 85   b $\frac{3}{5}$ of 85   c $\frac{7}{5}$ of 85

Method

Solution		Commentary
a $85 \div 5 = 17$ Therefore $\frac{1}{5}$ of 85 = 17	85 ↔ | 17 | 17 | 17 | 17 | 17 |	Find one-fifth of a number by dividing the number into five equal parts.
b $17 \times 3 = 51$ Therefore $\frac{3}{5}$ of 85 = 51	| 17 | 17 | 17 | 17 | 17 |	Three-fifths is three of the equal parts, so multiply $\frac{1}{5}$ by 3
c $17 \times 7 = 119$ Therefore $\frac{7}{5}$ of 85 = 119	| 17 | 17 | 17 | 17 | 17 | | 17 | 17 |	Seven-fifths is seven of the equal parts, so multiply $\frac{1}{5}$ by 7 The answer is greater than 85, because $\frac{7}{5}$ is greater than 1 whole.

Practice (A)

 1 Work out:

    a $\frac{1}{4}$ of 12   **b** $\frac{1}{3}$ of 21   **c** $\frac{1}{6}$ of 42

    d $\frac{1}{5}$ of 60   **e** $\frac{1}{7}$ of 84   **f** $\frac{1}{4}$ of 88

 **2** Work out:

    a $\frac{3}{4}$ of 12   **b** $\frac{3}{7}$ of 21   **c** $\frac{5}{6}$ of 42

    d $\frac{5}{8}$ of 64   **e** $\frac{11}{12}$ of 72   **f** $\frac{7}{10}$ of 130

3 Calculate:

    a $\frac{1}{2}$ of 376   **b** $\frac{1}{4}$ of 944   **c** $\frac{3}{4}$ of 944

    d $\frac{3}{8}$ of 1000   **e** $\frac{1}{6}$ of 255   **f** $\frac{3}{5}$ of 261

4 Rob has £48

He spends $\frac{1}{6}$ of his money on a book. He spends $\frac{3}{8}$ of his money on a computer game.

How much does Rob spend in total?

5 There are 36 counters in a bag.

$\frac{1}{3}$ of the counters are blue.

$\frac{7}{12}$ of the counters are red.

The rest of the counters are yellow.

How many yellow counters are there?

6 Write <, > or = to complete the comparisons.

    a $\frac{1}{5}$ of 35 $\bigcirc$ $\frac{2}{7}$ of 35   **b** $\frac{3}{4}$ of 24 $\bigcirc$ $\frac{2}{3}$ of 27

    c $\frac{8}{9}$ of 56 $\bigcirc$ $\frac{5}{6}$ of 56   **d** $\frac{5}{8}$ of 56 $\bigcirc$ $\frac{3}{10}$ of 120

7 At a match, 240 people supported the red team and 270 supported the blue team.

$\frac{2}{3}$ of the people who supported the red team were adults.

$\frac{5}{9}$ of the people who supported the blue team were adults.

The rest of the people at the match were children.

    a Did more adults support the red team or the blue team?

    b Did more children support the red team or the blue team?

8 Faith has £360 in her bank account.

She spends $\frac{3}{4}$ of the money on a new phone.

She spends $\frac{5}{6}$ of the remaining money on some new shoes.

How much money does Faith have left?

What do you think? 💭

1 The cost of a train ticket is £48

The cost of the ticket increases by $\frac{1}{4}$

The following month, the new cost of the ticket decreases by $\frac{1}{4}$

Sven says, "The cost of the ticket is now the same as it was at the beginning."

Do you agree with Sven? Explain your answer.

2 A coat costs £200

It is reduced in price by $\frac{1}{10}$ every day in a sale.

Huda has £140

How many days will Huda have to wait before she can afford the coat?

Are you ready? (B)

1 Write down a number to make each statement correct.

a There are ☐ fifths in 1 whole.

b There are ☐ eighths in 1 whole.

c There are ☐ quarters in 1 whole.

d There are ☐ halves in 1 whole.

2 Work out the missing numbers.

a $12 \div 4 =$ ☐ $4 \times$ ☐ $= 12$

b $24 \div 6 =$ ☐ $6 \times$ ☐ $= 24$

c $35 \div 5 =$ ☐   ☐ $\times 5 = 35$

What do you notice? Discuss with a partner.

3 Work out:

a $345 \div 5 \times 2$ **b** $1498 \div 7 \times 3$ **c** $1850 \div 8 \times 3$

4 Work out:

a $\frac{1}{3}$ of 27   **b** $\frac{1}{5}$ of 65   **c** $\frac{2}{3}$ of 27   **d** $\frac{5}{8}$ of 56

Example 1

Amina spends three-fifths of her money on a magazine costing £6

How much money does she have left?

Method

Solution	Commentary
$\frac{3}{5}$ = £6 Amina's money £6 spent $\frac{1}{5}$ = £6 ÷ 3 = £2	Three of the five equal parts is £6 So each one of the parts must be £6 ÷ 3
£2 £2 £2 £2 £2 amount left Amina has $\frac{2}{5}$ left, which is £2 × 2 = £4	The bar shows there are two equal parts remaining. Each equal part is £2, so Amina has 2 × £2 left.

Example 2

$\frac{2}{3}$ of a number is 80. What is $\frac{3}{5}$ of the number?

Method

Solution	Commentary
40 40 40 80 The number is 40 × 3 = 120 $\frac{1}{5}$ of the number = 120 ÷ 5 = 24 24 24 24 24 24 $\frac{3}{5}$ of the number = 24 × 3 = 72	Two of the three equal parts is 80, so each one of the parts must be 80 ÷ 2 = 40 To find $\frac{3}{5}$ of the number, multiply the result by 3

Practice (B)

1 Work out the number if $\frac{1}{5}$ of the number is 6

?

6				

2 Work out the whole if:

    a $\frac{1}{4}$ of the whole is 3   **b** $\frac{1}{8}$ of the whole is 3

    c $\frac{1}{3}$ of the whole is 12   **d** $\frac{1}{6}$ of the whole is 13

3 Work out the unit fractions of the amount.

    a If $\frac{3}{4}$ of a number is 9, what is $\frac{1}{4}$ of the number?

    b If $\frac{5}{6}$ of a number is 30, what is $\frac{1}{6}$ of the number?

    c If $\frac{4}{7}$ of a number is 32, what is $\frac{1}{7}$ of the number?

4 Work out the number if $\frac{3}{5}$ of the number is 9

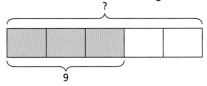

5 Work out the whole if:

    a $\frac{3}{4}$ of the whole is 9   **b** $\frac{3}{4}$ of the whole is 15

    c $\frac{2}{3}$ of the whole is 24   **d** $\frac{4}{7}$ of the whole is 16

6 Tiff spends $\frac{4}{7}$ of her money on a new game.

    The game costs £48

    a How much money did Tiff have to begin with?

    b How much money does Tiff have left?

7 Work out the following.

    a If $\frac{2}{5}$ of a number is 8, what is $\frac{1}{4}$ of the number?

    b If $\frac{2}{5}$ of a number is 8, what is $\frac{3}{4}$ of the number?

    c If $\frac{4}{5}$ of a number is 24, what is $\frac{5}{6}$ of the number?

    d If $\frac{5}{8}$ of a number is 25, what is $\frac{7}{10}$ of the number?

8 Work out the missing number.

    $\frac{3}{4}$ of 16 = $\frac{2}{3}$ of $\boxed{}$

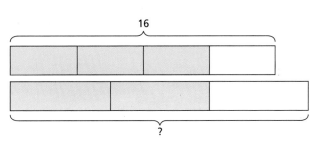

9 Work out the missing numbers.

    a $\frac{2}{5}$ of 15 = $\frac{2}{3}$ of $\boxed{}$ **b** $\frac{5}{8}$ of 32 = $\frac{4}{5}$ of $\boxed{}$ **c** $\frac{5}{7}$ of $\boxed{}$ = $\frac{7}{9}$ of 45

Consolidate – do you need more?

1 Work out:

a $\frac{1}{6}$ of 48

b $\frac{1}{9}$ of 63

c $\frac{2}{3}$ of 18

d $\frac{3}{8}$ of 48

e $\frac{7}{10}$ of 120

f $\frac{5}{12}$ of 96

2 Calculate:

a $\frac{1}{3}$ of 477

b $\frac{1}{8}$ of 992

c $\frac{3}{5}$ of 885

d $\frac{3}{7}$ of 2548

e $\frac{1}{4}$ of 657

f $\frac{4}{5}$ of 373

3 Junaid has £72

He spends $\frac{3}{8}$ of his money on a birthday present for his sister.

He spends $\frac{1}{3}$ of his money on some clothes.

How much does Junaid spend in total?

4 There are 32 children in a class.

$\frac{1}{2}$ of the children walk to school.

$\frac{3}{8}$ are driven to school.

The rest of the children cycle to school.

How many children cycle to school?

5 Write <, > or = to complete the comparisons.

a $\frac{1}{4}$ of 48 $\bigcirc$ $\frac{5}{6}$ of 12

b $\frac{5}{8}$ of 64 $\bigcirc$ $\frac{2}{3}$ of 45

c $\frac{3}{7}$ of 63 $\bigcirc$ $\frac{4}{9}$ of 63

d $\frac{7}{10}$ of 240 $\bigcirc$ $\frac{4}{5}$ of 190

6 Work out the whole if:

a $\frac{1}{3}$ of the whole is 7

b $\frac{3}{5}$ of the whole is 18

c $\frac{3}{4}$ of the whole is 27

d $\frac{2}{7}$ of the whole is 16

7 Faith spends $\frac{5}{8}$ of her money on some trainers.

The trainers cost £55

How much money does Faith have left?

8 Work out the following.

a If $\frac{3}{4}$ of a number is 12, what is $\frac{5}{8}$ of the number?

b If $\frac{2}{3}$ of a number is 10, what is $\frac{3}{5}$ of the number?

c If $\frac{5}{6}$ of a number is 20, what is $\frac{11}{12}$ of the number?

d If $\frac{4}{9}$ of a number is 36, what is $\frac{2}{3}$ of the number?

9 Ed and Flo both buy a copy of the same book.

Ed has £48 and spends $\frac{3}{8}$ of his money on the book.

Flo spends $\frac{2}{3}$ of her money on the book.

How much money did Flo have before she bought the book?

Stretch – can you deepen your learning?

1 Amina's monthly pay is £240

Her monthly pay is increased by $\frac{5}{12}$

How much is Amina's new monthly pay?

2 Ali's monthly pay is £220

His monthly pay is increased by $\frac{3}{11}$

Ali's new monthly pay is now $\frac{4}{5}$ of Abdullah's monthly pay.

How much is Abdullah's monthly pay?

3 Bev is paid £10.50 per hour.

She works 4 hours each day for three days in a week.

She spends $\frac{1}{4}$ of her weekly pay on a meal.

She spends $\frac{2}{5}$ of her weekly pay on travel costs.

How much money does Bev have left?

Are you ready? (A)

1 Convert each percentage to a fraction in its simplest form.

    a 50%   **b** 25%   **c** 10%   **d** 1%   **e** 75%   **f** 20%

2 Work out:

    a $\frac{1}{2}$ of 400   **b** $\frac{1}{4}$ of 400   **c** $\frac{1}{10}$ of 400

    d $\frac{1}{100}$ of 400   **e** $\frac{3}{4}$ of 400   **f** $\frac{1}{5}$ of 400

3 Convert each percentage to a fraction.

    a $33\frac{1}{3}$%   **b** $66\frac{2}{3}$%

4 What percentage is represented by each bar model?

    a

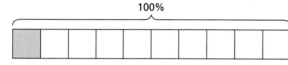

    b

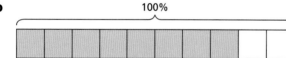

    c

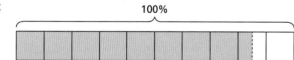

You need to know how to work out a percentage of an amount. Chapter 5.2 showed how to calculate fractions of amounts and you can use this knowledge alongside the equivalence of simple fractions and percentages.

For example, 25% is equivalent to $\frac{1}{4}$

To find 25% of 300, you would find one-quarter of 300, or 300 ÷ 4, which equals 75

So 25% of 300 is 75

It sometimes helps to break a percentage down into simpler percentages. For example, you could find 45% of an amount by calculating with simpler percentages such as:

- 25% + 10% + 10%
- 10% × 4 + 5%
- 50% − 5%
- 1% × 45

Example 1

Work out:

a 50% of 210   **b** 25% of 210   **c** 75% of 210

d $33\frac{1}{3}$% of 210   **e** $66\frac{2}{3}$% of 210

Method

Solution	Commentary
a 50% of 210 = 210 ÷ 2 = 105	Use the knowledge that 50% is $\frac{1}{2}$, and to find $\frac{1}{2}$ you divide by 2
b 25% of 210 = 210 ÷ 4 = 52.5	Use the knowledge that 25% is $\frac{1}{4}$, and to find $\frac{1}{4}$ you divide by 4 You could also find 25% by 'halving and halving again'. 210 ÷ 2 = 105, 105 ÷ 2 = 52.5
c 75% of 210 = 25% of 210 × 3 = 52.5 × 3 = 157.5	Use the knowledge that 75% is equal to 3 × 25% or $\frac{3}{4}$, so multiply the answer to part **b** by 3
d $33\frac{1}{3}$% of 210 = 210 ÷ 3 = 70	Use the knowledge that $33\frac{1}{3}$% is $\frac{1}{3}$, and to find $\frac{1}{3}$ you divide by 3
e $66\frac{2}{3}$% of 210 = $33\frac{1}{3}$% of 210 × 2 = 70 × 2 = 140	Use the knowledge that $66\frac{2}{3}$% is equal to 2 × $33\frac{1}{3}$% or $\frac{2}{3}$, so multiply the answer to part **d** by 2

Example 2

Work out:

a 10% of 80   **b** 30% of 80   **c** 15% of 44

d 95% of 70   **e** 27% of 3000

Method

Solution	Commentary
a 10% of 80 = 80 ÷ 10 = 8	Use the knowledge that 10% is $\frac{1}{10}$, and to find $\frac{1}{10}$ you divide by 10
b 30% of 80 = 3 × 10% of 80 = 3 × 8 = 24	Use the answer to part **a** and the fact that 30% is 3 × 10%.

c 10% of 44 = 4.4 5% of 44 = 2.2 15% of 44 = 4.4 + 2.2 = 6.6	Partition 15% into 10% and 5%. Divide by 10 to find 10%. Halve your answer to find 5%. Add your answers to find 15%.

d 10% of 70 = 7
 5% of 70 = 3.5
 90% of 70 = 63
 95% of 70 = 63 + 3.5
        = 66.5
	Divide by 10 to find 10%.
	 Halve your answer to find 5%.
	 90% of 70 = 9 × 10% of 70
	        = 9 × 7 = 63
	 Add your answers to find 95%.

e 1% of 3000 = 30 20% of 3000 = 600 7% of 3000 = 210 27% of 3000 = 600 + 210 = 810	Find 1% by dividing by 100 Find 20% either by multiplying 1% (30) by 20 or by finding 10% (300) and multiplying by 2 Find 7% by multiplying 1% (30) by 7 Add your answers to find 27%.

Practice (A)

1 Work out:

    a 50% of 24   **b** 50% of 300   **c** 25% of 24   **d** 25% of 360

    e 10% of 240   **f** 10% of 24   **g** 1% of 300   **h** 1% of 240

2 Work out:

    a 75% of 24   **b** 75% of 92   **c** $33\frac{1}{3}$% of 60

    d $33\frac{1}{3}$% of 240   **e** $66\frac{2}{3}$% of 60   **f** $66\frac{2}{3}$% of 240

3 Work out:

    a 10% of 60   **b** 30% of 60   **c** 90% of 60

    d 70% of 60   **e** 20% of 60   **f** 1% of 60

4 Work out:

    a 10% of 200   **b** 1% of 200   **c** 5% of 200

    d 15% of 200   **e** 45% of 200   **f** 47% of 200

5 There are 1240 students in a school. 30% of the students are left-handed.

    a How many students are left-handed?

    b How many students are right-handed?

6 600 people attend a football match.

55% of the people are supporters of the home team.

A ticket to the match costs £12

What is the total cost of tickets for supporters of the home team?

7 Write <, > or = to complete the comparisons.

a 50% of 36 ◯ 25% of 76

b 10% of 270 ◯ $33\frac{1}{3}$% of 93

c 1% of 2680 ◯ 5% of 520

d 75% of 480 ◯ 90% of 400

8 Mario has to pay 25% tax on all his earnings over £12 000

He earns £18 500

Work out how much tax Mario has to pay.

What do you think? (A)

1 Amina and Beca are working out 95% of 500

Amina says, "I am going to find 10% and multiply it by 9 then find 5% and add them together."

Beca says, "I am going to find 5% and subtract it from the whole."

a Use both methods to calculate 95% of 500

Discuss with a partner which method you prefer.

b How many other ways can you find to calculate 95%?

Are you ready? (B)

1 Convert the percentages to decimals.

a 63%   **b** 89%   **c** 91%   **d** 4%   **e** 40%   **f** 109%

2 Work out:

a 1200 × 0.46   **b** 1200 × 0.4   **c** 1200 × 0.06

d 0.86 × 48   **e** 0.8 × 48   **f** 0.06 × 48

Using your calculator

You can use your calculator to work out percentages of amounts.

To work out 27% of 85, type **2** **7** **SHIFT** **%** **×** **8** **5**

Although you can work out most percentages mentally by building up from multiples of 10% and 1%, it can be more efficient to use a calculator when numbers become more difficult.

Using decimal and percentage equivalences, you can find a percentage of an amount by multiplying by a decimal. For example, 87% is equal to 0.87 so to find 87% of 132 you can multiply 132 by 0.87 using a calculator, giving 114.84. In this case, 0.87 is the **decimal multiplier**.

Example 1

Work out 16% of 6500

Method

Solution	Commentary
16% of 6500 = 0.16 × 6500 = 1040	First convert 16% to a decimal: $16\% = \frac{16}{100} = 0.16$ Then enter 0.16 × 6500 into your calculator.

Example 2

Ron scores 15 marks on a maths test.

The maximum mark available on the test was 25

What percentage did Ron score?

Method

Solution	Commentary
15 ÷ 25 = 0.6 0.6 × 100 = 60% Ron scored 60%.	To work out the percentage, divide the marks Ron scored by the total number of marks available. Write the decimal as a percentage by multiplying by 100

Practice (B)

1 Write down the decimal multiplier you would use to find each percentage.

    a 54%   **b** 76%   **c** 8%   **d** 60%   **e** 42.8%   **f** 104%

2 Work out:

    a 32% of 27 500   **b** 54% of 420   **c** 27% of 508

    d 87% of 68   **e** 93% of 88   **f** 8% of 3.6

3 Write each quantity as a percentage of the other.

    a 12 marks as a percentage of 20 marks

    b £8 as a percentage of £10

    c 7 days as a percentage of 10 days

    d 45p as a percentage of £1

4 A laptop costs £1200

    In a sale, it is reduced to 78% of the original price.

    What is the cost of the laptop in the sale?

5 A restaurant adds a 13% service charge on top of the price of the food.

The cost of Tiff's food is £45

What service charge will be added to her bill?

6 A company donates 8% of its annual profit to charity.

a The company's annual profit last year was £823 000

Calculate the amount that the company donated to charity last year.

b This year, the company's profit was £48 000 greater than last year.

Calculate the amount that the company donated to charity this year.

7 Filipo, Ed and Jakub are competing in a long jump competition.

Filipo jumps exactly 4 m.

Ed jumps 75% of the distance Filipo jumped.

Jakub jumps 81% of Ed's distance.

How far did Jakub jump?

Are you ready? (C)

1 Convert each percentage to a decimal.

    a 76%   **b** 70%   **c** 6%

2 Convert each percentage to a decimal.

    a 104%   **b** 125%   **c** 180%

3 71% of Earth's surface is covered by water.

What percentage of Earth's surface is **not** covered by water?

4 Work out the missing decimal in each calculation.

    a 0.7 + ☐ = 1   **b** 0.82 + ☐ = 1   **c** 0.09 + ☐ = 1

You can use a range of different methods to calculate percentage increases and decreases, including the use of multipliers. Bar models can help to visualise percentage increases and decreases.

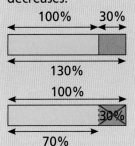

This bar model shows a percentage **increase**.

This bar model shows a percentage **decrease**.

Example 1

A shop has a sale. All the prices are reduced by 15%.

Work out the sale price of an item that normally costs £280

Method A – Calculation and subtraction

Solution	Commentary
10% of £280 = £28 5% of £280 = £14 15% of £280 = £28 + £14 = £42	You can work out the **reduction** in price using the non-calculator methods you looked at earlier in the chapter, or by entering 0.15 × £280 on a calculator.
Sale price = £280 – £42 = £238	You can then find the sale price by subtracting the reduction from the normal price.

Method B – The multiplier method

Solution	Commentary
100% 15% 85% 100% – 15% = 85% 85% as a decimal is 0.85 0.85 × £280 = £238	If you decrease an amount by 15%, there will be 85% of the original amount remaining. Use your knowledge of converting between percentages and decimals. To find 85%, you can multiply by 0.85 0.85 is the decimal **multiplier** used to find a 15% reduction.

Example 2

Emily earns £34 000. This year, her salary will increase by 7%.

Work out Emily's new salary.

Method A

Solution	Commentary
1% of £34 000 £34 000 ÷ 100 = £340 7% of £34 000 = £340 × 7 = £2380	You can find 1% of any amount by dividing it by 100 To find 7% of an amount, work out 1% and then multiply this by 7. Or use your calculator and enter 0.07 × 34 000
100% 7% 34 000 2380 Emily's new salary = £34 000 + £2380 = £36 380	Find Emily's new salary by adding the increase of £2380 to her original salary.

Method B

Solution	Commentary
100% 7% 107%	
100% + 7% = 107% 107% as a decimal is 1.07 1.07 × £34 000 = £36 380	When an amount is increased by 7%, it becomes 107% of the original amount. To find 107% of an amount, you can multiply by 1.07

Practice (C)

1 The cost of a television was £480

In a sale, the price is reduced by 20%.

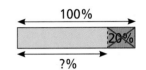

  a What percentage of the original price of the television is the sale price?

  b Write the decimal equivalent of your answer to part **a**.

  c Use the decimal equivalent from part **b** to find the sale price of the television.

2 Write down the decimal multiplier you would use to **reduce** a quantity by each percentage.

  a 40%   **b** 25%   **c** 75%   **d** 51%   **e** 7%   **f** 95%

3 A pair of trainers costs £80. The trainers are reduced by 28% in a sale.

Rhys says, "The cost of the trainers is £80 × 0.28 = £22.4"

  a Explain the two mistakes that Rhys has made.

  b What is the sale price of the trainers?

4 A house was valued at £241 000 last year.

The value of the house has decreased by 16% this year.

What is the value of the house now?

5 A company employs 240 people.

The number of employees increases by 20%.

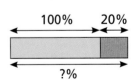

  a What percentage of the original number of employees does the company have now?

  b Write the decimal equivalent of your answer to part **a**.

  c Use the decimal equivalent to find how many employees the company has now.

6 Write down the decimal multiplier you would use to **increase** a quantity by each percentage.

  a 40%   **b** 25%   **c** 7%   **d** 51%   **e** 7.5%   **f** 50.2%

7 A stadium can hold 64 000 people.

The capacity of the stadium is increased by 26%.

How many people can the stadium hold now?

8 The decimal multiplier 1.23 works out an increase of 23%.

Describe the effect of these decimal multipliers.

   a 1.08   **b** 0.87   **c** 1.4   **d** 0.7   **e** 1.65   **f** 0.04

What do you think? (C)

1 In January, the cost of a train ticket is £12.50

In April, the train company increases the price of the ticket by 20%.

In November, the price of the ticket is reduced by 20%.

Benji says, "The price of a ticket will return to £12.50 in November."

Do you agree with Benji? Explain your answer.

Consolidate – do you need more?

1 Work out:

   a 50% of 56   **b** 25% of 132   **c** 10% of 2050   **d** 1% of 4000

2 Work out:

   a 75% of 36   **b** 70% of 400   **c** $33\frac{1}{3}$% of 960

   d 40% of 540   **e** $66\frac{2}{3}$% of 360   **f** 60% of 60

3 There are 24 000 fans at a rugby match.

40% of the fans support the away team.

   a How many fans support the away team?

   b How many fans support the home team?

4 Write each quantity as a percentage of the other.

   a 92 marks as a percentage of 100 marks

   b 18 m as a percentage of 20 m

   c 130 as a percentage of 200

   d 40 cm as a percentage of 1 m

5 Work out:

   a 27% of 34 600   **b** 46% of 360   **c** 76% of 410

   d 93% of 82   **e** 81% of 64   **f** 6% of 2068

6 A car cost £8400 two years ago.

The value of the car is now 65% of the original price.

What is the value of the car now?

7 Write down the decimal multiplier you would use to change a quantity by each percentage.

 a Increase a quantity by: **i** 20% **ii** 45% **iii** 9%

 b Decrease a quantity by: **i** 37% **ii** 8% **iii** 60%

8 A house was valued at £164 500 last year.

The value of the house has increased by 21% this year.

What is the value of the house now?

9 A music website had 31 800 subscribers.

The number of subscribers has decreased by 32%.

How many subscribers does the website have now?

10 In January, the cost of a flight to New York was £650

In June, the price increased by 30%.

In November, the price was reduced by 15%.

What was the cost of the flight in November?

Stretch – can you deepen your learning?

1 The length and the width of a rectangle are both increased by 10%.

Bev says, "The area will also increase by 10%."

Do you agree with Bev? Explain your answer.

2 A ball is dropped from a height of 4 metres.

After each bounce, it reaches 75% of the previous height.

After how many bounces would it reach a height less than 1 metre?

3 Ali works three days a week from 8:30 am to 4:00 pm.

He is paid £18 per hour.

At the end of the week, he is given a 20% bonus.

He is taxed 25% of all his earnings.

How much money does Ali earn in one week after tax?

Are you ready? (A)

1 Work out:

   a 20% of 90   **b** 35% of 80   **c** 2% of 300   **d** 27% of 800

2 Write down the decimal multiplier equivalent to:

   a a 30% increase   **b** a 40% decrease   **c** an 8% increase

   d a 4% decrease   **e** a 27% decrease   **f** a 32% increase.

3 Increase 600 by:

   a 10%   **b** 6%   **c** 18%   **d** 70%

4 Work out:

   a £24 × 4   **b** £15 × 5   **c** £96 ÷ 3   **d** £124 ÷ 4

Using your calculator 📱

You can use the power key on your calculator to work out repeated multiplications. For example, if you want to multiply a quantity by 1.05, and then 1.05 again, and then 1.05 again, you could multiply it by 1.05^3

On a calculator, you would use the key that looks like this, $x^■$.
Enter **1** **.** **0** **5** first, then $x^■$, then **3** . You should get 1.157625

Money invested in a savings account *earns* **interest**, which can be paid monthly or **annually**. You *pay* interest on money borrowed from a bank or other source of finance. The rate of interest is often given **per annum**, which means every year.

In this chapter, you will learn about two different types of interest: simple and compound.

Simple interest is paid on savings but does not earn interest itself. For example, if you had £100 and earned £5 interest in one year, you would now have £105. If you did not **deposit** or withdraw any money from the account, in the following year, you would earn interest only on the £100, not the extra £5

Example

Work out the interest earned on £750 invested for 3 years in an account earning 4% simple interest per year.

> The initial amount invested is called the **principal**.

Method

Solution	Commentary
£750 × 0.04 = £30.00 in 1 year	First find 4% of £750. Remember, to find 4% of an amount you can multiply by 0.04
£30.00 × 3 = £90.00 The interest earned would be £90.00	The amount of interest earned does not change (unless money is added/removed), so multiply by 3 because the money is invested for 3 years.

Practice (A)

1 Samira invests £400 in an account paying simple interest at 5% per annum.

    a Work out 5% of £400

    b Work out how much money Samira will have in her account after:

        i 1 year   **ii** 2 years   **iii** 3 years.

2 Rob invests £800 in an account earning simple interest at 3% per annum.

    Work out the total amount of money in Rob's account after 4 years.

3 Tiff invests £3500 in an account earning 8% simple interest per annum.

    Work out how much interest Tiff will have earned after 5 years.

4 A bank pays 4% simple interest per annum.

    a Rhys invests £2400 for 3 years.

        Work out the total amount of money in Rhys's account after 3 years.

    b Junaid invests £3200 for 3 years.

        Work out how much interest Junaid will earn.

5 Faith buys a new television costing £1200. She pays a 15% deposit.

    She pays 5% simple interest per annum on the remaining amount.

    Faith pays off the amount over one year.

    How much will Faith pay in total?

6 Zach buys a new television costing £1800. He pays a 10% deposit.

    He pays 5% simple interest per annum on the remaining amount.

    Zach pays off the amount over 2 years.

    How much will Zach pay in total?

What do you think? (A) 💭

1 Ed invests £2800 in an account earning 7.5% simple interest per annum.

Work out the total amount of money in Ed's account after 3 years.

2 Amina invests £600 for 4 years in a bank account paying simple interest.

At the end of 4 years, Amina has earned £48 interest.

Work out the annual rate of interest.

3 Lida invests £500 for 3 years in a bank account paying simple interest.

At the end of 3 years, Amina has £590 in her account.

Work out the annual rate of interest.

Are you ready? (B)

1 Write down the decimal multiplier equivalent to:

a a 60% increase **b** a 10% decrease **c** a 3% increase

d a 2% decrease **e** a 19% decrease **f** a 21% increase.

2 Work out the following. Round your answers to 2 decimal places.

a 1.2^2 **b** 1.05^3 **c** 1.42^5

3 Work out the following. Round your answers to 2 decimal places.

a $1.2^2 \times £200$ **b** $£400 \times 1.05^3$ **c** $1.42^5 \times £100$

4 Kath invests £2200 in an account earning 7% simple interest per annum.

Work out how much interest Kath will have earned after 4 years.

Simple interest is not the only kind of interest.

Compound interest is added to your savings each year and you will then earn interest on the total amount in future years. For example, if you have £100 and earned £5 in interest in 1 year, you would now have £105. The following year, you would earn interest on the full £105 in your account, i.e. £105 × 1.05 = £110.25. This is more than would be in the account if it was earning simple interest over the same period (£110).

Note that interest (compound or simple) applies to borrowing as well as saving. So if you take out a loan, you pay interest rather than earning it.

Also, interest isn't always calculated annually, but could be monthly, or even daily. Many banks will use compound interest calculated on a daily basis.

Example

Mario invests £1900 in a savings account earning 4% compound interest per year.

How much money will he have in his account after 3 years?

Method

Solution	Commentary
100% + 4% = 104% = 1.04	First work out the required decimal multiplier.
After 1 year: £1900 × 1.04 = £1976	Remember, the multiplier for a 4% increase is 1.04
After 2 years: £1976 × 1.04 = £2055.04 After 3 years: £2055.04 × 1.04 = £2137.2416	Then work out the total value of Mario's savings after each year for 3 years. Multiply the total amount in the account at the end of each year by 1.04
This rounds to £2137.24	Round your answer to 2 decimal places for money.

On your calculator, you would enter 1900 × 1.04 × 1.04 × 1.04. Alternatively, you could enter 1900×1.04^3

Practice (B)

1. Mario invests £400 at 2% compound interest per annum.

   a Work out 2% of £400

   b How much money will Mario have in his account after 1 year?

   c How much interest will he earn in the second year?

   d How much money will Mario have in his account after 2 years?

2. Emily invests £2600 at 5% compound interest per annum.

   How much money will Emily have in her account after 3 years?

3. Filipo invests £3000 at 8% compound interest per annum.

   a How much will Filipo have in his account after 2 years?

   b How much will Filipo have in his account after 3 years?

   c How much will Filipo have in his account after 6 years?

4. Rob buys a car worth £20 000

   The value of the car **depreciates** by 12% each year for the first 4 years.

   What is the value of the car at the end of the 4 years?

   > Depreciation is a loss in the value of goods over time.

5 Abdullah buys an antique painting for £3800

Each year the value of the painting **appreciates** by 13%.

What will the value of the painting be after 10 years?

> Appreciation is a gain in the value of goods over time.

6 When a ball is dropped, it bounces to 70% of the previous height each time.

The ball is dropped from 3 metres.

Work out the height the ball will reach after the third bounce.

7 Samira invests £800 at 7.5% compound interest per annum.

Each year Samira takes £100 out of the account.

How much will Samira have in her account after 2 years?

What do you think? (B)

1 A bath contains 90 litres of water.

When the plug is pulled out, 15% of the water is lost every minute.

After how many minutes will the volume of water be less than half the original volume?

2 Chloe invests £4000 at 12% compound interest rate per annum.

How many years will it take Chloe to reach £6000, assuming that she does not withdraw any money?

Consolidate – do you need more?

1 Huda invests £800 in an account paying simple interest at 8% per annum.

  a Work out 8% of £800

  b Work out how much money Huda will have in her account after:

   i 1 year **ii** 3 years **iii** 6 years.

2 Benji invests £1400 in an account earning simple interest at 4% per annum.

Work out the total amount of money in Benji's account after 5 years.

3 Kath invests £2400 in an account earning 12% simple interest per annum.

Work out how much interest Kath will have earned after 4 years.

4 Marta buys a new television costing £1500. She pays a 20% deposit.

She pays 9% simple interest per annum on the remaining amount.

Marta pays off the amount over 1 year.

How much will Marta pay in total?

5 Lida invests £3000 at 10% compound interest per annum.

How much money will Lida have in her account after 6 years?

6 Sven invests £500 at 6% compound interest per annum.

    a How much will Sven have in his account after 2 years?

    b How much will Sven have in his account after 4 years?

    c How much will Sven have in his account after 8 years?

7 Zach buys a car worth £16 500

The value of the car depreciates by 11% each year for the first 3 years.

What is the value of the car at the end of the 3 years?

8 Jakub invests £300 in an account earning 7% compound interest per annum.

How long will it take for Jakub to double the amount of money in the account?

Stretch – can you deepen your learning?

1 A bank offers two different savings accounts.

Super Saver

Simple interest
3.5% per annum

Gold Saver

Compound interest
3% per annum

Ali wants to invest £12 000 in one of the accounts.

    a Which account will earn more interest if he invests for 2 years?

    b Which account will earn more interest if he invests for 12 years?

2 Beca invests £500 in an account that earns 5% interest per annum.

After 4 years Beca has £600 in the account.

Does Beca's account earn simple or compound interest?

3 A bank is offering a new savings account.

> Earn 8% simple interest per annum for the first two years and
> 2.5% compound interest per annum for any additional years

Junaid invests £6400 in the account.

How much will Junaid have in his account after 7 years?

Are you ready?

1 Convert each percentage to a decimal.

    a 70%   **b** 6%   **c** 82%   **d** 74.5%   **e** 130%   **f** 108.5%

2 Write the decimal multiplier equivalent to:

    a an increase by 80%   **b** a decrease by 80%   **c** a decrease by 5%

    d an increase by 3%   **e** an increase by 13.8%   **f** a decrease by 8.5%.

3 Work out the missing numbers.

    a $36 \div 1.2 = \boxed{}$ $1.2 \times \boxed{} = 36$

    b $36 \div 0.8 = \boxed{}$ $0.8 \times \boxed{} = 36$

    c $609 \div 1.45 = \boxed{}$ $1.45 \times \boxed{} = 609$

    d $50.4 \div 0.84 = \boxed{}$ $0.84 \times \boxed{} = 50.4$

When a value has been increased by 20%, the new value is 120% of the **original value**.

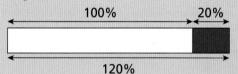

When a value has been decreased by 20%, the new value is 80% of the original value.

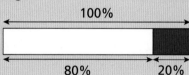

Sometimes you will want to find the original amount prior to a percentage change.

These are often called **reverse percentage** problems. They can be solved using calculator and non-calculator methods.

Example 1

The price of a calculator is reduced by 20% to £6.40. Work out its original price.

Method A

Solution	Commentary
100% − 20% = 80% 100% [bar model: 80% with 20% crossed out] 80%	Work out what percentage of the original price is left after the reduction.
80% of the original price is £6.40 10% would be £6.40 ÷ 8 = £0.80	80% = £6.40 so 10% = 80% ÷ 8
The original price was 10 × £0.80 = £8.00	The original price is 100%, which is 10 × 10%.

Method B

Solution	Commentary
80% = £6.40 ÷ 80 ⤸ ÷ 80 1% = £0.08 × 100 ⤸ × 100 100% = £8	£6.40 is equal to 80% of the original price. You need to find 100%. To do this, divide by 80 to find 1% and then multiply by 100 to find 100%.
So the original price was £8.00	State the final answer.

Example 2

Beca's rent increases by 30%. After the increase, her rent is £780 per month.

Work out the cost of her rent before the increase.

Method

Solution	Commentary
100% + 30% = 130% 130% 100%　30%	Work out what percentage of the original rent the increased rent is.
130% of the original rent is £780 So 10% would be £780 ÷ 13 = £60	130% = £780 so 10% = 130% ÷ 13
The original rent was 10 × £60 = £600 You could have worked this out by finding 1% and then multiplying by 100 to find 100%. 130% = £780 ÷ 130 ⤸ ÷ 130 1% = £6 × 100 ⤸ × 100 100% = £600	The original rent is 100%, which is 10 × 10%. State the final answer.

Example 3

Flo earns 3% interest on her savings each year. At the end of the first year, she has a total of £1648 in her account.

Work out how much she deposited before the interest was added.

Method

Solution	Commentary
100% + 3% = 103%	Work out the new percentage after the increase of 3%
	Write this as the equivalent decimal multiplier.
103% = 1.03	The decimal multiplier for a 3% increase is 1.03
 $103\% = £1648$ ÷ 103 ÷ 103 $1\% = £16$ × 100 × 100 $100\% = £1600$	
So the original amount was £1600	State the final answer.

Practice

 1   25% of a number is 12

25%	25%	25%	25%

12

Work out:

a   100% of the number   **b**   50% of the number   **c**   75% of the number.

 2   30% of a number is 24

10%	10%	10%	10%	10%	10%	10%	10%	10%	10%

24

Work out:

a   10% of the number   **b**   100% of the number

c   70% of the number   **d**   15% of the number.

 3   110% of a number is 44

100%

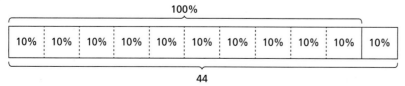

10%	10%	10%	10%	10%	10%	10%	10%	10%	10%	10%

44

Work out:

a   100% of the number   **b**   40% of the number

c   150% of the number   **d**   85% of the number.

4 40% of the students in Year 10 wear glasses.

88 students wear glasses.

How many students are there in Year 10?

5 After a 30% pay rise, Benji earns £390 per week.

Work out Benji's weekly salary before the pay rise.

6 Bev buys some trainers costing £69

This cost includes tax at a rate of 15%.

What is the price of the trainers without tax?

7 Huda buys a tablet device in a sale.

The sale price is 80% of the original price.

The sale price is £460

Work out the original price of the tablet.

8 The price of a jumper in a 15% off sale is £40.80

What was the original price of the jumper?

9 Sven is given a 13% pay rise.

His new salary is £22 600

What was Sven's salary before the pay rise?

What do you think?

1 Copy and complete the table.

Original amount	Percentage change	New amount
32	Increase by 10%	
40		37.6
	Decrease by 40%	66
12	Decrease by 17%	
16		25.6

Consolidate – do you need more?

1 20% of a number is 34

20%	20%	20%	20%	20%

34

Work out:

a   100% of the number   b   10% of the number   c   60% of the number.

2 75% of a number is 69

25%	25%	25%	25%

69

Work out:

a 100% of the number

b 50% of the number

c 10% of the number

d 35% of the number.

3 120% of a number is 96

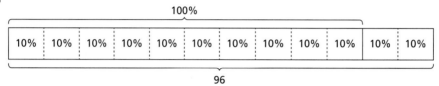

Work out:

a 100% of the number

b 30% of the number

c 170% of the number

d 45% of the number.

4 60% of the counters in a bag are red.

There are 42 red counters in the bag.

How many counters are there in total in the bag?

5 After a 30% reduction in price, the cost of a car is £14 000

What was the value of the car before the reduction?

6 A restaurant adds a 15% service charge on top of the bill.

Rob pays £26

What was Rob's bill before the service charge was added?

7 Samira sold her house for £225 000

The value of her house had reduced by 7.5% since she bought it.

Work out how much Samira paid for her house.

8 Ali bought a car two years ago.

The value of the car decreased by 20% in the first year.

The value decreased by a further 15% in the second year.

Now the car is valued at £12 580

How much did Ali pay for the car?

9 The price of a train ticket has increased by 12.5%.

The new price of the train ticket is £13.50

What was the price of the train ticket before the increase?

Stretch – can you deepen your learning?

1 The cost of a train ticket from Manchester to Leeds decreases by 20%.

The cost of a train ticket from Manchester to Liverpool increases by 20%.

The train tickets for both journeys now cost £12

What was the original difference in price between a ticket from Manchester to Leeds and a ticket from Manchester to Liverpool?

2 The cost of a concert ticket has increased by £24

The ticket now costs £106

What is the percentage increase?

3 In the year after being built, the value of a house increased by 20% to £144 000

The following year, the value of the house increased by the same amount of money.

What was the value of the house at the end of that two-year period?

4 In the year after being built, the value of a house increased by 15% to £184 000

The following year, the value of the house increased by the same percentage.

What was the difference in the value of the house between the start and end of that two-year period?

Percentages: exam practice

1 Complete a copy of this table of equivalent values.

Decimal	0.25		0.75
Fraction		$\frac{1}{2}$	
Percentage	25%		

[3 marks]

2 Work out 10% of 40 [1 mark]

3 Work out $\frac{3}{5}$ of 50 [2 marks]

4 Sophie earns £2000 per month.

She saves 30%.

How much money does she save each month? [2 marks]

5 What is $\frac{1}{8}$ as a decimal? [2 marks]

6 A hotel has 60 guests.

45 of the guests are adults.

(a) Work out 45 out of 60 as a percentage. [2 marks]

20% of the 45 adults wear glasses.

(b) Write the number of adults who wear glasses as a fraction of the 60 guests.

Give your answer in its simplest form. [3 marks]

7 Ron invests £4500 for 3 years in a savings account.

He is paid 4% per annum compound interest.

How much does Ron have in his savings account after 3 years? [3 marks]

8 In a sale, normal prices are reduced by 15%.

The sale price of a tablet is £102

Work out the normal price of the tablet. [3 marks]

6 Accuracy

In this block, we will cover...

6.1 Decimal places

Example 1

Round each number to the given degree of accu

a 4.68 to 1 decimal place **b** 7.

Method

Solution	Comment
a 4.68	You can c
4.6 4.65 4.7	is closer to
4.68 = 4.7 to 1 decimal place	to 1 decim
b 7.397	

6.2 Significant figures

Practice

1   **a**   In each number, which digit is the 1st s

    i   2451   **ii**   4 809 173

   b   What is the place value of each of thos

2   **a**   In each number, which digit is the 2nd

    i   2451   **ii**   4 809 173

   b   What is the place value of each of thos

3   Write each number correct to 1 significant

   a   457   **b**   28.87   **c**   103.21

6.3 Estimation

Consolidate – do you need more

1   **a**   Copy and complete these statements t

    37 rounded to 1 significant figure = ☐

    88 rounded to 1 significant figure = ☐

    So 37 × 88 ≈ ☐ × ☐ = ☐

    Is your answer an overestimate or und

   b   Work out 37 × 88

2   **a**   Estimate the answer to each calculatio

    i   5276 – 891   **ii**   652 + 193 + 728

6.4 Limits of accuracy

Stretch – can you deepen your le

1   There are 40 sweets in a bag correct to 1 s

   4 of the sweets are given out.

   a   What is the greatest number of sweets

   b   What is the smallest number of sweets

2   There are 60 pencils in box A correct to 1 s

   There are 90 pencils in box B rounded to t

   a   Work out the greatest possible differe

   b   Work out the smallest possible total of

6.1 Decimal places

Are you ready?

1 What is the value of the digit 5 in each number?

    a 2560   **b** 25.6   **c** 2.56   **d** 0.256

2 Round 3752 to: **a** the nearest 10   **b** the nearest 100   **c** the nearest 1000

3 Write down the middle value of each pair of numbers.

4 a Estimate the number each arrow is pointing to.

    b For each number found in part **a**, write down which integer it is closest to.

Decimal numbers can be rounded to a number of decimal places. This can make them easier to understand, and can also make calculations simpler to perform.

Rounding to different degrees of accuracy, such as 'to the nearest integer', 'to 1 decimal place' or 'to 2 d.p.' will affect how precise a number is.

For example, 7.893 71 rounded to 2 decimal places is 7.89

Rounded to 3 decimal places, it is 7.894

Example 1

Round each number to the given degree of accuracy.

a 4.68 to 1 decimal place   **b** 7.397 to 2 decimal places

Method

Solution	Commentary
a 4.68 4.6 4.65 4.7 4.68 = 4.7 to 1 decimal place	You can clearly see from the number line that 4.68 is closer to 4.7 than to 4.6, therefore 4.68 rounded to 1 decimal place is 4.7
b 7.397 7.39 7.395 7.40 7.397 = 7.40 to 2 decimal places	The next number with 2 decimal places after 7.39 is 7.40

Example 2

$\pi = 3.14159\ldots$

Round π to: **a** 1 decimal place   **b** 2 decimal places   **c** 3 decimal places.

Method

Solution	Commentary
a 3.14159… 3.1 3.15 3.2 π = 3.1 to 1 decimal place	π is greater than 3.1, so use a number line from 3.1 to 3.2 You can see that π is closer to 3.1 than 3.2
b 3.14159… 3.14 3.145 3.15 π = 3.14 to 2 decimal places	π is greater than 3.14, so use a number line from 3.14 to 3.15 You can see that π is closer to 3.14 than 3.15
c 3.14159… 3.141 3.1415 3.142 π = 3.142 to 3 decimal places	π is greater than 3.141, so use a number line from 3.141 to 3.142 You can see that π is closer to 3.142 than 3.141

Practice

1 **a** Estimate the number each arrow is pointing to.

5.0 5.1 5.2 5.3 5.4 5.5 5.6 5.7 5.8 5.9 6.0

b Round each number found in part **a** to 1 decimal place.

2 **a** Explain the mistake that has been made in each working out.

i 7.86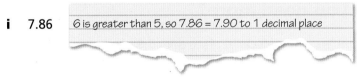
6 is greater than 5, so 7.86 = 7.90 to 1 decimal place

ii 1.326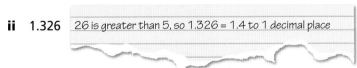
26 is greater than 5, so 1.326 = 1.4 to 1 decimal place

iii 4.24
4 is less than 5, so 4.24 = 4.1 to 1 decimal place

b Round each number in part **a** to 1 decimal place.

3 Round each number to 1 decimal place.

a	2.34	**b**	10.48	**c**	7.15	**d**	0.13
e	4.261	**f**	6.03	**g**	6.95	**h**	9.972

4 **a** Estimate the number each arrow is pointing to.

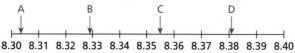

b Round each number found in part **a** to 2 decimal places.

5 Round each number to 2 decimal places.

a	2.341	**b**	2.345	**c**	8.026	**d**	0.149
e	0.194	**f**	0.196	**g**	10.055	**h**	9.976

6 Copy and complete the table.

	Rounded to nearest integer	Rounded to 1 d.p.	Rounded to 2 d.p.	Rounded to 3 d.p.
5.1783				
11.2056				
3.0095				
8.9999				

7 Work out the following. Give each answer correct to 2 decimal places.

a 2.47×4.518 **b** $8.43 \div 3.65$ **c** $\sqrt{6.84 + 2.91}$ **d** 3.147^4

8 The price of a litre of petrol is 153.5p

a Round the price to the nearest penny.

b Round the price to the nearest pound.

Consolidate – do you need more?

1 **a** Which two numbers with 1 decimal place does 7.81 lie between?

| 7 and 8 | | 7.8 and 7.9 | | 7.1 and 7.2 |

b What is 7.81 rounded to 1 decimal place?

c Round the numbers below to 1 decimal place.

i 7.18   **ii** 7.45   **iii** 7.96   **iv** 7.01

2 Round each number to 1 decimal place.

a	3.12	**b**	20.83	**c**	6.45	**d**	0.74
e	5.737	**f**	4.04	**g**	11.97	**h**	0.981

3 **a** Which two numbers with 2 decimal places does 4.836 lie between?

| 4.8 and 4.9 | 4 and 5 | 4.83 and 4.84 |

b What is 4.836 rounded to 2 decimal places?

c Round the numbers below to 2 decimal places.

i 4.861   **ii** 4.825   **iii** 4.803   **iv** 4.897

4 Round each number to 2 decimal places.

a 4.561   **b** 1.972   **c** 9.036   **d** 0.155

e 0.298   **f** 0.291   **g** 100.005   **h** 19.995

5 Copy and complete the table.

	Rounded to nearest integer	Rounded to 1 d.p.	Rounded to 2 d.p.	Rounded to 3 d.p.
9.908 98				
105.1055				
$\frac{3}{8}$				
π				

6 Work out the following. Give each answer correct to 2 decimal places.

a 3.147×2.59 **b** $12.654 \div 2.13$ **c** $\sqrt{9.876 - 1.001}$ **d** 4.203^5

7 The distance from town A to town B is 14.7518 km.

Round the distance to the nearest:

a kilometre   **b** metre   **c** 10 metres   **d** 100 metres.

Stretch – can you deepen your learning?

1 A number rounds to 8.5 to 1 decimal place.

The same number rounds to 8 to the nearest integer.

What could the number be? Write three possible answers.

2 Amina makes a number using four of these cards. Each digit is different.

| 0 | 1 | 2 | 3 | 4 | 5 | 6 | 7 | 8 | 9 |

Amina's number rounds to 10.4 to 1 decimal place.

Write down all the possible numbers she could have made.

3 Two websites correctly report the population of France to different degrees of accuracy.

Website A: 67.7 million   **Website B:** 67. ☐ ☐ million

a What could the missing digits from Website B be?

b Write your answer as an integer.

Are you ready?

1 Round each number to the nearest 10

   a 256.06   **b** 25.6   **c** 2.56   **d** 0.256

2 Round each number to the nearest integer.

   a 256.06   **b** 25.6   **c** 2.56   **d** 0.256

3 Round each number to 1 decimal place.

   a 256.06   **b** 25.62   **c** 2.56   **d** 0.256

4 Round each number to 3 decimal places.

   a 0.256256   **b** 25.61256   **c** 2.5605   **d** 0.256999

5 Beca is working out a calculation.

Here is her calculator answer:

Write Beca's answer to:

   a the nearest hundred   **b** the nearest integer   **c** 2 decimal places.

You can round, say, to the nearest 10 or 100, or to the nearest integer. You can also round to a number of decimal places. There is another type of rounding which is particularly important in ensuring that answers are given to an appropriate accuracy, and this is called **significant figures**.

Significant figures are the most important digits in a number and they give you an idea of its size.

For example, the number of views of an online video might be 198 753 762, which could be rounded to 200 000 000 to 1 significant figure.

Similarly, the measurement 0.000 058 197 metres could be rounded to 0.000 06 to 1 significant figure. It can be helpful to round to significant figures rather than decimal places in situations like this, where if you round to 3 decimal places the answer would be 0.000 metres.

Example 1

Write down the 1st significant figure in each of these numbers.

a 972 863   **b** 0.007 9324

Method

Solution	Commentary
a 9 is the first significant figure	1st non-zero digit from left $\downarrow$ 972 863 $\uparrow$ 2nd significant figure
b 7 is the first significant figure	1st non-zero digit from left. $\downarrow$ 0.0079324 $\uparrow$ $\uparrow$ 2nd significant figure 3rd significant figure

Example 2

Round the number 26 719 to 1 significant figure.

Method A

Solution	Commentary
②6 719 The 1st significant figure is 2 This is in the 10 000s column, so rounding to 1 significant figure will be the same as rounding to the nearest 10 000 $\downarrow$ 20 000 ———— 25 000 ———— 30 000 26 719 is closer to 30 000, so 26 719 is 30 000 to 1 s.f.	First identify the 1st significant figure. Notice the approximation must be of the same **order of magnitude** as the number itself – it must represent the original number.

Method B

Solution	Commentary
26 719 The first non-zero digit is the 2, therefore 2 is the first significant figure.	Locate the significant figure for the degree of accuracy required. The first non-zero digit from the left is the first significant figure.
2⸽6 719 6 is more than 5	Look at the next digit to the right. Is it 5 or more?
As 6 is more than 5, round up. Adding 1 to the 2 gives 3. The first significant figure represents ten thousands, so four zeros must be added to make it the correct size. 26 719 is 30 000 to 1 s.f.	If it is 5 or more, round up by adding 1 to the previous digit. If it is less than 5, round by keeping the previous digit the same. If the degree of accuracy is 10 or more, fill in zeros to make the number the correct size.

Example 3

Round 0.006 9563 to 2 significant figures.

Method

Solution	Commentary
0.006 9¦563	The 1st significant figure is 6
	The 2nd significant figure is 9
	You need to look at the number after 9
	This is 5 so round 9 to the next ten-thousandth.
To 2 significant figures 0.006 9563 is 0.0070	So to 2 significant figures 0.006 9563 is 0.0070
	Remember to include the zero after the 7 to show that the number has been rounded to 2 significant figures and not just 1 significant figure.

Practice

1 **a** In each number, which digit is the 1st significant figure?

    i 2451   **ii** 4809 173   **iii** 0.004 361   **iv** 6.000 184

  b What is the place value of each of those digits?

2 **a** In each number, which digit is the 2nd significant figure?

    i 2451   **ii** 4809 173   **iii** 0.004 361   **iv** 6.000 184

  b What is the place value of each of those digits?

3 Write each number correct to 1 significant figure.

  a 457   **b** 28.87   **c** 103.21   **d** 0.086   **e** 0.005 09   **f** 0.097

4 Round each number to 2 significant figures.

  a 452   **b** 30.91   **c** 403.28   **d** 0.0235   **e** 0.003 08   **f** 0.0796

5 There were 97 503 fans at a football match.

  a Round the number of fans to 1 significant figure.

  b At the match, 10 853 scarves were sold.

    Write the number of scarves sold to 2 significant figures.

6 Copy and complete the table.

	Rounded to 1 s.f.	Rounded to 2 s.f.	Rounded to 3 s.f.
23 763			
1 056 198			
0.2076			
0.000 023 55			

7 Work out the following. Give each answer correct to 2 significant figures.

  a 1425 × 809

  b 14.261 ÷ 1.85

  c $\sqrt{2.45 + 3.784}$

  d 0.013^3

What do you think? 💭

1 Sven is rounding the number 3.04 to 2 significant figures.

He says, "The answer is 3"

Do you agree with Sven? Explain your answer.

2 A box contains 200 nails correct to 1 significant figure.

  a What is the lowest possible number of nails in the box?

  b What is the greatest possible number of nails in the box?

3 The number of people who watched a TV programme was 240 000 rounded to 2 significant figures.

  a What is the greatest possible number of people who watched the TV programme?

  b What is the lowest possible number of people who watched the TV programme?

Consolidate – do you need more?

1 **a** In each number, which digit is the 1st significant figure?

  i 3589   **ii** 2 309 865   **iii** 0.020 56   **iv** 1.050 817

  b In each number above, which digit is the 3rd significant figure?

2 Write each number to 1 significant figure.

  a 261   **b** 35.12   **c** 205.89   **d** 0.019   **e** 0.009 07   **f** 0.092

3 Round each number to 2 significant figures.

  a 261   **b** 40.56   **c** 908.16   **d** 0.0506   **e** 0.006 907   **f** 0.0996

4 A student measures the length of a table and obtains a measurement of 4.58 metres.

Round the measurement to 2 significant figures.

5 The mass of a small object is 0.008 63 kilograms.

Round the measurement to 1 significant figure.

6 A scientist measures the temperature of a liquid and records it as 24.95°C.

Write the temperature correct to 2 significant figures.

7 The volume of a gas is measured as 0.006 715 litres.

Round this measurement to 3 significant figures.

8 Copy and complete the table.

	Rounded to 1 s.f.	Rounded to 2 s.f.	Rounded to 3 s.f.
105 672			
23 091 124			
0.3089			
0.000 093 95			

9 Work out the following. Give each answer correct to 2 significant figures.

a $1.3^2 + 2.6^2$

b $\dfrac{3.4 \times 4.5}{6.43 - 2.67}$

c $\sqrt{6.35 - 3.194}$

d 0.0206^4

Stretch – can you deepen your learning?

1 The length and the width of the rectangle have been rounded to 1 significant figure.

4 cm

6 cm

What is the smallest possible area of the rectangle?

2 Samira is rounding 0.3519 to 1 significant figure and also to 1 decimal place.

She says, "The answer to both is 0.4. When you round decimal numbers to 1 significant figure and 1 decimal place, the answer is always the same."

Do you agree with Samira? Explain your answer.

3 A number rounded to 1 significant figure is 30 000

The same number rounded to 2 significant figures is 30 000

The same number rounded to 3 significant figures is 30 000

What could the number be?

6.3 Estimation

Are you ready?

1 Round each number to 1 significant figure.

   a 3289   **b** 417   **c** 182   **d** 45   **e** 0.24   **f** 0.58

2 Work out:

   a 400 + 200   **b** 3000 − 50   **c** 0.2 + 0.6   **d** 3000 + 400 + 50 + 200

3 Work out:

   a 200 × 50   **b** 400 ÷ 200   **c** 0.2 × 0.6   **d** 50 × 0.2

Estimating means to work out something close to the correct answer. It is easy to make mistakes when working out calculations either mentally, using a written method or using a calculator, and estimation is a great way to sense-check your answers. Rounding numbers, especially to 1 significant figure, will give estimated numbers that you can mentally calculate with.

An estimate is often called an **approximation**.

Example 1

Estimate the answer to each calculation. Then work out the actual answer.

   a 5320 + 996   **b** 627 × 6

Method

Solution	Commentary
a Estimate: 5000 + 1000 = 6000	5320 rounded to 1 significant figure is 5000
	996 rounded to 1 significant figure is 1000
	The total of 5000 and 1000 is 6000, therefore 5320 + 996 ≈ 6000

Actual answer:
 5320 + 996 = 5320 + 1000 − 4
         = 6320 − 4
         = 6316
	The actual answer should be greater than 6000 because 5320 > 5000, and even though 996 < 1000, it is only 4 less.
	 First add 1000 and then subtract 4 from your answer.
	 6316 is close to 6000 but greater, therefore the estimate supports your answer.

b Estimate: 600 × 6 = 3600	627 rounded to 1 significant figure is 600 so 627 × 6 ≈ 3600
Actual answer: 627 × 3 × 2 = 1881 × 2 = 3762	The actual answer should be greater than 3600 as 627 > 600 6 = 3 × 2 so use factors to break the calculation down. The actual answer is 3762, and the estimate supports this.

Example 2

A Christmas song was streamed 913 837 times in December.

Estimate the mean number of times the song was streamed per day in the month.

Method

Solution	Commentary
913 837 = 900 000 to 1 significant figure	Round both numbers to 1 significant figure.
	9ᵎ13 837 1 is less than 5, so you round to 900 000
31 = 30 to 1 significant figure	There are 31 days in December.
	3ᵎ1 1 is less than 5, so you round to 30
900 000 ÷ 30 = 30 000	The calculation for the exact mean is 913 837 ÷ 31

Example 3

Estimate the answer to $\dfrac{4.27 \times 17.65}{0.209}$

Method

Solution	Commentary
$\dfrac{4.27 \times 17.65}{0.209} \approx \dfrac{4 \times 20}{0.2}$	Round each number to 1 significant figure. The symbol ≈ means 'approximately equal to'.
$\dfrac{4 \times 20}{0.2} = \dfrac{80}{0.2} = \dfrac{800}{2} = 400$	You can multiply the numerator and denominator of the fraction by 10 and not change its value.
$\dfrac{4.27 \times 17.65}{0.209} \approx 400$	The answer to the calculation is approximately 400

Practice

1 Choose the most appropriate estimate for each multiplication.

Select from the four options given.

a	**37 × 62**	30 × 60	40 × 70	40 × 60	35 × 65
b	**224 × 486**	220 × 480	200 × 400	300 × 500	200 × 500
c	**82 × 314**	80 × 300	100 × 300	80 × 310	80 × 320
d	**4.5 × 18**	4 × 20	5 × 20	5 × 18	0 × 20

2 Copy and complete these statements to estimate the answer to 41 × 75

41 rounded to 1 significant figure = ☐

75 rounded to 1 significant figure = ☐

So 41 × 75 ≈ ☐ × ☐ = ☐

3 Copy and complete these statements to estimate the answer to 0.875 + 0.015 926

0.875 rounded to 1 significant figure = ☐

0.015 926 rounded to 1 significant figure = ☐

So 0.875 + 0.015 926 ≈ ☐ + ☐ = ☐

4 **a** Round each number to 1 significant figure.

    i 956   **ii** 231

  b Use your answers to part **a** to estimate the answer to each calculation.

    i 956 − 231   **ii** 231 + 956   **iii** 956 × 231   **iv** 956 ÷ 231

5 **a** Estimate the answer to each calculation.

    i 3624 − 986   **ii** 231 + 425 + 519   **iii** 372 × 581   **iv** 2184 ÷ 529

  b Work out each calculation above, rounding the answer to 2 decimal places if appropriate.

    Were your estimates greater or less than the actual answers?

    Why do you think that is the case? Discuss with a partner.

6 **a** Estimate the answer to each calculation.

    i 4.817 − 1.938   **ii** 0.32 + 0.55 + 0.78   **iii** 6.28 × 0.34   **iv** 8.42 ÷ 0.24

  b Work out each calculation above, rounding the answer to 2 decimal places where appropriate.

7 28 students are going on a school trip.

The school buys each student an apple and a carton of juice.

  a Estimate the total cost the school pays for the items.

  b Work out the total cost the school pays for the items.

8 Estimate the answer to each calculation.

  a $21.8 + \dfrac{48.1}{5.2}$ **b** $\dfrac{21.8}{5.2} + 48.1$ **c** $\dfrac{21.8 + 48.1}{5.2}$ **d** $\dfrac{21.8 + 48.1}{0.52}$

What do you think? 💭

1 Ali and Junaid are estimating the answer to 18 × 287

Ali says, "I am going to round each number to the nearest hundred."

Junaid says, "I am going to round each number to 1 significant figure."

Whose method do you think is more appropriate? Explain your reasons.

Consolidate – do you need more?

1 a Copy and complete these statements to estimate the answer to 37 × 88

37 rounded to 1 significant figure = ☐

88 rounded to 1 significant figure = ☐

So 37 × 88 ≈ ☐ × ☐ = ☐

Is your answer an overestimate or underestimate?

b Work out 37 × 88

2 a Estimate the answer to each calculation.

    i 5276 − 891   **ii** 652 + 193 + 728   **iii** 513 × 684   **iv** 4092 ÷ 365

b Work out each calculation above, rounding the answer to 2 decimal places where appropriate.

3 a Estimate the answer to each calculation.

    i 8.197 − 2.719   **ii** 0.465 + 0.134 + 0.861

    iii 9.08 × 0.44   **iv** 9.29 ÷ 0.31

b Work out each calculation above, rounding the answer to 2 decimal places where appropriate.

4 Estimate the answer to: $\dfrac{322 \times 74}{98 \times 31}$

5 Estimate the answer to each calculation.

    a $78.5 + \dfrac{15.8}{3.92}$ **b** $\dfrac{78.5}{3.92} + 15.8$ **c** $\dfrac{78.5 + 15.8}{3.92}$

6 A pair of jeans costs £42

A T-shirt costs £18.50

A pair of trainers costs £54

A pair of socks costs £4.20

Estimate the total cost of buying 2 pairs of jeans, 4 T-shirts, 1 pair of trainers and 5 pairs of socks.

7 A gardening company is laying a patio.

Each square metre requires 38 bricks.

The total area of the garden is 92 square metres.

a Estimate the total number of bricks required for the patio.

b The bricks are sold in packs of 500. Each pack costs £142

Estimate the total cost of the bricks to make the patio.

8 A car's fuel efficiency is approximately 18.1 miles per litre of petrol.

The car is travelling a total distance of 364 miles on a journey.

Estimate the number of litres of petrol the car will need for the journey.

9 A school buys 38 new tablets that cost £295.45 each and 21 new screens that cost £1038.40 each.

Estimate the total cost.

Stretch – can you deepen your learning?

1 Estimate the area of each shape.

a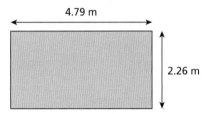

4.79 m

2.26 m

b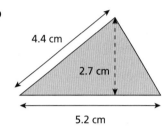

4.4 cm

2.7 cm

5.2 cm

c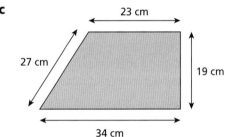

23 cm

27 cm

19 cm

34 cm

d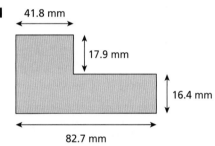

41.8 mm

17.9 mm

16.4 mm

82.7 mm

2 Estimate the mean mass of the boxes.

2.8 kg   7.15 kg   6.4 kg   9.82 kg   1.71 kg

3 Samira has not got her calculator. She needs to estimate the square root of 29

Copy and complete her working out.

29 > 25, so $\sqrt{29}$ > ☐   29 < 36, so $\sqrt{29}$ < ☐

So $\sqrt{29}$ is between ☐ and ☐

4 a Work out two consecutive integers between which each of these roots lie.

'Consecutive' means next to each other.

For each pair of integers, suggest which integer you think the root is closer to.

   i $\sqrt{72}$ **ii** $\sqrt{51}$ **iii** $\sqrt{94}$ **iv** $\sqrt{139}$

b Work out the square root of each number. Round your answers to 2 decimal places.

   i $\sqrt{72}$ **ii** $\sqrt{51}$ **iii** $\sqrt{94}$ **iv** $\sqrt{139}$

Are you ready?

1 Round each number to the nearest hundred.

    a 374   **b** 347   **c** 3347

2 Round each number to 2 decimal places.

    a 4.167   **b** 0.035   **c** 10.2091

3 Round each number to 1 significant figure.

    a 4074   **b** 0.474   **c** 40.74

4 Which of the numbers below round to 2000 to 1 significant figure?

    219   1512   2764   2018   2399

5 Which of the numbers below round to 0.4 to 1 significant figure?

    4.41   0.39   0.415   4.39   0.351

When you work with rounded numbers, your answers are less accurate than working with exact numbers. Similarly, when you measure a length, your measurement is less accurate than the actual length. It is useful to know how far measurements could be from the actual values.

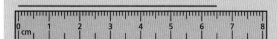

Error intervals are the limits of accuracy when a number has been rounded or **truncated**. They are the range of possible values that a number or a quantity could actually be.

Here are some possible error intervals: $35 \leqslant x < 45$ $15.2\,\text{cm} \leqslant h < 15.3\,\text{cm}$

Example

A stick is measured as 17 cm long to the nearest centimetre.

a Which of these values could be the actual length of the stick?

| 16.36 cm | 16.52 cm | 16.601 cm | 17.01 cm | 17.49 cm | 17.5 cm |

b Write the error interval for the length, l, of the stick in centimetres.

Method

Solution	Commentary
a 16.52 cm, 16.601 cm, 17.01 cm, 17.49 cm	All of these numbers round to 17 to the nearest integer. Notice that 16.36 would round to 16 and 17.5 would round to 18
b 16.5 cm $\leqslant l <$ 17.5 cm	You read this as 'l is greater than or equal to 16.5 and less than 17.5'
	17.5 is **not** included in the error interval as 17.5 rounds to 18 to the nearest integer.
	16.5 **is** included in the error interval as 16.5 rounds to 17 to the nearest integer.

Practice

1 **a** Which of the numbers below round to 5 to the nearest integer?

| 5.2 | 4.51 | 5.5 | 5.09 | 4.897 | 4.499 |

b Which is the correct error interval for a number, x, that rounds to 5 to the nearest integer?

| $4.5 < x \leqslant 5.5$ | $4.5 < x < 5.5$ | $4.5 \leqslant x \leqslant 5.5$ | $4.5 \leqslant x < 5.5$ |

2 A number, a, is given as 13 rounded to the nearest integer.

Copy and complete the error interval for a.

$12.5 \leqslant a < \boxed{}$

3 **a** Which of the numbers below round to 3.2 to 1 decimal place?

| 3.1 | 3.09 | 3.15 | 3.25 | 3.167 | 3.249 |

b Which is the correct error interval for a number, y, that rounds to 3.2 to 1 decimal place?

| $3.15 < y < 3.25$ | $3.15 \leqslant y < 3.25$ | $3.15 \leqslant y \leqslant 3.25$ | $3.15 < y \leqslant 3.25$ |

4 A number, b, is given as 22.6 rounded to 1 decimal place.

Copy and complete the error interval for b.

$\boxed{} \leqslant b < \boxed{}$

5 The number of people on a train is 80, correct to the nearest ten.

   a Write the lowest possible number of people on the train.

   b Write the greatest possible number of people on the train.

6 Mario is thinking of a number with 2 decimal places.

He says, "When I round my number to the nearest integer, the answer is 7"

   a Write the smallest possible number Mario could be thinking of.

   b Write the greatest possible number Mario could be thinking of.

7 The length of a fence, f, is given as 60 m.

Write the error interval for f if the measurement is:

   a correct to the nearest metre

   b correct to 1 significant figure.

8 A number, n, is 500 when rounded.

Write the error interval for n if it has been rounded to:

   a 1 significant figure   **b** 2 significant figures   **c** 3 significant figures.

Consolidate – do you need more?

1 **a** Which of the numbers below round to 6000 to 1 significant figure?

      5498   6013.5   6499   5500.5   6500

   b Which is the correct error interval for a number, x, that rounds to 6000 to 1 significant figure?

      $5500 \leqslant x \leqslant 6500$ $5500 \geqslant x > 6500$ $5500 < x \leqslant 6500$ $5500 \leqslant x < 6500$

2 **a** Which of the numbers below round to 0.25 to 2 decimal places?

      0.2   0.251   0.15   0.245   0.2549   0.255

   b Which is the correct error interval for a number, y, that rounds to 0.25 to 2 decimal places?

      $0.245 < y < 0.255$ $0.245 \leqslant y < 0.255$ $0.245 \leqslant y \leqslant 0.255$ $0.245 \geqslant y < 0.255$

3 A number, c, is given as 40 rounded to the nearest ten.

Copy and complete the error interval for c.

   $\boxed{} \leqslant c < \boxed{}$

4 The population of a country is 68 000 000 correct to 2 significant figures.

   a What is the smallest possible population of the country?

   b What is the greatest possible population of the country?

5 Faith is thinking of a number with 3 decimal places.

She says, "When I round my number to 1 significant figure, the answer is 0.8"

a Write the smallest possible number Faith could be thinking of.

b Write the greatest possible number Faith could be thinking of.

6 The mass of a box, b, is given as 10 kg.

Write the error interval for b if the measurement is:

a correct to the nearest kilogram

b correct to 1 significant figure.

Stretch – can you deepen your learning?

1 There are 40 sweets in a bag correct to 1 significant figure.

4 of the sweets are given out.

a What is the greatest number of sweets that could be left in the bag?

b What is the smallest number of sweets that could be left in the bag?

2 There are 60 pencils in box A correct to 1 significant figure.

There are 90 pencils in box B rounded to the nearest ten.

a Work out the greatest possible difference in the number of pencils in each box.

b Work out the smallest possible total of pencils.

3 A ticket to the theatre costs £28

The number of people attending the theatre is 300 to the nearest hundred.

What is the difference between the greatest and smallest possible amount of money the theatre could have raised from ticket sales?

Accuracy: exam practice

1 Round each number to the nearest 100   **1–3**

   (a) 479   **[1 mark]**

   (b) 10 630   **[1 mark]**

   (c) 3012   **[1 mark]**

2 Round 31.652 to:

   (a) 1 decimal place   **[1 mark]**

   (b) 2 significant figures   **[1 mark]**

3 Round 3.849 to 1 significant figure.   **[1 mark]**

4 Use your calculator to work out the value of: **3–5**

$$\frac{7.95 + \sqrt{8.74}}{2.03 \times 1.49}$$

   (a) Write down all the figures on your calculator display.   **[2 marks]**

   (b) Write down your answer to part **(a)** correct to 3 significant figures.   **[1 mark]**

5 Work out an estimate for:

$$\frac{3112 \times 13}{98}$$ **[3 marks]**

6 Work out an estimate for:

$$\frac{43 \times 12}{0.51}$$ **[3 marks]**

7 The mass of a sack of potatoes is 25 kg, correct to the nearest kilogram.

   (a) Write down the smallest possible mass of the sack of potatoes.   **[1 mark]**

   (b) Write down the largest possible mass of the sack of potatoes.   **[1 mark]**

Number: exam practice

1 **(a)** Work out the sum of 548 and 3082 [1 mark]

(b) Work out the difference between 983 and 47 [1 mark]

(c) Work out the product of 31 and 56 [3 marks]

2 Write $\frac{11}{20}$ as a decimal. [2 marks]

3 **(a)** Work out $0.65^2 \times \sqrt{43}$

Write down all the figures on your calculator display. [2 marks]

(b) Round your answer to part **(a)** to 2 significant figures. [1 mark]

4 232 people were on a train.

95 people got off.

57 people got on.

How many people are now on the train? [2 marks]

5 Work out 15% of £24 [2 marks]

6 Write 67% as a fraction. [1 mark]

7 Amir runs 3 km.

How many metres does Amir run? [1 mark]

8 At 7 pm the temperature is –1°C.

By 10 pm the temperature has fallen by 3°C.

What is the temperature at 10 pm? [1 mark]

1–3

9 Amy has £230 in the bank.

She receives 2.5% simple interest on the money.

How much interest does she receive? **[2 marks]**

10 At a school, there are 120 students in Year 11

$\frac{1}{4}$ of the students study art.

20% of the students study dance.

The rest of the students study PE.

How many students study PE? **[3 marks]**

11 Work out:

(a) −2 × 3 **[1 mark]**

(b) −24 ÷ −6 **[1 mark]**

12 Two glasses contain water.

A 500 ml glass is 30% full.

A 400 ml glass is half full.

Which glass contains less water? **[3 marks]**

13 Write 250 as a product of its prime factors.

Give your answer in index form. **[3 marks]**

14 Work out $1\frac{2}{3} + 2\frac{4}{5}$

Write your answer as a mixed number. **[3 marks]**

15 Write 27 million in standard form. **[2 marks]**

16 The population of a city is 190 000 correct to 3 significant figures.

(a) Write down the lowest possible population. **[1 mark]**

(b) Write down the greatest possible population. **[1 mark]**

17 A car is bought for £12 000

Its value depreciates by 15% each year for the first three years.

What is the value of the car at the end of the three years? **[3 marks]**

7 Understanding algebra

In this block, we will cover...

7.1 Notation and simplifying

Example 2

Pencils cost w pence and pens cost x pence.

Write an expression for the cost of:

a 4 pencils   **b** 7 pens   **c** 3 p

Method

Solution	Commentary
a $4w$	Pencils cost w pence so 4 pencils as $4w$.
b $7x$	Pens cost x pence so 7 pens cost
c $3w + 11x$	Use + to represent 'and' in the q

7.2 Index laws

Practice (A)

1. Simplify the expressions.

   a $g^2 \times g^3$ **b** $h^5 \times h^5$

   d $m^2 \times m^3 \times m^4$ **e** $n^2 \times n^5 \times n$

2. Simplify:

   a $b^5 \div b^3$ **b** $c^8 \div c^2$

   d $e^9 \div e^3$ **e** $\dfrac{f^7}{f^2}$

3. Simplify, leaving your answers in index for

   a $(h^3)^2$ **b** $(k^2)^5$

7.3 Substitution

Consolidate – do you need more

1. Work out the value of each of these expre

   a $d + 3$ **b** $\dfrac{d}{3}$ **c** $3d$

2. Substitute $y = 15$ to find the value of each

   a $3y + 5$ **b** $3y - 5$ **c** $\dfrac{y}{3} +$

3. Substitute $x = 12$ and $y = 30$ to find the val

   a $x + y$ **b** $y - x$ **c** xy

   e $\dfrac{x}{y}$ **f** $3x - y$ **g** $y -$

7.4 Algebraic structures

Stretch – can you deepen your le

1. Write a simplified expression for the area

   a   **b**

   k cm $3g$

2. Given the identity $3(x + 2y) + 4(5x + by) \equiv a$

Are you ready? (A)

1   Write each term without mathematical operations.

    a $4 \times y$
    b $y \times 4$
    c $4 \div y$

    d $y \div 4$
    e $y \times y$
    f $a \times y$

2   Write each of these as fractions.

    a One-half
    b One-third
    c Three-quarters

Algebra involves using letters and numbers. You need to be familiar with some of the language associated with algebra.

A **variable** is an unknown quantity, usually depicted by a letter such as x or y.

A **term** is a number or a variable, or a product of both, such as:

7 $4a$ x^2 $-7a$ $3a$ $5xy$

In the term $-7a$ above, a is the variable and -7 is called the **coefficient**. $-7a$ means $-7 \times a$.

In algebra, the multiplication sign is not shown. So $5xy$ means $5 \times x \times y$.

An **expression** is a collection of terms separated by plus or minus signs, such as:

$-x + y$ $3m - 7$ $4a - 3a + 18$ $\frac{p}{q} - 3z$

In the expression $\frac{p}{q} - 3z$ above, $\frac{p}{q}$ is how you write $p \div q$ using algebra.

Algebra tiles can be used to help build terms and expressions.

This example shows the expression $4x - 2y + 2x^2 + 1$

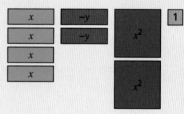

Terms are **like** if they have the same letter(s) and same power(s). For example, $12ab$ and $-7ab$ are like terms but $3x$ and x^3 are not like terms.

You can **simplify** expressions (make them simpler) by adding or subtracting the coefficients of any like terms.

Example 1

Represent each statement using objects or models and using symbols.

    a 2 more than a number
             b 3 less than a number

    c 1 more than double a number
       d double '1 less than a number'

Method

Solution	Commentary
a Let a cube represent the number. Let the number be k $k + 2$	You are adding 2 ones to the number.
b $k - 3$	Use a different representation for −1 than for +1
c $2k + 1$	For double the number, use two cubes. Remember you write $2k$, not $2 \times k$ or $k \times 2$
d $2k - 2$	One less than a number is Doubling gives two cubes and two −1s.

Example 2

Pencils cost w pence and pens cost x pence.

Write an expression for the cost of:

a 4 pencils **b** 7 pens **c** 3 pencils and 11 pens

Method

Solution	Commentary
a $4w$	Pencils cost w pence so 4 pencils cost 4 lots of w pence. You can write this as $4w$.
b $7x$	Pens cost x pence so 7 pens cost 7 lots of x pence. You can write this as $7x$.
c $3w + 11x$	Use + to represent 'and' in the question. Remember, you can't mix up variables. For example, it is incorrect to write this as $14wx$.

Practice (A)

1 p is a number.

Write an expression for:

a 6 more than p **b** 3 less than p **c** double p

d one-third of p **e** 2 less than half of p **f** double '5 more than p'.

2 Oranges cost m pence and apples cost r pence.

a Write an expression for the cost of:

   i 2 oranges   **ii** 5 apples   **iii** 2 oranges and 5 apples.

b Write an expression for the difference between the cost of 5 apples and 3 oranges.

3 **a** Amina is x years old.

Write an expression for how old she will be:

   i in 3 years' time   **ii** in k years' time   **iii** w years from now.

b Write an expression for Amina's age 6 years ago.

4 Mario gets £b pocket money each month. He saves £10 and spends the rest.

Write expressions for:

a how much money Mario spends each month

b the total amount of money he spends each year.

Mario's friend, Ed, gets half as much pocket money as him.

c Write an expression for how much pocket money Ed gets every month.

5 Write expressions for:

a the sum of h and g **b** 3 more than the sum of h and g

c the product of h and g **d** the quotient of h and g.

> The quotient is the result obtained by dividing one quantity by another.

Are you ready? (B)

1 Simplify:

a $y + y$ **b** $2y + y$ **c** $2y - y$ **d** $y + y + y$ y and $1y$ are the same.

2 Work out:

a $3 - 4$ **b** $-3 - 4$ **c** $-3 + 4$ **d** $3 - -4$ **e** $3 + -4$

3 Identify the terms that are 'like' $3w$.

3 $2w$ $-5w$ w $3p$ $4w^2$ pw

Example

Simplify the expressions.

a $3k + 4p + 2k$ **b** $5a - 3 + 2a + 5$ **c** $x^2 + 3x + x^2 - x$

Method

Solution	Commentary
a $3k + 4p + 2k \equiv 3k + 2k + 4p$ $\equiv 5k + 4p$	You cannot simplify further as $5k$ and $4p$ are unlike terms. You can show this using objects. 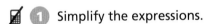
b $5a - 3 + 2a + 5 \equiv 5a + 2a - 3 + 5$ $\equiv 7a + 2$	Collect like terms first. Simplify by adding or subtracting the coefficients or numbers.
c $x^2 + 3x + x^2 - x \equiv x^2 + x^2 + 3x - x$ $\equiv 2x^2 + 2x$ Remember that $\equiv$ means 'is equivalent to'. A statement using $\equiv$ is true for all values of the variable or variables.	Collect like terms first. Simplify by adding or subtracting the coefficients. You cannot simplify further as x^2 and x are unlike.

Practice (B)

1 Simplify the expressions.

  a $2g + 4g$ **b** $5h + 3h + 6$ **c** $2k + 3p + 4k + 2p$

  d $3u + 4 + 1 + 4u$ **e** $2j + 7j + 3r + j$ **f** $e + 2a + e$

2 Simplify the expressions.

  a $3a + 5b + 2a + b$ **b** $3a + 5b - 2a + b$ **c** $3a + 5b + 2a - b$

  d $4c - 3d + 4d + 5c$ **e** $4c + 3d - 4d - 5c$ **f** $-4c + 3d - 4d + 5c$

3 Simplify the expressions.

  a $6e^2 + 5f + 4e^2 + 8f$ **b** $3e^2 + 8f - e^2$ **c** $7ef + 5f + 3ef$

  d $7x^2 + 3x + 2x^2 - 5x$ **e** $6x^2 + 2x - x^2 - 2x^2$ **f** $xy + 3x + 3y + yx$

4 Write a simplified expression for the perimeter of this rectangle.

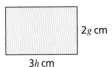

2g cm

3h cm

5 Simplify the expressions.

  a $2gh^2 + 5gh - gh^2$ **b** $2gh^2 + 5g^2h + gh^2$ **c** $2gh^2 - 5gh + gh + 5gh^2$

Consolidate – do you need more?

1 g is a number.

Write an expression for:

a 5 less than g **b** 8 more than g **c** 1 more than double g

d double '4 less than g'   **e** one-quarter of g **f** 3 less than one-third of g.

2 Write an expression for the number of days in n weeks.

3 Write expressions for how many chocolates there are altogether in y boxes if each box contains:

a 8 chocolates   **b** 25 chocolates   **c** x chocolates.

4 Simplify the expressions.

a $2m + 3k + m + 2k$ **b** $2m + 3k - m$ **c** $2m + 3k + m - 2k$

d $2m - 3k + 2k$ **e** $2m + 3k - 2k - m$ **f** $m - 5m - 3k - 5k$

g $m^2 - 3m + 3m^2$ **h** $m^2 - 4m - m$ **i** $3m^2 - 4m - 3m^2 + m$

Stretch – can you deepen your learning?

1 Write three expressions that simplify to $6x + 5$

2 Write simplified expressions for the area of each shape.

a
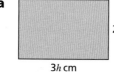
2g cm
3h cm

b

3y cm

3 Write simplified expressions for the perimeter of each shape.

a
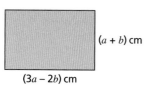
$(a + b)$ cm
$(3a - 2b)$ cm

b

5k cm

White Rose
MATHS

Are you ready? (A)

 1 Write each of these using powers.

    a $5 \times 5 \times 5$ **b** $h \times h$ **c** $6 \times 6 \times 6 \times 6 \times 6$ **d** $k \times k \times k \times k$

2 Simplify these expressions.

    a $5 \times u$ **b** $m \times p$ **c** $6 \times w \times g$ **d** $4h \times h$

| Base of the power | Index of the power | The plural of 'index' is 'indices'. |

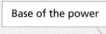

Coefficient $\rightarrow 3x^2$

Expanded form

Being able to represent numbers and letters involving powers in expanded form is essential for understanding the laws of indices.

2^4 in expanded form is $2 \times 2 \times 2 \times 2$ y^3 in expanded form is $y \times y \times y$

Note that any number or letter raised to the power of 0 is 1

Example

Simplify: **a** $h^5 \times h^3 \times h$ **b** $y^8 \div y^5$ **c** $(a^5)^3$

Method

Solution	Commentary
a $h^5 \times h^3 \times h$ $= h \times h \times h \times h \times h \times h \times h \times h \times h$ $= h^9$	Rewrite the calculation in expanded form. The base stays the same and the powers are added. Now that the calculation is written in expanded form, you are finding the result of 'h' multiplied by itself nine times, which can be simplified to h^9.
b $y^8 \div y^5$ $\dfrac{y^8}{y^5} = \dfrac{y \times y \times y \times y \times y \times y \times y \times y}{y \times y \times y \times y \times y}$ $= \dfrac{y \times y \times y \times \cancel{y} \times \cancel{y} \times \cancel{y} \times \cancel{y} \times \cancel{y}}{\cancel{y} \times \cancel{y} \times \cancel{y} \times \cancel{y} \times \cancel{y}}$ $= y \times y \times y$ $y^8 \div y^5 = y^3$	Write the calculation as a fraction. Then write both the numerator and the denominator in expanded form. The base stays the same and the powers are subtracted. Now cancel common factors. You are now left with the product of three ys, which is y^3 so $y^8 \div y^5 = y^3$
c $(a^5)^3 \equiv a^5 \times a^5 \times a^5$	Rewrite the calculation in expanded form.
$a^5 \times a^5 \times a^5 = a^{15}$	Use the addition law for indices to simplify the expression.
$(a^5)^3 \equiv a^{5 \times 3} = a^{15}$	Notice the connection between the powers in the question and the power in the answer. When finding powers of powers, you find the product of the powers: $5 \times 3 = 15$

Practice (A)

1 Simplify the expressions.

a $g^2 \times g^3$ b $h^5 \times h^5$ c $k^3 \times k$

d $m^2 \times m^3 \times m^4$ e $n^2 \times n^5 \times n$ f $p \times p^8 \times p$

2 Simplify:

a $b^5 \div b^3$ b $c^8 \div c^2$ c $d^9 \div d$

d $e^9 \div e^3$ e $\dfrac{f^7}{f^2}$ f $\dfrac{g^8}{g^4}$

3 Simplify, leaving your answers in index form.

a $(h^3)^2$ b $(k^2)^5$ c $(m^4)^4$ d $(p^5)^7$

What do you think? 💡

1 Emily says that $5y^3 \times 2y^5 = 7y^8$ because you need to add when two terms are being multiplied.

Do you agree with Emily?

Are you ready? (B)

1 Simplify: a $g^4 \times g$ b $3a \times 2a$ c $4b \times b$ d $15c \div 5$

2 Evaluate: a $\dfrac{1}{4^2}$ b $7 + {-4}$ c $7 - {-4}$ d $-7 - {-4}$

Example 1

Simplify: a $3a^2 \times 4a^5$ b $10w^6 \div 5w^2$ c $(2b^7)^4$

Method

Solution	Commentary
a $3a^2 \times 4a^5$	Rewrite the calculation in expanded form.
$= 3 \times a \times a \times 4 \times a \times a \times a \times a \times a$	The coefficients 3 and 4 can be multiplied to get 12
$= 3 \times 4 \times a \times a \times a \times a \times a \times a \times a$	
$= 12a^7$	The base 'a' is being multiplied by itself seven times, which can be simplified to a^7.
b $10w^6 \div 5w^2$	You can write the calculation as a fraction.
$\dfrac{10w^6}{5w^2} = \dfrac{10 \times w \times w \times w \times w \times w \times w}{5 \times w \times w}$	Then write both the numerator and denominator in expanded form.
$= \dfrac{10 \times w \times w \times w \times w \times \cancel{w} \times \cancel{w}}{5 \times \cancel{w} \times \cancel{w}}$	Divide the coefficients, 10 and 5, to get 2
$= 2 \times w \times w \times w \times w = 2w^4$	Cancelling common factors leaves the product of four ws, which is w^4.
	So $10w^6 \div 5w^2 = 2w^4$

c $(2b^7)^4 \equiv 2 \times b^7 \times 2 \times b^7 \times 2 \times b^7 \times 2 \times b^7$	
$(2b^7)^4 \equiv (2 \times 2 \times 2 \times 2) \times (b^7 \times b^7 \times b^7 \times b^7)$	Simplify the numerical part of the expression by multiplying the numbers.
$(2b^7)^4 \equiv 2^4 \times b^{28} \equiv 16b^{28}$	Use the addition law for indices to simplify the algebraic part of the expression.
$(2b^7)^4 \equiv 16b^{28}$	You raise the integer to the power outside the bracket: $2^4 = 16$
	When working out powers of powers, you find the product of the powers: $7 \times 4 = 28$

Example 2

Evaluate: **a** 5^0 **b** 3^{-2} **c** $a^5 \times a^{-3}$ **d** $a^5 \div a^{-3}$

Method

Solution	Commentary
a $5^0 = 1$	Any number or letter raised to the power of 0 is 1
b $3^{-2} = \dfrac{1}{3^2} = \dfrac{1}{9}$	Rewrite as the reciprocal with a positive power.
c $5 + -3 = 2$ So $a^5 \times a^{-3} = a^2$	The index laws work just the same for negative powers, so add the powers.
d $5 - -3 = 8$ So, $a^5 \div a^{-3} = a^8$	Again, the same index laws apply. As this involves division, subtract the powers. Be careful when subtracting -3 from 5

Practice (B)

 1 Simplify the expressions, giving your answers in index form.

     a $2x^3 \times x^5$ **b** $y^3 \times 4y^2$ **c** $2w^3 \times 3w^4$

     d $7u \times 5u^7$ **e** $3h^4 \times 5h^3$ **f** $6g^3 \times g^3$

2 Simplify:

     a $9b^6 \div 3b^4$ **b** $8c^7 \div 4c^2$ **c** $10d^5 \div 2d$

     d $12e^8 \div 3e^4$ **e** $\dfrac{18f^{10}}{3f^5}$ **f** $\dfrac{3g^{12}}{2g^4}$

3 Simplify, leaving your answers in index form.

     a $(3h^4)^2$ **b** $(4n^2)^3$ **c** $(2w^5)^5$ **d** $(5p^8)^3$

4 Evaluate:

     a 3^0 **b** y^0 **c** 5^{-2} **d** 2^{-3}

5 Simplify:

a $g^5 \times g^{-3}$ **b** $h^3 \times h^{-4}$ **c** $k^{-3} \times k$

d $a^{-3} \times a^5$ **e** $b^{-5} \times b^2$ **f** $x \times x^{-3}$

6 Simplify:

a $y^5 \div y^{-2}$ **b** $w^3 \div w^{-4}$ **c** $c^{-3} \div c^2$

d $d^{-5} \div d$ **e** $b^{-5} \div b^{-2}$ **f** $g \div g^{-4}$

Consolidate – do you need more?

1 Simplify the expressions.

a $g^3 \times g^4$ **b** $k^5 \times k$ **c** $m^5 \times m^2 \times m$

d $b^7 \div b^3$ **e** $c^9 \div c$ **f** $\dfrac{g^{10}}{g^5}$

2 Simplify, leaving your answers in index form.

a $(h^2)^3$ **b** $(k^7)^4$ **c** $(m^8)^5$

d $(4g^5)^2$ **e** $(3n^4)^3$ **f** $(2w^8)^4$

3 Simplify the expressions.

a $2m^4 \times 5m^3$ **b** $y^4 \times 5y^6$ **c** $3w^4 \times 5w^6$

d $6b^7 \div 3b^2$ **e** $12c^8 \div 4c^2$ **f** $\dfrac{9g^6}{g^3}$

4 Evaluate:

a 7^0 **b** p^0 **c** 8^{-2}

5 Simplify:

a $g^4 \times g^{-2}$ **b** $k^{-5} \times k$ **c** $a^{-4} \times a^3$

d $y^4 \div y^{-3}$ **e** $c^{-1} \div c^3$ **f** $b^{-6} \div b^{-4}$

Stretch – can you deepen your learning?

1 Simplify:

a $\dfrac{y^2 \times y^3}{y^4}$ **b** $\dfrac{x^4 \times x}{x^3}$ **c** $\dfrac{u^7}{u^3 \times u^2}$

d $\dfrac{w^3 \times w \times w^2}{w^2 \times w^3}$ **e** $\dfrac{3w^5 \times 8w^3}{6w^4}$ **f** $\dfrac{6g^4 \times 3g^3}{9g^5 \times 2g^4}$

2 Simplify:

a $6g^5 \times 4g^{-2}$ **b** $8y^{-5} \times 8y^{-3}$ **c** $15x^8 \div 5x^{-4}$ **d** $24h^{-2} \div 8h^{-3}$

3 Simplify:

a $(x^{-2})^{-3}$ **b** $(y^5)^{-4}$ **c** $(4w^{-3})^2$ **d** $(2u^4)^{-3}$

Are you ready?

 1 In words, explain the meaning of these expressions.

     a $5x$ **b** $5 + x$ **c** $\dfrac{x}{5}$ **d** $x - 5$ **e** $5 - x$

2 Explain the difference between $2x$ and x^2

3 Work out:

     a 3^2 **b** $(-3)^2$ **c** 4×-3 **d** -4×-3 **e** $(-4)^2 \times -3$

Using your calculator 🖩

If you are using a calculator when substituting into expressions, be careful with negative numbers. Ensure that you use brackets where required.

For example, to evaluate the expression a^2 when $a = -3$, you need to key in

      rather than

Example 1

Work out the value of these expressions when $a = 80$

a $2a + 9$ **b** $3a - 9$ **c** $\dfrac{a}{2} + 17$ **d** $500 - 3a$

Method

Solution	Commentary

a $2a + 9$
    $= 2 \times 80 + 9$
    $= 160 + 9$
    $= 169$
	Replace a with 80
	 Remember that $2a$ stands for 'a multiplied by 2'.
	
	 <table><tr><td>a</td><td>a</td><td>9</td></tr></table> <table><tr><td>80</td><td>80</td><td>9</td></tr></table>

b $3a - 9$
    $= 3 \times 80 - 9$
    $= 240 - 9$
    $= 231$
	<table><tr><td>a</td><td>a</td><td>a</td></tr></table> <table><tr><td colspan="2">$3a - 9$</td><td>9</td></tr></table> <table><tr><td>80</td><td>80</td><td>80</td></tr></table> <table><tr><td colspan="2">$240 - 9 = 231$</td><td>9</td></tr></table>

c $\dfrac{a}{2} + 17$
    $= \dfrac{80}{2} + 17$
    $= 40 + 17$
    $= 57$
	Remember that $\dfrac{a}{2}$ means 'a divided by 2' so $\dfrac{80}{2}$ means '80 divided by 2'.
	
	 <table><tr><td colspan="2">a</td><td></td></tr><tr><td>$\frac{a}{2}$</td><td></td><td></td></tr><tr><td>$\frac{a}{2}$</td><td>17</td><td></td></tr></table> <table><tr><td colspan="2">80</td></tr><tr><td>40</td><td></td></tr><tr><td>40</td><td>17</td></tr></table>

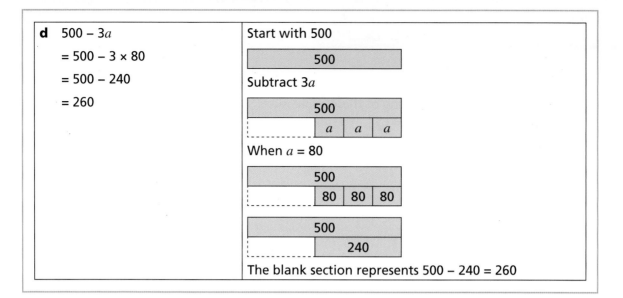

d $500 - 3a$

$= 500 - 3 \times 80$

$= 500 - 240$

$= 260$

Start with 500

500

Subtract $3a$

When $a = 80$

The blank section represents $500 - 240 = 260$

Example 2

Evaluate the formula $V = 5p^3$ when $p = 3$

Method

Solution	Commentary
When $p = 3$, $p^3 = 3 \times 3 \times 3 = 27$ So $5p^3 = 5 \times 27 = 135$ $V = 135$	$5p^3$ means $5 \times p^3$

Example 3

Work out the value of a^2b when:

a $a = 4$ and $b = 5$ **b** $a = -4$ and $b = 5$ **c** $a = -4$ and $b = -5$

Method

Solution	Commentary
a $a^2b = 4^2 \times 5$ $= 16 \times 5$ $= 80$	Work out the squared term first because of the order of operations.
b $a^2b = (-4)^2 \times 5$ $= 16 \times 5$ $= 80$	It is best to put -4 in brackets to help remember that the calculation is -4×-4 Remember, the product of two negative numbers is positive.
c $a^2b = (-4)^2 \times -5$ $= 16 \times -5$ $= -80$	Remember, the product of a negative number and a positive number is negative.

Practice

1 Work out the value of each of these expressions when $y = 30$

a $y + 5$ **b** $\dfrac{y}{5}$ **c** $5y$ **d** $y - 5$

2 Substitute $w = 12$ to work out the value of each expression.

a $3w + 4$ **b** $3w - 4$ **c** $\dfrac{w}{4} + 5$ **d** $5 + 4w$ **e** $50 - 3w$

3 Evaluate these expressions when $h = 4.6$

a $\dfrac{h}{10} + 8$ **b** $8 + 10h$ **c** $10h - 8$ **d** $8 - \dfrac{h}{10}$

e $\dfrac{h}{2} - 2$ **f** $h^2 - 20$ **g** $20 + h^2$ **h** $2 - \dfrac{h}{2}$

4 Substitute $m = 25$ and $p = 80$ to work out the value of:

a $m + p$ **b** $p - m$ **c** mp **d** $\dfrac{p}{m}$

e $\dfrac{m}{p}$ **f** $4m - p$ **g** $p - 3m$ **h** $2p + 5m$

5 **a** Work out the value of $5k + 4n$ when:

     i $k = 3$ and $n = 5$ **ii** $k = -3$ and $n = 5$

     iii $k = 3$ and $n = -5$ **iv** $k = -3$ and $n = -5$

   b Work out the value of $5k - 4n$ when:

     i $k = 3$ and $n = 5$ **ii** $k = -3$ and $n = 5$

     iii $k = 3$ and $n = -5$ **iv** $k = -3$ and $n = -5$

6 $p = 6$ and $r = -8$

Work out:

a $p + r$ **b** $r - p$ **c** pr **d** $p - 3r$

e $\dfrac{3r}{p}$ **f** $\dfrac{pr}{3}$ **g** $p - \dfrac{r}{2}$ **h** $\dfrac{p - r}{2}$

7 Evaluate the expressions when $x = 3$, $y = 4$ and $z = -5$

a $4x^2$ **b** xy^2 **c** x^2y **d** $3y^3$

e $4z^2$ **f** xz^2 **g** x^2z **h** $3z^3$

What do you think? 💡

1 $a = 8$ and $b = -3$

How many expressions involving both a and b can you find with a value of:

a -5 **b** 24?

2 **a** Work out $5e^2$ and $(5e)^2$ when:

     i $e = 10$ **ii** $e = 50$ **iii** $e = -2$

   b Write down a value of e for which $5e^2$ and $(5e)^2$ are equal.

Consolidate – do you need more?

1 Work out the value of each of these expressions when $d = 36$

   a $d + 3$ **b** $\dfrac{d}{3}$ **c** $3d$ **d** $d - 3$

2 Substitute $y = 15$ to work out the value of each expression.

   a $3y + 5$ **b** $3y - 5$ **c** $\dfrac{y}{3} + 5$ **d** $5 + 3y$ **e** $50 - 3y$

3 Substitute $x = 12$ and $y = 30$ to work out the value of:

   a $x + y$ **b** $y - x$ **c** xy **d** $\dfrac{y}{x}$

   e $\dfrac{x}{y}$ **f** $3x - y$ **g** $y - 3x$ **h** $3x - 2y$

4 $n = 8$ and $w = -10$

Work out:

   a $n + w$ **b** $w - n$ **c** nw **d** $3n - 3w$

   e $\dfrac{4w}{n}$ **f** $\dfrac{nw}{4}$ **g** $n - \dfrac{w}{2}$ **h** $\dfrac{n - w}{2}$

5 Evaluate the expressions when $a = 5$, $b = 6$ and $c = -4$

   a $3a^2$ **b** ab^2 **c** a^2b **d** $2b^3$

   e $3c^2$ **f** ac^2 **g** a^2c **h** $2c^3$

Stretch – can you deepen your learning?

1 Work out the value of each expression when $g = 3$ and $w = 4$

   a g^w **b** w^g **c** $g^{\sqrt{w}}$ **d** $\sqrt{w^g}$

2 Substitute $e = 100$ and $f = 64$ to work out the value of each expression.

   a $f(e - f)$ **b** $2f(e + f)$ **c** $\sqrt{f}\,(e - f)$

   d $\sqrt{f}\left(\sqrt{e} - \sqrt{f}\right)$ **e** $\sqrt{ef}\,(e - f)$ **f** $\dfrac{e + f}{\sqrt{f}}$

7.4 Algebraic structures

Are you ready?

1 Simplify each expression.

    a $2x + 8 + 6x - 2$ **b** $12 + 8w - 8 + 10w$

2 State whether each of the following is **true** or **false**.

    a $a \times a = 2a$ **b** $b + b = 2b$

    c $b \times a \times a = a^2b$ **d** $6ab^2 = 3a \times 2b$

3 Work out the area of each shape.

    a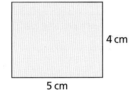

    b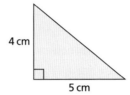

A **variable** is a numerical quantity that might change, often denoted by a letter.

Variables can be used to make **expressions**, **equations**, **identities** and **formulae**.

In this chapter you will use algebraic symbols to represent different types of relationships.

$3x + 5$	is an **expression**. It is a collection of terms. You can work out the value of the expression from different values of x.
$3(x + 5) = 36$	is an **equation**. You can solve this to find the value of x.
$3(x + 5) \equiv 3x + 15$	is an **identity**. It is true for all values of the letter x.
$A = bh$	is a **formula**. It is the formula for the area of a parallelogram of base b and perpendicular height h. You can substitute values for b and h to work out the value of A.

Example 1

Match each of the following pieces of algebra to one of the words.

$15 = 2a + 8$

$x^2 - x - 20$

$5(2a - 3) \equiv 10a - 15$

$A = \frac{1}{2}(a + b)$

equation

expression

formula

identity

Method

Solution	Commentary
$15 = 2a + 8$ is an equation.	You can solve this to find the value of a. So it is an equation.
$x^2 - x - 20$ is an expression.	This is a collection of terms and you don't know the value of x. So it is an expression.
$5(2a - 3) \equiv 10a - 15$ is an identity.	This is true for all values of a. So it is an identity.
$A = \frac{1}{2}(a + b)h$ is a formula.	This is the formula for the area of a trapezium. Substitute in values of a, b and h to work out the value of A.

Example 2

Write a simplified expression for the area of the rectangle.

g cm

$2h$ cm

Method

Solution	Commentary
Area = length × width	This is the formula for the area of a rectangle.
$= 2h \times g$	The length is $2h$ and the width is g so multiply these together.
$= 2hg$ cm²	Remember that you don't use the × symbol when writing expressions in their simplest form. The units for area are cm².

Practice

1 Match each of the following to the correct word.

$2w + 5a^2 - 3$

$V = lwh$

$6 + 5h = 21$

$15 + 10y \equiv 10y + 15$

equation

expression

formula

identity

2 One pencil costs p pence. One ruler costs r pence.

Write down an expression for the cost of 3 pencils and 5 rulers.

3 Copy and complete the identities.

a $5y - 2y \equiv \boxed{}$
b $4x + \boxed{} \equiv 7x$

c $3w \times \boxed{} \equiv 24wg$
d $6h + 3h \equiv 5h + \boxed{}$

4 **a** Write down a simplified expression for the perimeter of the rectangle.

b Write down a simplified expression for the area of the rectangle.

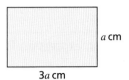

a cm

$3a$ cm

Consolidate – do you need more?

1 Match each of the following to the correct word.

| $5g \times 9h \equiv 45gh$ | $a^2 + b^2 = c^2$ | $5x + 20 = 3x + 22$ | $5g + 2g^2$ |

| equation | expression | formula | identity |

2 Given $w = 6$ and $p = -3$, work out the value of each expression.

a $5w + 2p$
b $5w - 2p$
c $2wp$
d $2w^2p$

3 Write an expression for the volume of the cuboid.

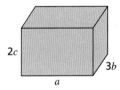

$2c$

$3b$

a

Stretch – can you deepen your learning?

1 Write a simplified expression for the area of each circle.

a

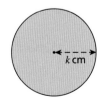

k cm

b

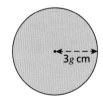

$3g$ cm

c

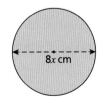

$8x$ cm

2 Given the identity $3(x + 2y) + 4(5x + by) \equiv ax + 18y$, work out the values of a and b.

Understanding algebra: exam practice

White Rose MATHS

1 Simplify:

(a) $y + y + y + y$ [1 mark]

(b) $5a + 3b + 2a - b$ [2 marks]

(c) $4t^2 + 7t - t^2$ [1 mark]

2 Write an expression for 6 less than x. [1 mark]

3 One table costs t pounds.

One chair costs c pounds.

Write an expression for the cost of 4 tables and 12 chairs. [2 marks]

4 If $a = 3$ and $b = 10$, evaluate $5a + 8b$. [2 marks]

5 Simplify:

(a) $d \times d \times d$ [1 mark]

(b) $4f \times 5g$ [1 mark]

6 A square has side lengths x.

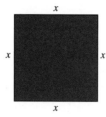

(a) Write an expression for the perimeter of the square. [1 mark]

(b) Write an expression for the area of the square. [1 mark]

7 The cost C, in pounds, of a holiday cottage can be found using the formula

$C = 100d + 25p$

where d = number of days and p = number of people.

How much does the holiday cottage cost for 6 people for 1 week? [3 marks]

8 Simplify:

(a) $m^9 \div m^3$ [1 mark]

(b) $5x^4y^3 \times 3x^2y$ [2 marks]

1–3

3–5

8 Working with brackets

In this block, we will cover...

8.1 Expanding and factorising a single bracket

Example 1

a Multiply out the brackets in the expression 5(

b By substituting $a = 100$, work out 5×98

Method

Solution	Commenta
a $5(a - 2) \equiv 5a - 10$	Multiply ea $5 \times a = 5a$ a
b When $a = 100$ $5(100 - 2) = 500 - 10 = 490$	If $a = 100$, Alternative

8.2 Expanding and simplifying double brackets

Practice

1. Expand and simplify:

   a $(x + 3)(x + 4)$ **b** $(x + 5)(x +$

   d $(y + 1)(y + 4)$ **e** $(w + 7)(w +$

2. Expand and simplify:

   a $(x + 5)(x - 3)$ **b** $(x - 3)(x +$

   d $(p + 3)(p - 5)$ **e** $(g - 7)(g +$

3. **a** What is the same and what is different

8.3 Factorising quadratic expressions

Consolidate – do you need more?

1. Factorise these quadratic expressions comp

   a $x^2 + 8x + 7$ **b** $y^2 + 9y + 14$

   d $y^2 + 11y + 28$ **e** $w^2 + 21w + 20$

2. Factorise these expressions completely.

   a $x^2 + 7x - 8$ **b** $x^2 + 6x - 16$

   d $y^2 + 7y - 30$ **e** $w^2 - w - 30$

   g $g^2 - 5g + 36$ **h** $p^2 - 4p + 45$

8.4 Identities

Stretch – can you deepen your le

1. Copy and complete the expressions to crea

   a $4(x + 3) \equiv 4x + \boxed{}$

   b $\boxed{}(y + 7) \equiv \boxed{}y + 21$

   c $3(\boxed{}w - 4) + 6(2w - \boxed{}) \equiv 18w - 24$

2. Use the diagram to create an identity.

8.1 Expanding and factorising a single bracket

Are you ready? (A)

1 Simplify:

    a $2a + 2a + 2a$ **b** $2b \times 3$ **c** $3c \times 2$

    d $5d \times -2$ **e** $-3 \times -2e$ **f** $3f \times 5f$

2 Simplify by collecting like terms.

    a $2a + 5 + 3a + 4$ **b** $4c + 8 + 2c - 1$ **c** $3f + 2 - f + 1$ **d** $6 + 2g - 1 - 3g$

Multiplying out a single bracket means to multiply the term directly in front of the bracket by all the terms inside. The command words used on exam papers are '**Expand…**' or '**Multiply out…**'.

Here are some ways of showing the expansion of $3(x + 2)$:

 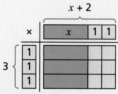

$$3(x + 2) \equiv 3x + 6$$

If you are asked to '**Expand and simplify**', you need to multiply out several single brackets and then simplify by collecting like terms.

Factorising is the inverse of the process of multiplying out a bracket. Look for common factors to write the expression in its bracketed form. The command word used on exam papers is '**Factorise…**'.

Shown right is an illustration using algebra tiles to show an expression being arranged in factorised form.

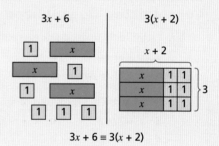

$$3x + 6 \equiv 3(x + 2)$$

If you are asked to '**Factorise fully**', you must find the highest common factor (HCF) of the terms in the expression. For $8p + 12pq$ this would be $4p$, so $8p + 12pq \equiv 4p(2 + 3q)$

Example 1

a   Multiply out the brackets in the expression $5(a - 2)$.

b   By substituting $a = 100$, work out 5×98

Method

Solution	Commentary
a $5(a - 2) \equiv 5a - 10$	Multiply each term in the bracket by 5 $5 \times a = 5a$ and $5 \times -2 = -10$
b When $a = 100$ $5(100 - 2) = 500 - 10 = 490$	If $a = 100$, then $a - 2 = 98$ Alternatively, you could substitute $a = 100$ into $5a - 10$

Example 2

Expand the brackets and simplify the answers if possible.

a $p(p + 5)$ **b** $3(m + 2n) - 2(4n - 3m)$

Method

Solution	Commentary
a $p(p + 5) \equiv p^2 + 5p$	Remember $p \times p$ is written p^2, not pp.
b $3(m + 2n) - 2(4n - 3m)$ $\equiv 3m + 6n - 8n + 6m$ $\equiv 9m - 2n$	First, expand both sets of brackets. Be careful with the second bracket: $-2 \times 4n$ gives $-8n$ but $-2 \times -3m$ gives $+6m$. Then collect the like terms, being careful with any negative terms.

Practice (A)

1 Multiply out:

   a $3(a + 4)$ **b** $5(c - 2)$ **c** $4(3 + e)$

   d $2(3g + 4)$ **e** $6(2h - 1)$ **f** $5(4 - 3j)$

   g $7(3x + 5)$ **h** $8(2 - 4k)$ **i** $9(10 + 9y)$

2 Expand:

   a $2(b + 3)$ **b** $-2(b + 3)$ **c** $-2(b - 3)$

   d $3(4g - 2)$ **e** $-3(4g + 2)$ **f** $-3(4g - 2)$

3 Multiply out and simplify:

   a $4(g + 5) + 2g$ **b** $5(h + 1) - 3h$ **c** $4(2 + r) + 4r$

   d $3(k - 3) - 1$ **e** $5(2y + 3) + y + 4$ **f** $6(4 + 2p) - 4 - 10p$

4 Expand and simplify:

   a $5(c + 2) + 3(c + 6)$ **b** $3(x - 2) + 7(x + 1)$ **c** $2(6e + 7) + 5(e - 2)$

   d $3(7t + 10) - 3(4t + 2)$ **e** $3(2d + 7) + 2(d - 3)$ **f** $3(2n - 1) - 2(3n - 4)$

5 Multiply out:

   a $x(x + 8)$ **b** $g(3g - 2)$ **c** $2k(5 + k)$

   d $t(t^2 + g)$ **e** $w^2(3w + 2y)$ **f** $ef(2e - 3f^2)$

Are you ready? (B)

1 Work out the highest common factor of each set of numbers.

   a 8 and 12   **b** 36 and 42   **c** 35 and 63   **d** 16, 24 and 40

2 Work out the highest common factor of each set.

   a $9x$ and $12y$ **b** $9xy$ and $12yz$ **c** $9x^2$ and $12x$ **d** $9x^2$ and $12xy$

Example

Factorise these expressions.

a $3x + 6$ **b** $10y + 5$ **c** $6p + 15q + 9$

d $8a + 12$ **e** $3xy + x^2$ **f** $5ab + 15bc + 10b^2$

Method

Solution	Commentary
a $3x + 6 \equiv 3(x + 2)$	The terms $3x$ and 6 have a common factor of 3 so $3x + 6 \equiv 3(\boxed{} + \boxed{})$ "What do I need to multiply 3 by to get $3x$?"; the answer is x. **You could also think of this as $3x \div 3$ and $6 \div 3$, as the inverse of multiplication is division.** "What do I need to multiply 3 by to get 6?"; the answer is 2
b $10y + 5 \equiv 5(2y + 1)$	The common factor is 5
c $6p + 15q + 9 \equiv 3(2p + 5q + 3)$	There are three terms here but the process is the same. You need a factor that is common to all three.
d $8a + 12 \equiv 4(2a + 3)$	Although 2 is a factor of both $8a$ and 12, their HCF is 4, so that is the factor to choose.
e $3xy + x^2 \equiv x(3y + x)$	The common factor can be a variable.
f $5ab + 15bc + 10b^2 \equiv 5b(a + 3c + 2b)$	Both 5 and b are factors of all three terms. The highest common factor is $5b$.

Practice (B) ✍

1   Factorise the expressions.

   a $2x + 8$ **b** $4h - 16$ **c** $24y + 8$

   d $8p - 8$ **e** $6 + 3k$ **f** $12 - 6t$

   g $4m + 14$ **h** $15 + 12w$ **i** $6u - 28$

2   Factorise:

   a $12x + 16y + 20$ **b** $15h - 20 + 35g$ **c** $16 - 40w - 24g$

3   Factorise fully:

   a $x^2 + x$ **b** $y - y^2$ **c** $3f + f^2$

   d $5g^2 + 2g$ **e** $8u - 3u^2$ **f** $4n^3 + 3n$

4   Factorise fully:

   a $5 + 15p^2$ **b** $2n + 4n^2$ **c** $2x^2 - 8x$

   d $14y + 7y^2$ **e** $5g^2 + 10g$ **f** $6w^3 - 2w^2$

> To factorise fully, make sure you take out the HCF.

5   Factorise fully:

   a $8h^2 + 16hf$ **b** $6ab^2 + 24a^2b$

   c $12x^2y + 20xy + 8x^3$ **d** $15cd^2 + 10c^2d^2 + 25cd$

What do you think? 💭

1 Kath and Tiff attempt to fully factorise the expression $14xy^2 - 28x^2y$.

Kath's solution is $7xy(2y - 4x)$.

Tiff's solution is $14xy(y - 2x)$.

Whose solution is correct? What mistake has the other person made?

2 Explain why $10x - 7y$ cannot be factorised.

3 An expression expands to give $16a + 10ab$.

What could the expression have been if it involved:

a one bracket? **b** two single brackets?

Consolidate – do you need more?

1 Multiply out:

a $4(b + 3)$ **b** $4(2b + 3)$ **c** $4(2b - 3)$

d $4(2b + 3c)$ **e** $4(2b - 3c + 5)$ **f** $-4(b + 3)$

g $b(b + 3)$ **h** $b(2b - 5)$ **i** $4b(5 - b^2)$

2 Expand and simplify:

a $4(g + 2) + 5(g + 6)$ **b** $2(h - 1) + 6(h + 3)$ **c** $3(5w + 4) + 3(w - 2)$

d $2(5y + 6) - 3(3y + 2)$ **e** $2(3p + 4) + 3(4p - 1)$ **f** $8(3k - 2) - 4(2k - 3)$

3 Factorise the expressions.

a $3c + 9$ **b** $15 - 5d$ **c** $7e - 7$

d $6f + 16$ **e** $12g + 15h + 27$ **f** $18 - 42j - 24k$

4 Factorise fully:

a $m^2 + m$ **b** $8n + 3n^2$ **c** $4 + 12p^2$

d $3r + 9r^2$ **e** $10u + 6u^2$ **f** $14wy - 6w^2$

Stretch – can you deepen your learning?

 **1** Work out:

a $35 \times 45 + 35 \times 55$ **b** $47 \times 56 - 47 \times 46$ **c** $27 \times 71 + 29 \times 27$

2 Write a simplified expression for the perimeter of the triangle.

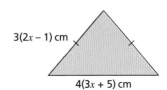

$3(2x - 1)$ cm

$4(3x + 5)$ cm

White Rose MATHS

Are you ready?

1 Multiply out:

a $2(y + 3)$ **b** $y(y + 3)$ **c** $2(y - 3)$ **d** $y(y - 3)$

2 Simplify the expressions.

a $5y + 6y$ **b** $5y - 6y$ **c** $-5y + 6y$ **d** $-5y - 6y$

3 Write as a single term without mathematical operations.

a $-2 \times y$ **b** $-2 \times -y$ **c** $-2 \times -3y$ **d** $2 \times -3y$

4 Simplify the expressions.

a $x^2 + 2x + 5x + 10$ **b** $x^2 + 2x - 5x - 10$ **c** $x^2 - 2x + 5x - 10$

Expanding a pair of brackets means to multiply them together. Using an area model can help.

Expand $(x + 2)(x + 3)$

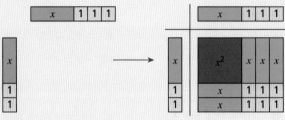

The area is $x^2 + 5x + 6$

So $(x + 2)(x + 3) \equiv x^2 + 5x + 6$

Example

Expand and simplify: **a** $(x + 2)(x + 5)$ **b** $(x + 2)(x - 3)$

Method

Solution	Commentary
a $(x + 2)(x + 5) \equiv x^2 + 2x + 5x + 10$ $\equiv x^2 + 7x + 10$	As $2x$ and $5x$ are like terms they can be added to give $7x$ as the 'middle' term in the final answer. You could also use the area model or algebra tiles to show this expansion. Again there is one x^2 tile, seven x tiles and ten 1s in the expansion.

b $(x + 2)(x - 3) \equiv x^2 - 3x + 2x - 6$

$\equiv x^2 - x - 6$

Be careful with positive and negative numbers.

Simplifying $-3x + 2x = -1x$, which you write as $-x$

Using algebra tiles you can see

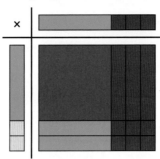

You can simplify by making zero pairs.

The answer is $x^2 - x - 6$

Practice

1. Expand and simplify:

   a $(x + 3)(x + 4)$ **b** $(x + 5)(x + 2)$ **c** $(y + 3)(y + 5)$

   d $(y + 1)(y + 4)$ **e** $(w + 7)(w + 5)$ **f** $(w + 1)(w + 9)$

2. Multiply out and simplify:

   a $(x + 5)(x - 3)$ **b** $(x - 3)(x + 6)$ **c** $(p + 7)(p - 5)$

   d $(p + 3)(p - 5)$ **e** $(g - 7)(g + 6)$ **f** $(g + 1)(g - 8)$

3. **a** What is the same and what is different about the expansions of
   $(y + 7)(y + 3)$ and $(y - 7)(y - 3)$?

   b Expand and simplify:

   i $(x - 3)(x - 2)$ **ii** $(h - 4)(h - 7)$ **iii** $(g - 5)(g - 8)$

4. Multiply out and simplify:

   a $(x - 2)(x + 2)$ **b** $(x - 7)(x + 7)$ **c** $(x + 1)(x - 1)$

   What do you notice?

5. **a** Benji thinks that $(y + 5)^2 \equiv y^2 + 25$

   Explain why Benji is incorrect and write the correct expansion of $(y + 5)^2$

   b Expand and simplify:

   i $(y + 3)^2$ **ii** $(x + 1)^2$ **iii** $(w - 4)^2$ **iv** $(r - 8)^2$

6 **a** Marta is expanding $(x + 4)(x + 3)$ and $(2x + 4)(x + 3)$.

Marta knows $(x + 4)(x + 3) \equiv x^2 + 7x + 12$

So she thinks $(2x + 4)(x + 3) \equiv 2x^2 + 14x + 24$

Explain why Marta is wrong and work out the correct expansion of $(2x + 4)(x + 3)$.

b Multiply out and simplify:

  i $(2y + 1)(y + 3)$ **ii** $(x + 2)(3x + 4)$ **iii** $(2w + 5)(3w + 1)$

7 Expand and simplify:

  a $(3w + 2)(w - 4)$ **b** $(p - 5)(5p + 2)$ **c** $(4y + 3)(3y - 1)$

  d $(3g - 2)(2g - 3)$ **e** $(2x + 3)^2$ **f** $(5h - 7)^2$

Consolidate – do you need more?

1 Expand and simplify:

  a $(g + 3)(g + 1)$ **b** $(x + 4)(x + 6)$ **c** $(y - 2)(y + 6)$

  d $(p + 2)(p - 1)$ **e** $(h - 4)(h + 1)$ **f** $(k + 3)(k - 7)$

2 Multiply out and simplify:

  a $(x - 4)(x - 2)$ **b** $(u - 4)(u - 5)$ **c** $(y - 3)(y - 1)$

  d $(t - 3)(t + 3)$ **e** $(w + 9)(w - 9)$ **f** $(n + 11)(n - 9)$

3 Expand and simplify: **a** $(x + 1)^2$ **b** $(x - 10)^2$

4 Multiply out and simplify:

  a $(2x + 3)(x + 4)$ **b** $(3u + 5)(u - 1)$ **c** $(y - 4)(4y + 3)$

  d $(2h - 4)(4h - 2)$ **e** $(3w + 2)^2$ **f** $(6g - 1)^2$

Stretch – can you deepen your learning?

1 Write an expression for the area of each shape.

  a
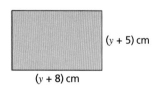
  $(y + 5)$ cm

  $(y + 8)$ cm

  b
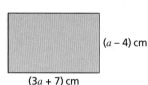
  $(a - 4)$ cm

  $(3a + 7)$ cm

2 Multiply out $(g + h)(k + m)$.

3 All these expressions are equivalent.

  $(x + 3)(x + 4)$ $(x + 2)(x + 5) + a$ $(x + 2)(x + 6) + bx$ $(x + 3)^2 + cx + d$

Work out the values of the letters a, b, c and d.

4 Expand and simplify: **a** $(e + 2)(e - 2)$ **b** $(r - 5)(r + 5)$ **c** $(b + 9)(b - 9)$

Can you generalise your findings?

5 Work out the value of: **a** $62^2 - 38^2$ **b** $84^2 - 16^2$ **c** 32×28

8.3 Factorising quadratic expressions

Are you ready?

1 Factorise fully: **a** $4g + 12$ **b** $g^2 - 4g$ **c** $4g^2 - 8g$

2 Write down the factors of each number.

    a 10 **b** 12 **c** 20 **d** 24

3 Write down a pair of numbers that have:

    a a sum of 5 and a product of 6   **b** a sum of 3 and a product of –10

    c a sum of –3 and a product of –18   **d** a sum of –9 and a product of 20

Factorising a quadratic expression is the opposite of expanding brackets.

A quadratic expression in the form of $x^2 + bx + c$ requires two brackets $(x + d)(x + e)$ when factorised.

This technique can be used to solve quadratic equations.

Example

Factorise: **a** $p^2 + 8p + 12$ **b** $q^2 + 4q - 12$

> You can check your answers by expanding and simplifying to see if you get the expression you started with.

Method

Solution	Commentary
a $p^2 + 8p + 12$ <table><tr><td>×</td><td>p</td><td></td></tr><tr><td>p</td><td>p^2</td><td></td></tr><tr><td></td><td></td><td>12</td></tr></table> $2 \times 6 = 12$ and $2 + 6 = 8$	Use a table, filling in the p^2 and constant term. The side lengths can be partly filled in with p as you know $p \times p = p^2$
<table><tr><td>×</td><td>p</td><td>6</td></tr><tr><td>p</td><td>p^2</td><td>$6p$</td></tr><tr><td>2</td><td>$2p$</td><td>12</td></tr></table>	You now need to find a pair of values which multiply to make the constant (12) and sum to make the coefficient of p (8). The only possible values are 2 and 6
So $p^2 + 8p + 12 \equiv (p + 2)(p + 6)$	Write the quadratic as a product of the two factors.
b $q^2 + 4q - 12$ <table><tr><td>×</td><td>q</td><td></td></tr><tr><td>q</td><td>q^2</td><td></td></tr><tr><td></td><td></td><td>–12</td></tr></table> $\longrightarrow$ <table><tr><td>×</td><td>q</td><td>6</td></tr><tr><td>q</td><td>q^2</td><td>$6q$</td></tr><tr><td>–2</td><td>$-2q$</td><td>–12</td></tr></table> $-2 \times 6 = -12$ and $-2 + 6 = 4$	Use a table, filling in the q^2 and constant term, and look for a pair of values which multiply to make the constant (–12) and sum to make the coefficient of q (4).
So $q^2 + 4q - 12 \equiv (q - 2)(q + 6)$	Write the quadratic as a product of the two factors.

Factorising a quadratic expression such as $(x^2 - 9)$ is a special case, often known as the **difference of two squares**.

The model here shows why $(x^2 - 9)$ factorised is $(x + 3)(x - 3)$.

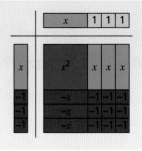

Example

Factorise $y^2 - 16$

Method

Solution	Commentary
$y^2 - 16 \equiv y^2 + 0x - 16$	$y^2 - 16$ is a case of the **difference of two squares**. You can think of it as $y^2 + 0x - 16$
$4 \times -4 = -16$ $4 + -4 = 0$	You now need two values with a product of -16 and a sum of 0 4 and -4 are the only possible options.
$y^2 - 16 \equiv (y + 4)(y - 4)$	

Practice

1. Factorise these quadratic expressions.

   a $x^2 + 6x + 5$ b $x^2 + 7x + 10$ c $x^2 + 8x + 15$

   d $y^2 + 7y + 12$ e $w^2 + 12w + 20$ f $p^2 + 9p + 20$

2. Factorise these expressions completely.

   a $x^2 + 6x - 7$ b $x^2 + 4x - 32$ c $x^2 + 3x - 10$

   d $g^2 + 2g - 15$ e $k^2 + 3k - 18$ f $h^2 + 7h - 18$

3. Factorise these expressions completely.

   a $x^2 - 2x - 3$ b $x^2 - 5x - 6$ c $x^2 - 2x - 15$

   d $y^2 - 4y - 12$ e $a^2 - 6a - 16$ f $c^2 - c - 20$

4. a Chloe is factorising $x^2 - 9x + 20$

   She says $4 \times 5 = 20$ and $4 + 5 = 9$ so $x^2 - 9x + 20 \equiv (x + 4)(x + 5)$

   Explain why Chloe is wrong and correctly factorise $x^2 - 9x + 20$

   b Factorise:

   i $x^2 - 9x + 18$ ii $y^2 - 10y + 24$ iii $w^2 - 11w + 30$ iv $u^2 - 14u + 24$

5 Factorise these expressions completely.

   a $x^2 + 8x + 16$ **b** $y^2 + 10y + 25$ **c** $w^2 - 4w + 4$ **d** $p^2 - 12p + 36$

   What do you notice about your answers?

6 Factorise:

   a $x^2 - 9$ **b** $y^2 - 49$ **c** $w^2 - 100$

   What do you notice about your answers?

7 Factorise these expressions completely.

   a $p^2 + 11p + 24$ **b** $q^2 + 13q + 36$ **c** $r^2 - 7r + 12$

   d $s^2 - s - 6$ **e** $u^2 - 8u - 20$ **f** $v^2 - 16v + 64$

   g $w^2 - 7w - 30$ **h** $x^2 - 11x + 30$ **i** $y^2 + y - 30$

Consolidate – do you need more?

1 Factorise these quadratic expressions completely.

   a $x^2 + 8x + 7$ **b** $y^2 + 9y + 14$ **c** $x^2 + 10x + 21$

   d $y^2 + 11y + 28$ **e** $w^2 + 21w + 20$ **f** $p^2 + 12p + 36$

2 Factorise these expressions completely.

   a $x^2 + 7x - 8$ **b** $x^2 + 6x - 16$ **c** $x^2 + 4x - 21$

   d $y^2 + 7y - 30$ **e** $w^2 - w - 30$ **f** $a^2 - 10a - 24$

   g $g^2 - 5g + 36$ **h** $p^2 - 4p + 45$ **i** $u^2 - 18u + 81$

3 Factorise:

   a $x^2 - 1$ **b** $g^2 - 64$ **c** $k^2 - 144$

Stretch – can you deepen your learning?

1 The expression for the area of the rectangle is $(x^2 - 5x - 50)\,\text{cm}^2$.

$(x - 10)$ cm

Write an expression for the unknown side length.

2 Work out the unknown values a, b, c and d.

$x^2 + 5x + a \equiv (x + b)(x + 3)$

$x^2 - x + c \equiv (x + 7)(x - d)$

3 Factorise $64x^2 - 9y^2$

4 If $(x + 5)(x - 5) = 24$, show that $(x + 7)(x - 7) = 0$

8.4 Identities

White Rose MATHS

Are you ready?

1 Simplify:

a $2y + y$ **b** $2y^2 + y + y^2$ **c** $2y - y$ **d** $3y + w - y + 3w$

2 Expand and simplify:

a $3(2y + 1) + 2(y + 4)$ **b** $3(x - 3) + 4(x - 2)$

3 Factorise:

a $6y + 18$ **b** $24x - 16$ **c** $15w^2 - 35w$

Remember that the sign $\equiv$ means 'is equivalent to' or 'identical to'.

For example:

$x + x \equiv 2x$ $x \times 4 \equiv 4x$ $a + b \equiv b + a$

The expressions on either side of the sign are always equal, no matter what the value of the variable(s).

Example 1

Are the following statements **true** or **false**?

a $t + 3t \equiv 4t$ **b** $5 \times t \times 2 \equiv 10t$ **c** $t^2 - t \equiv 2$

Method

Solution	Commentary
a $t + 3t \equiv 4t$ is true.	You can prove this using manipulatives or pictures and verify by checking with numbers.
b $5 \times t \times 2 \equiv 10t$ is true.	Work from left to right. $5 \times t \equiv 5t$ $5t \times 2 \equiv 10t$
c $t^2 - t \equiv 2$ is false.	t^2 and t are unlike terms, so you can't subtract directly. Check with a value, say $t = 10$ $10^2 - 10 = 100 - 10 = 90$, not 2 So the statement is false.

Example 2

Show that $3(2 + 3x) + 3x \equiv 6(2x + 1)$

Method

Solution	Commentary
Left-hand side: $3(2 + 3x) + 3x \equiv 6 + 9x + 3x$	Expand the brackets on the left-hand side of the identity.
$6 + 9x + 3x \equiv 12x + 6$	Collect like terms.
$12x + 6 \equiv 6(2x + 1)$ Same as the right-hand side of the identity.	Now fully factorise this expression to show it is the same as the right-hand side.

Practice

1 Are these identities **true** or **false**? Explain your answers.

a $5x + x \equiv 5x^2$ **b** $5x + 4y \equiv 9xy$ **c** $5x + 4x \equiv 9x$

d $5x + y + 4x \equiv 9x + y$ **e** $5 \times y \equiv y + y + y + y + y$ **f** $5y - y \equiv 5$

g $5xy - y \equiv 5x$ **h** $5x + 4y - x \equiv 4x + 4y$ **i** $5x + 4y + x \equiv 10xy$

2 Copy and complete the identities.

a $5g + 2g \equiv \boxed{}$ **b** $10h - 3h \equiv \boxed{}$

c $9k + 2k - k \equiv \boxed{}$ **d** $6p^2 + 2p + 3p^2 \equiv \boxed{}$

e $5w - m + 5w + 2m \equiv \boxed{}$ **f** $\boxed{} + \boxed{} - \boxed{} \equiv 4f + 4n$

3 **a** Show that $8(2u + 3) \equiv 2(12 + 8u)$

    b Show that $3(2y + 5) - 3 \equiv 6(y + 2)$

    c Show that $3(4 + 2h) + 8(h - 1) \equiv 2(7h + 2)$

4 Work out the values of a and b such that these identities are true.

    a $3(y + 1) + 4(y + 2) \equiv ay + b$

    b $4(h - 3) + 3(h + 5) \equiv ah + b$

    c $6(w + a) + 2(w + 7) \equiv bw + 20$

    d $x(x + a) + 3(x + b) \equiv x^2 + 5x + 18$

Consolidate – do you need more?

1 Which of the expressions are equivalent to $3p$?

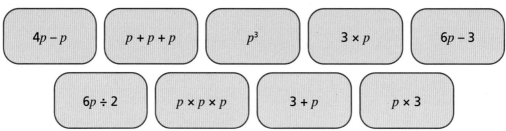

$4p - p$ $p + p + p$ p^3 $3 \times p$ $6p - 3$

$6p \div 2$ $p \times p \times p$ $3 + p$ $p \times 3$

2 Are these identities **true** or **false**? Explain your answers.

a $3m + 3m \equiv 6m^2$
b $2m + 3k + 4m + 5k \equiv 14mk$
c $3m^2 + k + k \equiv 3m^2 + 2k$
d $9m + k + m \equiv 10m + k$
e $4m - 4 \equiv m$
f $7m - k - 2k \equiv 7m - 3k$

3 Copy and complete the identities.

a $5t - t \equiv \boxed{}$
b $7g^2 + 2g + 11g^2 \equiv \boxed{}$
c $3f - e + 2f - 5f \equiv \boxed{}$
d $\boxed{} + \boxed{} - \boxed{} \equiv 3a - 2b$

4 Show that $6(3x + 10) \equiv 6 + 9(2x + 6)$

5 Work out the values of a and b such that these identities are true.

a $5(g + 2) + 4(g - 2) \equiv ag + b$
b $3(a + y) + 4(y + 8) \equiv by + 35$

Stretch – can you deepen your learning?

1 Copy and complete the expressions to create identities.

a $4(x + 3) \equiv 4x + \boxed{}$
b $\boxed{}(y + 7) \equiv \boxed{}y + 21$
c $3(\boxed{}w - 4) + 6(2w - \boxed{}) \equiv 18w - 24$

2 Use the diagram to create an identity.

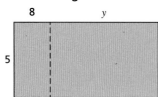

3 There are two values of y for which the expressions $2y$ and y^2 are equivalent.

Work out the two values.

Working with brackets: exam practice

1 Multiply out:

   (a) $2(a + 6)$ **[2 marks]**

   (b) $y(y - 4)$ **[2 marks]**

   (c) $4(6 + 5x)$ **[2 marks]**

2 Factorise:

   (a) $3w - 6$ **[1 mark]**

   (b) $b^2 + 7b$ **[1 mark]**

3 Factorise fully:

   $16m + 12n$ **[2 marks]**

4 Write an expression for the area of the rectangle.

                                                          [1 mark]

5 Expand and simplify:

   (a) $5(t + 7) + 2(t + 3)$ **[2 marks]**

   (b) $8(y - 3) - 7(y - 2)$ **[2 marks]**

6 Multiply out and simplify:

   (a) $(x + 6)(x - 9)$ **[2 marks]**

   (b) $(3x - 2)(x - 1)$ **[2 marks]**

7 Work out the values of a and b.

   $y^2 + 5y + a \equiv (y + b)(y + 3)$ **[2 marks]**

In this block, we will cover...

9.1 Inputs, outputs and inverses

Example

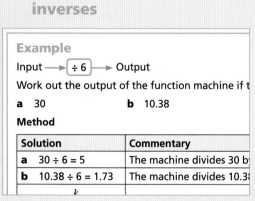

Input ⟶ $\div 6$ ⟶ Output

Work out the output of the function machine if t

a 30   **b** 10.38

Method

Solution	Commentary
a $30 \div 6 = 5$	The machine divides 30 by
b $10.38 \div 6 = 1.73$	The machine divides 10.38

9.2 Linear equations

Practice (B)

🖩 **1** Solve the equations.

     a $5y + 2 = 27$ **b** $3p - 20 = 1($

     d $5 + 6u = 11$ **e** $51 = 2e - 7$

2 Solve the equations.

     a $5q + 8 = 3$ **b** $17 + 4x = 5$

     d $2w - 9 = -1$ **e** $-19 = 6r - 1$

3 Solve these equations.

     a $3(x + 2) = 15$ **b** $8(r - 5) = 24$

9.3 Unknowns on both sides

Consolidate – do you need more

1 Solve the equation $2y + 11 = 4y + 3$

You can use the bar model to help you.

2 **a** Explain what your first step would be i

     i $7a + 4 = 3a + 24$

     iii $6 + 9m = m + 54$

     b Solve the equations in part **a**.

3 Solve these equations.

     a $3b - 5 = 2b + 4$ **b** $12e - 5 = 5$

9.1 Inputs, outputs and inverses

Are you ready? (A)

1 Simplify:

   a $3 \times y$ **b** $p \times p$ **c** $h \div 4$ **d** $7 \div w$

2 Write these using algebraic notation.

   a Double y **b** y divided by 5   **c** 5 more than y **d** 5 less than y

---

You can use 'function machine' diagrams to model **functions**.

A function is a mathematical relationship with an input and an output.

Function machines	Two-step function machines
Input ⟶ $\boxed{\times 5}$ ⟶ Output	Input ⟶ $\boxed{\times 2}$ ⟶ $\boxed{+ 3}$ ⟶ Output

---

Example

Input ⟶ $\boxed{\div 6}$ ⟶ Output

Work out the output of the function machine if the input is:

a 30   **b** 10.38   **c** k **d** $5m$

Method

Solution	Commentary
a $30 \div 6 = 5$	The machine divides 30 by 6
b $10.38 \div 6 = 1.73$	The machine divides 10.38 by 6
c $k \div 6 = \dfrac{k}{6}$	The machine divides k by 6, so the output is $k \div 6$, which you write as $\dfrac{k}{6}$
d $5m \div 6 = \dfrac{5m}{6}$	The machine divides $5m$ by 6

Practice (A)

1 Work out the output when you input 60 into each of these function machines.

   a Input ⟶ $\boxed{+ 10}$ ⟶ Output   **b** Input ⟶ $\boxed{\times 10}$ ⟶ Output

   c Input ⟶ $\boxed{- 10}$ ⟶ Output   **d** Input ⟶ $\boxed{\div 10}$ ⟶ Output

   e Input ⟶ $\boxed{+ 0.6}$ ⟶ Output   **f** Input ⟶ $\boxed{\times 0.6}$ ⟶ Output

   g Input ⟶ $\boxed{- 0.6}$ ⟶ Output   **h** Input ⟶ $\boxed{\div 0.6}$ ⟶ Output

2 Calculate the output when you input each of the following into these function machines.

a 18.2   **b** 281

  i   Input $\longrightarrow$ [× 6.5] $\longrightarrow$ Output   **ii**   Input $\longrightarrow$ [+ 13.9] $\longrightarrow$ Output

 iii   Input $\longrightarrow$ [− 21.4] $\longrightarrow$ Output   **iv**   Input $\longrightarrow$ [+ 3500] $\longrightarrow$ Output

  v   Input $\longrightarrow$ [subtract from 400] $\longrightarrow$ Output

3 Work out the output when you input y into each of these function machines.

 a   Input $\longrightarrow$ [× 8] $\longrightarrow$ Output   **b**   Input $\longrightarrow$ [+ 8] $\longrightarrow$ Output

 c   Input $\longrightarrow$ [− 8] $\longrightarrow$ Output   **d**   Input $\longrightarrow$ [÷ 8] $\longrightarrow$ Output

 e   Input $\longrightarrow$ [subtract from 8] $\longrightarrow$ Output

4 Work out the output when you input each of the following into these function machines.

a 7   **b** h **c** $5h$

  i   Input $\longrightarrow$ [− 6] $\longrightarrow$ Output   **ii**   Input $\longrightarrow$ [× 6] $\longrightarrow$ Output

 iii   Input $\longrightarrow$ [÷ 6] $\longrightarrow$ Output   **iv**   Input $\longrightarrow$ [× g] $\longrightarrow$ Output

  v   Input $\longrightarrow$ [× h] $\longrightarrow$ Output   **vi**   Input $\longrightarrow$ [÷ h] $\longrightarrow$ Output

Are you ready? (B)

1 Write each term in the simplest way.

 a $6 \times e$ **b** $c \times 5$ **c** $7 \div f$

 d $g \div 8$ **e** $h \times h$ **f** $j \times k$

2 Write down the inverse of these operations.

 a   × 3   **b**   + 4   **c**   ÷ 5   **d**   − 6

3 What is the inverse of squaring?

Example

The output of each of these function machines is 10

Work out the input for each machine.

a   Input $\longrightarrow$ [− 20] $\longrightarrow$ 10   **b**   Input $\longrightarrow$ [× 20] $\longrightarrow$ 10

c   Input $\longrightarrow$ [÷ 20] $\longrightarrow$ 10   **d**   Input $\longrightarrow$ [− p] $\longrightarrow$ 10

Method

Solution	Commentary
a $10 + 20 = 30$	The inverse of 'subtract 20' is 'add 20'.
b $10 \div 20 = 0.5$	The inverse of 'multiply by 20' is 'divide by 20'.
c $10 \times 20 = 200$	The inverse of 'divide by 20' is 'multiply by 20'.
d $10 + p$	The inverse of 'subtract p' is 'add p'.

Practice (B)

1 The output for each function machine is 60

Work out the inputs.

a $? \longrightarrow \boxed{+6} \longrightarrow 60$ **b** $? \longrightarrow \boxed{-6} \longrightarrow 60$

c $? \longrightarrow \boxed{\times 6} \longrightarrow 60$ **d** $? \longrightarrow \boxed{\div 6} \longrightarrow 60$

2 Calculate the input for each function machine.

a $? \longrightarrow \boxed{+3.5} \longrightarrow 145.5$ **b** $? \longrightarrow \boxed{-5.82} \longrightarrow 24.28$

c $? \longrightarrow \boxed{\div 5.7} \longrightarrow 1.89$ **d** $? \longrightarrow \boxed{\times 4.8} \longrightarrow 74.88$

e $? \longrightarrow \boxed{-83.23} \longrightarrow 0$ **f** $? \longrightarrow \boxed{\times 24.53} \longrightarrow 0$

3 The output for each function machine is w.

Work out the inputs, giving your answers in correct algebraic notation.

a $? \longrightarrow \boxed{+9} \longrightarrow w$ **b** $? \longrightarrow \boxed{\times 9} \longrightarrow w$

c $? \longrightarrow \boxed{\div 9} \longrightarrow w$ **d** $? \longrightarrow \boxed{-9} \longrightarrow w$

4 The output for each function machine is $8g$.

Work out the inputs, giving your answers in correct algebraic notation.

a $? \longrightarrow \boxed{\times 4} \longrightarrow 8g$ **b** $? \longrightarrow \boxed{-4} \longrightarrow 8g$

c $? \longrightarrow \boxed{+4} \longrightarrow 8g$ **d** $? \longrightarrow \boxed{\div 4} \longrightarrow 8g$

What do you think? 💬

1 The output for each function machine is $m + 7$

Work out the inputs.

a $? \longrightarrow \boxed{+5} \longrightarrow m + 7$ **b** $? \longrightarrow \boxed{-5} \longrightarrow m + 7$

c $? \longrightarrow \boxed{\div 5} \longrightarrow m + 7$ **d** $? \longrightarrow \boxed{\times 5} \longrightarrow m + 7$

2 $\text{Input} \longrightarrow \boxed{\text{square}} \longrightarrow \text{Output}$

Work out the input of this function machine if the output is:

a 36   **b** 2.25   **c** y^2 **d** $4y^2$

Are you ready? (C)

1 Here is a number puzzle: "Think of a number. Double it. Add 4."

a What is the answer to the puzzle if the number you think of is 10?

b What is the number you think of if the answer to the puzzle is 10?

2 Write these expressions in the simplest way.

a $3 \times a - 5$ **b** $(b + 3) \times 5$ **c** $c \div 3 + 5$ **d** $(d + 5) \div 3$

Example 1

Work out the output for each of these two-step function machines if the input is 24

a Input $\longrightarrow$ $\boxed{\times 2}$ $\longrightarrow$ $\boxed{+ 2}$ $\longrightarrow$ Output   **b** Input $\longrightarrow$ $\boxed{+ 2}$ $\longrightarrow$ $\boxed{\times 2}$ $\longrightarrow$ Output

c Input $\longrightarrow$ $\boxed{\times 2}$ $\longrightarrow$ $\boxed{+ a}$ $\longrightarrow$ Output   **d** Input $\longrightarrow$ $\boxed{+ a}$ $\longrightarrow$ $\boxed{\times 2}$ $\longrightarrow$ Output

Method

Solution	Commentary
a $24 \longrightarrow \boxed{\times 2} \longrightarrow 48 \longrightarrow \boxed{+ 2} \longrightarrow 50$	First multiply by 2 then add 2 to the result.
b $24 \longrightarrow \boxed{+ 2} \longrightarrow 26 \longrightarrow \boxed{\times 2} \longrightarrow 52$	
c $24 \longrightarrow \boxed{\times 2} \longrightarrow 48 \longrightarrow \boxed{+ a} \longrightarrow 48 + a$	
d $24 \longrightarrow \boxed{+ a} \longrightarrow 24 + a \longrightarrow \boxed{\times 2} \longrightarrow 2(24 + a)$ or $48 + 2a$	There are two ways to write the answer. For the second operation, you could multiply each term in the expression $24 + a$ by 2 to get $48 + 2a$, or you could write $2(24 + a)$, which means '2 times the value of $24 + a$'.

Example 2

The output of each of these two-step function machines is 10

Work out the input in each case.

a Input $\longrightarrow$ $\boxed{+ 4}$ $\longrightarrow$ $\boxed{\div 2}$ $\longrightarrow$ Output   **b** Input $\longrightarrow$ $\boxed{\times 2}$ $\longrightarrow$ $\boxed{+ 3}$ $\longrightarrow$ Output

Method

Solution	Commentary
a $? \longrightarrow \boxed{+ 4} \longrightarrow \boxed{\div 2} \longrightarrow 10$ $16 \longleftarrow \boxed{- 4} \longleftarrow 20 \longleftarrow \boxed{\times 2} \longleftarrow 10$ The input is 16	The inverse of '+ 4' is '– 4' and the inverse of '÷ 2' is '× 2'. Working backwards gives the inverse function machine.
b $? \longrightarrow \boxed{\times 2} \longrightarrow \boxed{+ 3} \longrightarrow 10$ $3.5 \longleftarrow \boxed{\div 2} \longleftarrow 7 \longleftarrow \boxed{- 3} \longleftarrow 10$ The input is 3.5	The inverse of '× 2' is '÷ 2' and the inverse of '+ 3' is '– 3'. Working backwards gives the inverse function machine.

You can make your answer the input to the original function machine to check that it gives the correct output.

$16 + 4 = 20$, $20 \div 2 = 10$. Correct.

$3.5 \times 2 = 7$, $7 + 3 = 10$. Correct.

Practice (C)

1 Work out the output when you input 8 into each of these two-step function machines.

a Input →| + 4 |→| ÷ 2 |→ Output

b Input →| ÷ 2 |→| + 4 |→ Output

c Input →| − 2 |→| × 4 |→ Output

d Input →| ÷ 4 |→| + 2 |→ Output

2 I think of a number. I multiply my number by 8 and then subtract 5

a Draw a function machine to show the number puzzle.

b Work out the output of the function machine if the input is:

i 7   **ii** 16.2   **iii** k

3 Marta thinks that if you input the same number into these two function machines you will always get the same output.

Input →| × 5 |→| + 2 |→ Output

Input →| × 2 |→| + 5 |→ Output

Input 10 into each function machine to show that Marta is wrong.

4 Work out the output when you input p into each of these function machines.

a Input →| × 3 |→| + 4 |→ Output

b Input →| + 3 |→| × 4 |→ Output

c Input →| − 3 |→| ÷ 4 |→ Output

d Input →| ÷ 4 |→| + 3 |→ Output

5 The output of each of these function machines is 8

Work out the inputs.

a Input →| + 4 |→| ÷ 2 |→ Output

b Input →| ÷ 2 |→| + 4 |→ Output

c Input →| − 2 |→| × 4 |→ Output

d Input →| ÷ 4 |→| + 2 |→ Output

6 I think of a number, subtract 7 and then multiply the result by 4

a Show this sentence using a two-step function machine.

b If my answer is 40, what number did I start with?

c If my answer is 4, what number did I start with?

d If my answer is 21, what number did I start with?

7 Input →| × 3 |→| + 8 |→ Output

Work out the input to this function machine if the output is:

a 44   **b** 245   **c** 8

d $3y + 8$ **e** $45y + 8$ **f** $12y + 11$

Consolidate – do you need more?

1 Work out the output when you input 50 into each of these function machines.

a Input → $+5$ → Output
b Input → $\times 5$ → Output
c Input → -5 → Output
d Input → $\div 5$ → Output
e Input → $+0.5$ → Output
f Input → $\times 0.5$ → Output
g Input → -0.5 → Output
h Input → $\div 0.5$ → Output

2 Work out the output when you input the following into each of the function machines below.

a t　　　　**b** $7t$

i Input → -7 → Output
ii Input → $+7$ → Output
iii Input → $\times 7$ → Output
iv Input → $\div 7$ → Output
v Input → subtract from 7 → Output
vi Input → $\times t$ → Output

3 The output of each function machine is 50

Work out the inputs.

a ? → $+5$ → 50
b ? → -5 → 50
c ? → $\times 5$ → 50
d ? → $\div 5$ → 50

4 Work out the input when the output of each of these function machines is:

a t　　　　**b** $7t$

i ? → $+7$ → Output
ii ? → $\div 7$ → Output
iii ? → -7 → Output
iv ? → $\times 7$ → Output

5 **a** Work out the output when you input 30 into each of these function machines.

i Input → $+10$ → $\div 2$ → Output
ii Input → $\div 2$ → $+10$ → Output
iii Input → -2 → $\times 10$ → Output
iv Input → $\div 10$ → $+2$ → Output

b Work out the input if the output of each of the function machines above is 30

6 I think of a number. I divide my number by 10 and then add 17

a Draw a two-step function machine to show the number puzzle.

b Work out the output of the function machine if the input is:

i 80　　　　**ii** 143　　　　**iii** f

c Work out the input of the function machine if the output is:

i 80　　　　**ii** 143　　　　**iii** f

7 **a** Work out the output when you input w into each of these function machines.

  i Input ⟶ ×5 ⟶ +2 ⟶ Output

  ii Input ⟶ +5 ⟶ ×2 ⟶ Output

  iii Input ⟶ −5 ⟶ ÷2 ⟶ Output

  iv Input ⟶ ÷5 ⟶ +2 ⟶ Output

  b Work out the input to each function machine above if the output is $10y + 22$

Stretch – can you deepen your learning?

1 Work out inputs for which both operations in each pair give:

  i the same output   **ii** different outputs.

  a Input ⟶ −20 ⟶ Output

    Input ⟶ subtract from 20 ⟶ Output

  b Input ⟶ ÷2 ⟶ Output

    Input ⟶ square root ⟶ Output

2 Work out the missing values or expressions.

  a ☐ ⟶ ×3 ⟶ + ☐ ⟶ $3e + 4$

  b ☐ ⟶ ×3 ⟶ + ☐ ⟶ $6f + 5$

  c ☐ ⟶ ×g ⟶ − ☐ ⟶ $g^2 - 8$

3 Copy and complete the function machines.

  a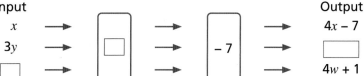

Input				Output
x	⟶	☐ ⟶ −7 ⟶		$4x - 7$
$3y$				☐
☐				$4w + 1$

  b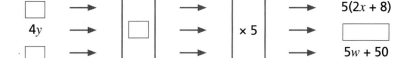

Input				Output
☐	⟶	☐ ⟶ ×5 ⟶		$5(2x + 8)$
$4y$				☐
☐				$5w + 50$

Are you ready? (A)

1 Write down the inverse of each operation.

    a $- 5$ **b** $+ 5$ **c** $\div 5$ **d** $\times 5$

2 Work out the value of each expression when $x = 20$

    a $50 + x$ **b** $50 - x$ **c** $50x$

    d $\dfrac{50}{x}$ **e** $\dfrac{x}{50}$

3 I think of a number and multiply it by 5

    The result is 35

    What number was I thinking of?

An **equation** is a mathematical statement that includes an equals sign to show that two expressions are equal.

When you **solve** an equation, you work out the value of the unknown letter. If you use a balancing method, you must do the **same to both sides** of the equation to keep it balanced.

You can represent an equation using a bar model. This might help you to see which operations to do to work out the value of each letter.

$$x + 5 = 12 \qquad\qquad \frac{y}{3} = 5 \qquad\qquad 2q = 18$$

x	5
12	

y		
5		

q	q
18	

Solutions to equations can be integers, negative numbers, fractions or decimals.

The balancing method will be used in the examples.

Example 1

Solve the equations.

a $x + 5 = 17$ **b** $\dfrac{y}{7} = 1$ **c** $0.5c = 9$

Method

Solution	Commentary
a $\quad x + 5 = 17$ $-5 \Big(\quad\quad\Big) -5$ $\quad\quad x = 12$	The aim of solving an equation is to get the unknown on its own. To do this, you can apply the inverse to 'undo' everything that is around the unknown. It is important to keep the equation balanced, so whatever you do to the left-hand side must also be done to the right-hand side.

b $\dfrac{y}{7} = 1$ $\times 7$ ⟍ ⟋ $\times 7$ $y = 7$	The inverse of division is multiplication, therefore multiply by 7 on both sides of the equation to get y on its own.
c $0.5c = 9$ $\div 0.5$ ⟍ ⟋ $\div 0.5$ $c = 18$	The inverse of multiplication is division, so divide both sides by 0.5 to get c on its own. Remember that dividing by 0.5 is the same as multiplying by 2

Some equations are more complex to solve, but you can't go wrong if you follow the correct rules of using the inverse.

Example 2

Solve the equations.

a $3 - w = 10$ **b** $\dfrac{20}{m} = 5$

Method

Solution	Commentary
a $3 - w = 10$ $+ w$ ⟍ ⟋ $+ w$ $3 = 10 + w$ $- 10$ ⟍ ⟋ $- 10$ $-7 = w$ $w = -7$	First you can add w to both sides to give $3 = 10 + w$ Now subtract 10 from both sides. You can leave your answer as $-7 = w$, however it is good practice to rewrite it with the letter on the left.
b $\dfrac{20}{m} = 5$ $\times m$ ⟍ ⟋ $\times m$ $20 = 5m$ $\div 5$ ⟍ ⟋ $\div 5$ $4 = m$ $m = 4$	First multiply by m so that it is no longer the denominator. Now divide by 5

Practice (A) 🖩

1 Solve the equations.

     a $g + 4 = 20$ **b** $4b = 20$ **c** $h - 4 = 20$ **d** $\dfrac{x}{4} = 20$

2 Solve each equation.

a $y + 72 = 118$ b $402 = w + 129$ c $87 + c = 291$

d $3.7 = 1.8 + r$ e $1201 + u = 4182$ f $0.29 = 0.124 + g$

3 Solve:

a $k - 19 = 31$ b $72 = d - 29$ c $c - 9.8 = 15.3$

4 Solve:

a $7a = 21$ b $\dfrac{a}{7} = 21$ c $18 = 10t$ d $18 = \dfrac{t}{10}$

e $24.8 = 4q$ f $24.8 = \dfrac{q}{4}$ g $\dfrac{u}{2.5} = 18$ h $18 = 2.5u$

5 Solve the equations.

a $w + 18 = 50$ b $w - 50 = 18$ c $50 - w = 18$

d $y - 19 = 5.6$ e $19 - y = 5.6$ f $6.5 = 91 - y$

6 Work out the values of the letters.

a $\dfrac{a}{12} = 6$ b $\dfrac{12}{a} = 6$ c $15c = 135$ d $\dfrac{c}{15} = 135$

e $21 = \dfrac{126}{w}$ f $\dfrac{35}{y} = 14$ g $\dfrac{200}{x} = 12.5$ h $91 = \dfrac{91}{f}$

What do you think? (A) 💀

1 Rhys says these equations are impossible to solve because the answer is the same as the starting number.

$75 + y = 75$ $75w = 75$

Do you agree with Rhys? Explain your answer.

2 a Explain the difference between solving the equations $p - 24 = 8$ and $24 - p = 8$

    b Explain the difference between solving the equations $\dfrac{p}{24} = 8$ and $\dfrac{24}{p} = 8$

Are you ready? (B)

1 Solve the equations.

a $3k = 21$ b $\dfrac{k}{3} = 21$ c $k - 3 = 21$ d $21 = k + 3$

2 Solve the equations.

a $a - 15 = 2$ b $15 - a = 2$ c $2 = \dfrac{a}{15}$ d $\dfrac{15}{a} = 2$

3 Multiply out:

a $3(x + 4)$ b $5(2x - 7)$

Example 1

Solve $25 = 5x + 10$

Method

Solution	Commentary
$25 = 5x + 10$ -10 $\quad\quad$ -10 $15 = 5x$ $\div 5$ $\quad\quad$ $\div 5$ $3 = x$ So $x = 3$	Subtract 10 from both sides of the equation. This is how you could represent solving the equation using bar models. $\begin{array}{ c }\hline 25 \\\hline \end{array}$ $\begin{array}{ c\| c }\hline 5x & 10 \\\hline\end{array}$ Divide both sides of the equation by 5 $\begin{array}{\|c\|}\hline 15 \\\hline 5x \\\hline\end{array}$ The solution to the equation is $x = 3$ $\begin{array}{\|c\|c\|c\|c\|c\|}\hline 3 & 3 & 3 & 3 & 3 \\\hline x & x & x & x & x \\\hline\end{array}$

Example 2

Solve $2(t - 2) = 13$

Method A

Solution	Commentary
$2(t - 2) = 13$ $2t - 4 = 13$ $+4$ $\quad\quad$ $+4$ $2t = 17$ $\div 2$ $\quad\quad$ $\div 2$ $t = \frac{17}{2} = 8.5$ So $t = 8.5$	Expand the bracket by multiplying everything inside it by 2 Add 4 to both sides of the equation. Divide by 2 on both sides of the equation. The solution to an equation is not always an integer. You can leave your answer as a fraction or change it to a decimal.

Method B

Solution	Commentary
$2(t - 2) = 13$ $\div 2$ $\quad\quad$ $\div 2$ $t - 2 = 6.5$ $t = 8.5$	You could divide both sides by 2 in the first step as an alternative method to expanding the bracket.

Example 3

Solve:

a $\dfrac{g}{4} - 2 = 9$

b $\dfrac{g - 2}{4} = 9$

Method

Solution	Commentary
a $\dfrac{g}{4} - 2 = 9$ $+2 \quad\quad\quad +2$ $\dfrac{g}{4} = 11$ $\times 4 \quad\quad\quad \times 4$ $g = 44$	Add 2 to both sides of the equation. Multiply both sides of the equation by 4
b $\dfrac{g - 2}{4} = 9$ $\times 4 \quad\quad\quad \times 4$ $g - 2 = 36$ $+2 \quad\quad\quad +2$ $g = 38$	Start by multiplying both sides of the equation by 4 Now add 2 to both sides of the equation.

Practice (B)

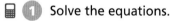 **1** Solve the equations.

    a $5y + 2 = 27$ **b** $3p - 20 = 10$ **c** $4k + 20 = 36$

    d $5 + 6u = 11$ **e** $51 = 2e - 7$ **f** $82 = 17 + 10w$

2 Solve the equations.

    a $5q + 8 = 3$ **b** $17 + 4x = 5$ **c** $1 = 3r + 10$

    d $2w - 9 = -1$ **e** $-19 = 6r - 1$ **f** $5g + 56 = 8$

3 Solve these equations.

    a $3(x + 2) = 15$ **b** $8(r - 5) = 24$ **c** $30 = 6(w - 8)$

    d $4(t + 3) = 8$ **e** $21 = 7(6 + y)$ **f** $5(p - 3) = -20$

4 Solve:

    a $\dfrac{w}{2} + 3 = 15$ **b** $\dfrac{p}{4} - 3 = 12$ **c** $7 + \dfrac{u}{3} = 11$ **d** $11 = \dfrac{r}{9} - 4$

5 Solve:

    a $\dfrac{w + 3}{2} = 15$ **b** $\dfrac{p - 3}{4} = 12$ **c** $\dfrac{7 + u}{3} = 11$ **d** $11 = \dfrac{r - 4}{9}$

6 Here are Mario's and Kath's methods for solving the equation $12 - 3x = 6$

Mario	Kath
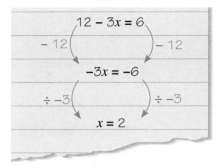	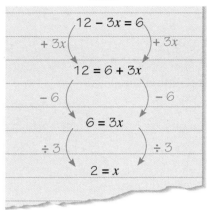

a Whose method do you prefer?

b Solve the equations.

    i $20 - 4x = 8$ **ii** $17 - 2x = 6$ **iii** $8 = 3 - 5x$

What do you think? (B) 💭

1 Huda thinks of a number.

She subtracts 5 and then divides by 3

Her result is 9

Lida thinks of a number.

She divides by 5 and then subtracts 3

Her result is 9

a Do you think Huda's and Lida's numbers are the same or different? Why?

b Write an equation to represent Huda's number puzzle and solve the equation.

c Write an equation to represent Lida's number puzzle and solve the equation.

Zach thinks of a number.

He adds 5 and then multiplies by 3

His result is 9

d Write an equation to represent Zach's number puzzle and solve the equation.

Consolidate – do you need more?

1 Solve the equations.

    a $p + 5 = 15$ **b** $5b = 15$ **c** $h - 5 = 15$ **d** $\frac{x}{5} = 15$

2 Solve each equation.

    a $y + 89 = 121$ **b** $k - 51 = 67$ **c** $392 = w + 208$

    d $75 + f = 261$ **e** $2.6 = 1.9 + r$ **f** $67.1 = h - 28.3$

3 Solve:

a $8x = 24$ **b** $\frac{x}{8} = 24$ **c** $155 = 2u$

d $155 = \frac{u}{2}$ **e** $30 - s = 19$ **f** $65 = 87 - y$

g $\frac{b}{18} = 9$ **h** $\frac{18}{b} = 9$ **i** $10.4 = \frac{26}{k}$

4 Solve the equations.

a $6y + 2 = 50$ **b** $4p - 20 = 8$ **c** $3r + 8 = 2$

d $14 + 6p = 23$ **d** $20 = 8w - 8$ **e** $108 = 56 + 10q$

f $17 - 5m = 7$ **g** $4 = 39 - 5b$ **h** $50 - 2x = 58$

5 Solve these equations.

a $5(w + 2) = 30$ **b** $7(p - 3) = 28$ **c** $48 = 6(v - 7)$

d $3(c + 6) = 12$ **e** $76 = 8(6 + m)$ **f** $8(a - 9) = -56$

6 Solve:

a $\frac{y}{8} + 9 = 12$ **b** $\frac{g + 9}{8} = 12$ **c** $\frac{k}{5} - 7 = 34$

d $\frac{m - 7}{5} = 34$ **e** $\frac{16 + y}{6} = 8$ **f** $16 + \frac{x}{6} = 8$

Stretch – can you deepen your learning?

1 Solve the equations.

a $\frac{p}{3} + 8 = 34$ **b** $\frac{2p}{3} + 8 = 34$

c $2\left(\frac{p}{3} + 8\right) = 34$ **d** $2\left(\frac{p}{3} - 8\right) = 34$

2 Solve the equations.

a $20 - 3p = 14$ **b** $20 + 3(p - 1) = 14$

c $20 - 3(p + 1) = 14$ **d** $20 - \frac{p + 1}{3} = 14$

3 Work out the values of the letters by forming and solving equations.

a
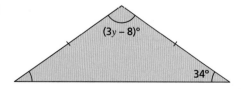

b
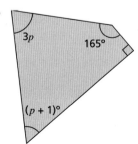

Are you ready?

1 Solve each equation.

   a $5y = 75$ **b** $108 = \frac{x}{3}$ **c** $m + 15 = 3$ **d** $17 = 25 - r$

2 Expand the brackets.

   a $5(a - 2)$ **b** $7(4 + y)$ **c** $4(8 - c)$ **d** $5(2g - 3)$

3 Solve each equation.

   a $2g + 7 = 15$ **b** $16 = \frac{w}{3} - 1$ **c** $8 - 5u = 3$ **d** $2(2x + 1) = 9$

To solve equations with the unknown on both sides of the equation, you should balance the equation so that the unknown is only on one side. This can be shown using algebra tiles:

$$2x + 3 = 5x + 1$$

 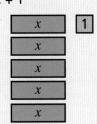

By subtracting $2x$ from both sides, the equation becomes $3 = 3x + 1$, which is much easier to solve.

Example 1

Solve the equation $4x + 2 = x + 17$

Method

Solution	Commentary
$4x + 2 = x + 17$ $-x$ ⟶ $-x$ $3x + 2 = 17$ -2 ⟶ -2 $3x = 15$ $\div 3$ ⟶ $\div 3$ $x = 5$	Represent the equation as a bar model. $4x + 2 = x + 17$ $\begin{array}{\|c\|c\|c\|c\|c\|} \hline x & x & x & x & 2 \\ \hline \end{array}$ $\begin{array}{\|c\|c\|} \hline x & 17 \\ \hline \end{array}$ Take x from both bars $3x + 2 = 17$ $\begin{array}{\|c\|c\|c\|c\|} \hline x & x & x & 2 \\ \hline \end{array}$ $\begin{array}{\|c\|} \hline 17 \\ \hline \end{array}$ Take 2 from both bars $3x = 15$ $\begin{array}{\|c\|c\|c\|} \hline x & x & x \\ \hline \end{array}$ $\begin{array}{\|c\|} \hline 15 \\ \hline \end{array}$ Divide into three equal parts $x = 5$ $\begin{array}{\|c\|} \hline x \\ \hline 5 \\ \hline \end{array}$

Example 2

Solve these equations.   **a** $3x = x + 6$ **b** $5y + 1 = 6y - 2$ **c** $z + 1 = 4 - 5z$

Method

Solution	Commentary
a $3x = x + 6$ $-x$ ⟶ $-x$ $2x = 6$ $\div 2$ ⟶ $\div 2$ $x = 3$	Subtract x from both sides of the equation. Divide both sides by 2
b $5y + 1 = 6y - 2$ $-5y$ ⟶ $-5y$ $1 = y - 2$ $+2$ ⟶ $+2$ $y = 3$	Subtract $5y$ from both sides of the equation. Find the value of y by adding 2 to both sides.
c $z + 1 = 4 - 5z$ $+5z$ ⟶ $+5z$ $6z + 1 = 4$ -1 ⟶ -1 $6z = 3$ $\div 6$ ⟶ $\div 6$ $z = 0.5$	Add $5z$ to both sides of the equation. This means the remaining term in z will have a positive coefficient. Take 1 from both sides of the equation to isolate the unknown. Divide both sides by 6

Practice

1 Solve the equation $5x + 4 = 2x + 19$

You can use the bar model to help you.

x	x	x	x	x	4

x	x	19			

2 Solve the equations.

a $6y + 5 = 3y + 20$ **b** $4w + 3 = w + 9$ **c** $2p + 15 = 8p + 3$

d $m + 31 = 7m + 1$ **e** $9t + 7 = 42 + 4t$ **f** $4k + 55 = 9k$

3 Solve these equations.

a $9y - 8 = 2y + 34$ **b** $7u - 3 = 5u + 2$ **c** $8p + 13 = 9p - 7$

4 Filipo is solving the equation $3x - 8 = 4 + 7x$

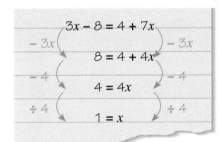

He checks his solution and discovers it is not correct.

$3 \times 1 - 8 = 3 - 8 = -5$ $4 + 7 \times 1 = 4 + 7 = 11$

$-5 \neq 11$

Identify Filipo's error and correct it.

5 Solve the equations.

a $7b - 11 = 2b - 1$ **b** $5k - 6 = 3k - 2$ **c** $m - 12 = 3m - 2$

d $5t + 6 = 2t - 3$ **e** $12a - 44 = 12 + 4a$ **f** $w - 4 = 6w - 6$

6 Solve these equations.

a $20 - 3p = 2p + 10$ **b** $5c + 9 = 30 - 2c$ **c** $8 - 2x = 3x - 7$

d $12 - 4y = y + 7$ **e** $10 - 2n = 16 + n$ **f** $20 - 4g = 13 - 2g$

What do you think? 💡

1 Jakub and Flo are solving the equation $4 - 6y = 16 - 10y$

Jakub says, "I'm going to add $6y$ to each side of the equation."

Flo says, "I'm going to add $10y$ to each side of the equation."

a Whose strategy is more sensible? Explain why.

b Solve the equation $4 - 6y = 16 - 10y$

Benji solves the equation by multiplying both sides by -1

$$4 - 6y = 16 - 10y$$
$\times -1$ $\times -1$
$$6y - 4 = 10y - 16$$

c Does Benji's method work?

2 Solve these equations.

  a $1 - 2u = 11 - 4u$ **b** $15 - 3e = 7 - e$ **c** $8 - 6m = 6 - 8m$

Consolidate – do you need more?

1 Solve the equation $2y + 11 = 4y + 3$

You can use the bar model to help you.

y	y		11		
y	y	y	y		3

2 **a** Explain what your first step would be in solving each of these equations.

    i $7a + 4 = 3a + 24$ **ii** $w + 26 = 7w + 8$

    iii $6 + 9m = m + 54$ **iv** $3x = 7x - 56$

  b Solve the equations in part **a**.

3 Solve these equations.

  a $3b - 5 = 2b + 4$ **b** $12e - 5 = 5e + 2$ **c** $3p + 54 = 13p - 6$

  d $7r - 11 = 3r - 3$ **e** $5g - 13 = g - 7$ **f** $9a - 44 = 15a - 20$

4 Solve these equations.

  a $25 - 4q = 2q + 1$ **b** $8y + 9 = 56 - 2y$ **c** $8 - 7e = 4e - 14$

  d $15 - 12h = 5 + 8h$ **e** $35 - 5m = 19 - m$ **f** $-4u = 10 - 2u$

Stretch – can you deepen your learning?

1 Solve the equations.

  a $3(y + 4) = 2y - 1$ **b** $3(y + 4) = 2(y - 1)$

  c $3(y + 4) = 1 - 2y$ **d** $3(4 - y) = 2(1 - y)$

2 The triangle and the square have the same perimeter.

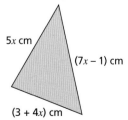
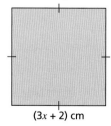

5x cm   (7x – 1) cm   (3 + 4x) cm   (3x + 2) cm

  a Work out the value of x.

  b Calculate the side length of the square.

Functions and linear equations: exam practice

White Rose
MATHS

1 Here is a function machine:

Input ⟶ | × 4 | ⟶ | + 5 | ⟶ Output

   (a) Work out the output when the input is 10   **[1 mark]**

   (b) Work out the input when the output is 9   **[1 mark]**

2 Solve:

   (a) $8y = 16$ **[1 mark]**

   (b) $m - 7 = 2$ **[1 mark]**

   (c) $\frac{x}{5} = 9$ **[1 mark]**

3 Solve:

   (a) $3 - p = -4$ **[1 mark]**

   (b) $\frac{1}{4}q = 20$ **[1 mark]**

4 Ben is x years old.

Amir is 5 years older than Ben.

Jack is three times Ben's age.

The sum of their ages is 100

How old is Ben?   **[3 marks]**

5 Solve:

   (a) $22 = 4a - 10$ **[2 marks]**

   (b) $5b - 3 = 2b + 8$ **[3 marks]**

6 Work out the perimeter of the parallelogram.

$(4x + 3)$ cm

x cm x cm

$(5x - 1)$ cm   **[4 marks]**

1–3

3–5

10 Formulae

White Rose
M▲THS

In this block, we will cover...

10.1 Using formulae

Example 1

Evaluate these formulae when $x = 2$, $y = 5$ and $z =$

a $G = 3x + 1$　　　**b** $G = y - 3x$　　　**c** G

Method

Solution	Commentary
a $G = 3x + 1$	Replace x with 2
$= 3 \times 2 + 1$	Remember $3x$ means '3
$= 6 + 1$	
$G = 7$	The value of G is 7
b $G = y - 3x$	Replace y with 5 and x w

10.2 Creating formulae

Practice

1. The total price of three pens and one ruler
   The price of one pen is £x.
   The price of one ruler is £1

   Which of these formulae represents the in

   $A = 3 + x$　　　$A = 3(x + 1)$

2. The cost of a helicopter ride is £78 plus £1(
   a Write a formula for the total cost (£C)

10.3 Rearranging formulae

Consolidate – do you need more

1. Rearrange each formula to make u the sub
   a $d = 3u$　　　**b** $d = \frac{u}{3}$
   e $d = u + e$　　　**f** $d = u - e$

2. Rearrange each formula to make t the sub
   a $p = 5t + k$　　　**b** $p = 5t - k$

   d $p = \frac{t}{5} - k$　　　**e** $p = k + \frac{t}{5}$

   Rearrange each formula to make u the sub

10.1 Using formulae

Are you ready? 🖩

1 Substitute $w = 16$ and $p = 24$ to work out the value of:

   a $w + p$ **b** $p - w$ **c** wp **d** $\dfrac{p}{w}$

   e $\dfrac{w}{p}$ **f** $4w - p$ **g** $p - 2w$ **h** $3p - 5w$

2 Evaluate the expressions when $x = 4$, $y = 5$ and $z = -3$

   a $3x^2$ **b** xy^2 **c** z^2 **d** xyz^2

A **formula** is an equation representing the relationship between two or more variables. It is often related to a real-life application.

For example, the conversion from degrees Celsius to degrees Fahrenheit can be calculated using the formula $F = \dfrac{9}{5}C + 32$, where F represents the temperature in degrees Fahrenheit and C the temperature in degrees Celsius.

To evaluate a formula, substitute numerical values for the variables and use the order of operations to work out the value.

Example 1

Evaluate these formulae when $x = 2$, $y = 5$ and $z = -3$

a $G = 3x + 1$ **b** $G = y - 3x$ **c** $G = \dfrac{xy}{20}$ **d** $G = 3y + 4z$

Method

Solution	Commentary

a $G = 3x + 1$
       $= 3 \times 2 + 1$
       $= 6 + 1$
    $G = 7$
	Replace x with 2
	
	 Remember $3x$ means '3 multiplied by x'.
	
	 The value of G is 7

b $G = y - 3x$
       $= 5 - 3 \times 2$
       $= 5 - 6$
    $G = -1$
	Replace y with 5 and x with 2
	
	 Work out 3×2 first to obey the order of operations.

c $G = \dfrac{xy}{20}$
       $= \dfrac{2 \times 5}{20} = \dfrac{10}{20}$
    $G = \dfrac{1}{2}$
	Replace x with 2 and y with 5
	
	 Remember that xy means 'x multiplied by y' and the fraction represents a division.
	
	 You can leave your answer as a fraction or write it as a decimal (0.5).

d $G = 3y + 4z$
       $= 3 \times 5 + 4 \times -3$
       $= 15 - 12$
    $G = 3$
	Replace y with 5 and z with -3
	
	 Work out each multiplication first to obey the order of operations. The product of a positive number and a negative number is negative.

Example 2

The formula for the area of a circle is $A = \pi r^2$

a Work out the area of a circle with diameter 12 cm. Give your answer in terms of π.

b The area of a circle is 100 cm².

Calculate the radius of the circle. Give your answer to 3 significant figures.

Method

Solution	Commentary
a If $d = 12$ cm, then $r = 12 \div 2$ $= 6$ cm	Remember, the radius of a circle is half of its diameter.
$A = \pi r^2$	Write down the formula for the area.
$A = \pi \times 6^2$	Substitute the value of r.
$A = 36\pi$ cm²	Leave your answer in terms of π, as requested.
b $A = \pi r^2$	Write down the formula for the area.
$100 = \pi r^2$ $\div \pi \quad\quad\quad \div \pi$ $\dfrac{100}{\pi} = r^2$	Substitute the value of A. You now have an equation in r. Start by dividing both sides by π.
$31.83\ldots = r^2$ $\sqrt{\quad} \quad\quad\quad \sqrt{\quad}$ $5.641\ldots = r$	Use your calculator to work out r^2. Do not round at this stage. Take the square root to find r.
$r = 5.64$ cm (3 s.f.)	Give your answer to 3 significant figures, as requested.

Practice

 1 **a** Using the formula $B = 3 + w$, calculate the value of B when $w = 12$

   b Using the formula $B = w - 3$, calculate the value of B when $w = 12$

   c Using the formula $B = 3w$, calculate the value of B when $w = 12$

   d Using the formula $B = \dfrac{w}{3}$, calculate the value of B when $w = 12$

2 Using these formulae, calculate the value of R when $y = 15$

   a $R = 2y + 7$ **b** $R = 4y - 3$ **c** $R = \dfrac{6y}{12}$ **d** $R = y^2$

3 Using the formula $A = 2m + 3n$, calculate the value of A for these values of m and n.

   a $m = 4, n = 7$ **b** $m = 0, n = 8$ **c** $m = -3, n = 5$ **d** $m = 4, n = -6$

4 Using these formulae, calculate the value of K when $a = 9$ and $b = 5$

   a $K = 2a + b$ **b** $K = 5a - b$ **c** $K = 3(a + b)$

   d $K = 5(3a - 4b)$ **e** $K = 2a^2$ **f** $K = ab^2$

5 The formula to convert temperatures in °C to temperatures in °F is $F = \frac{9}{5}C + 32$

Using the formula, convert a temperature of:

a 30°C to °F **b** 26°C to °F.

6 Using these formulae, calculate the value of T when $v = 36$

a $T = \frac{v}{4} - 8$ **b** $T = \frac{v - 8}{4}$ **c** $T = \sqrt{v}$

d $T = 3 + \sqrt{v}$ **e** $T = 3\sqrt{v}$ **f** $T = \frac{\sqrt{v}}{3}$

7 The area of a rectangle of length l and width w is given by the formula $A = lw$

Use the formula to:

a work out A when $l = 8\,mm$ and $w = 10\,mm$

b work out w when $A = 56\,mm^2$ and $l = 16\,mm$.

Consolidate – do you need more?

1 Using the formula $T = r - 8$, calculate the value of T for these values of r.

a $r = 10$ **b** $r = 50$ **c** $r = 8$ **d** $r = 2$

2 Using the formula $l = \frac{f}{4}$, calculate the value of l for these values of f.

a $f = 12$ **b** $f = 30$ **c** $f = 0$ **d** $f = -8$

3 Using these formulae, calculate the value of S when $g = 3$ and $h = 8$

a $S = 4h - g$ **b** $S = 4g + 3h$ **c** $S = 4(g + h)$

d $S = 4(5g - h)$ **e** $S = 5g^2$ **f** $S = 2gh^2$

4 Using these formulae, calculate the value of A when $m = 64$

a $A = \frac{m}{4} - 12$ **b** $A = \frac{m - 12}{4}$ **c** $A = \sqrt{m}$

d $A = 4 + \sqrt{m}$ **e** $A = 4\sqrt{m}$ **f** $A = \frac{\sqrt{m}}{4}$

5 The volume of a cylinder can be calculated using the formula $V = \pi r^2 h$, where V is the volume of the cylinder, r is the radius of the circular cross-section and h is the height of the cylinder.

a The radius of a cylinder is 5 cm.

The height of the cylinder is 10 cm.

Use the formula to calculate the volume of the cylinder.

b A cylinder has a volume of 5024 m³ and a radius of 10 m.

Use the formula to calculate the height of the cylinder.

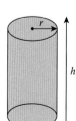

Stretch – can you deepen your learning?

1 The formula for the area of a trapezium is $A = \frac{1}{2}(a + b)h$.

   a Work out the value of A given $a = 8\,cm$, $b = 2\,cm$ and $h = 4\,cm$

   b Work out the value of h given $A = 56\,cm^2$, $a = 6\,cm$ and $b = 8\,cm$

   c Work out the value of b given $A = 84\,cm^2$, $a = 5\,cm$ and $h = 14\,cm$

2 The perimeter of a rectangle of length l and width w is given by the formula $P = 2(l + w)$.

   a Use the formula to:

     i work out P when $l = 8\,mm$ and $w = 10\,mm$

     ii work out w when $P = 46\,mm$ and $l = 16\,mm$.

   b If the perimeter of a rectangle is $50\,cm$, work out at least three sets of values for l and w.

3 Beca is substituting $u = 0$, $t = 4$ and $a = 2$ into the formula $s = ut + \frac{1}{2}at^2$

Here is her working.

$$s = 0 \times 4 + \frac{1}{2} \times 2 \times 4^2$$

$$= 4 + \frac{1}{2} \times 8^2$$

$$= 4 + \frac{1}{2} \times 16$$

$$= 4 + 8$$

$$s = 12$$

Explain the mistakes Beca has made and work out the correct solution for s.

4 An amount of money, M, is invested in a bank account.

The formula shows the value of the investment, V, after n years where y is the rate of compound interest per year.

$$V = M\left(1 + \frac{y}{100}\right)^n$$

Zach invests £1500 at an interest rate of 3.5%.

What is the value of his investment after:

   a 1 year   **b** 3 years   **c** 15 years?

10.2 Creating formulae

Are you ready?

1 Evaluate these formulae when $m = 8$

    a $A = 60m + 15$ **b** $A = 0.6m + 15$ **c** $A = 0.6m + 1.5$

2 Solve the equations.

    a $135 = 8n + 39$ **b** $10 = 0.45x + 1.9$

3 Write an expression for the perimeter of the rectangle.

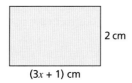

2 cm

$(3x + 1)$ cm

Example 1

An equilateral triangle has side length l.

a Write a formula for the perimeter, P, of the equilateral triangle in terms of l.

b Work out P when $l = 60$

c Work out l when $P = 60$

Method

Solution	Commentary
a $P = 3l$	Each side of the equilateral triangle has length l, so the perimeter P is given by $P = l + l + l$, which simplifies to $P = 3l$
b When $l = 60$, $P = 3 \times 60 = 180$	Substitute the value of l into the formula.
c When $P = 60$ $$60 = 3l$$ $\div 3 \Big(\quad \Big) \div 3$ $$20 = l$$ So $l = 20$	Substitute the value of P into the formula. Divide both sides by 3 to find l.

Example 2

An electrician charges £40 per hour plus a callout charge of £60

a Write a formula for the total cost (£C) of hiring the electrician for t hours.

b The electrician charges one client £220. For how many hours did they work?

Method

Solution	Commentary
a 1 hour costs £40	
So t hours costs $t \times 40 = £40t$	Multiply t by the cost per hour.
Total cost = charge for the hours worked + £60	Add the callout charge to the cost of t hours' work.
So $C = 40t + 60$	Write the formula in algebraic form. You could also have written $C = 60 + 40t$
b $220 = 40t + 60$ $-60 \quad\quad\quad -60$ $160 = 40t$ $\div 40 \quad\quad\quad \div 40$ $4 = t$	Substitute the values you know into the formula. Here you know that $C = 220$ Solve the equation using the methods you learned in Chapter 9.2
The electrician worked for 4 hours.	Write a statement to answer the question.

Practice

1 The total price of three pens and one ruler is £A.

The price of one pen is £x.

The price of one ruler is £1

Which of these formulae represents the information given?

$\boxed{A = 3 + x}$ $\boxed{A = 3(x + 1)}$ $\boxed{A = x + 1}$ $\boxed{A = 3x + 1}$

2 The cost of a helicopter ride is £78 plus £16 per mile.

a Write a formula for the total cost (£C) of using the helicopter for n miles.

b Sven is charged £478 for a helicopter ride.

For how many miles was the helicopter used?

3 A zoo charges £15 for a child ticket and £24 for an adult ticket.

Write a formula for the total cost, £T, of c child tickets and a adult tickets.

4 **a** Write a formula to work out the perimeter, P, of the rectangle.

b Use the formula to work out P when $y = 4$

9 cm

$(5y + 1)$ cm

5 A taxi driver charges a fare of £1.30, plus 85p for each mile travelled.

   a Which of these formulae represents the total cost, C, in pounds, of a journey of m miles?

$C = 1.3 + 85m$	$C = (1.3 + 0.85)m$	$C = 0.85m + 1.3$

   b Work out the cost of a 16-mile journey.

   c Beca has £20. How far can she travel in the taxi?

Consolidate – do you need more?

1 **a** Write a formula for the perimeter, P, of the pentagon in terms of x and y.

   b Work out the value of P when $x = 4\,\text{cm}$ and $y = 3\,\text{cm}$.

   c Work out the value of x when $y = 5\,\text{cm}$ and $P = 43\,\text{cm}$.

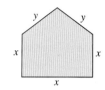

2 A plumber charges £30 per hour plus a callout charge of £75

   a Write a formula for the total cost ($£C$) of hiring the plumber for h hours.

   b Use the formula to calculate the total cost of hiring the plumber for 3 hours.

   c The plumber charges one client £240. For how many hours did the plumber work?

3 The cost of a taxi journey is 80p per mile plus a £5 standard charge.

   a Create a formula for the cost, £C, of a taxi journey in terms of the number of miles, m, travelled.

   b Use the formula to calculate the cost of a taxi journey of:

     i 5 miles   **ii** 26 miles.

   c A taxi journey costs £33. How long was the journey?

Stretch – can you deepen your learning?

1 Write a formula for the area of the trapezium, giving your answer in the simplest form.

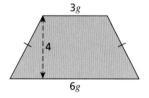

2 Speed is defined as distance divided by time.

   a Derive a formula connecting speed, s, distance, d, and time, t.

> Derive is another way of saying 'write' a formula.

   b Work out s when $d = 12$ miles and $t = 2$ hours.

   c Work out s when $d = 150$ metres and $t = 40$ seconds.

   d Work out d when $s = 30$ km/h and $t = 2$ hours.

Are you ready?

1 Write down the inverse of each operation.

     a $- 8$ **b** $\div 7$ **c** $\times 9$ **d** $+ 6$

2 The formula for the area of a triangle is $A = \frac{1}{2}bh$

     a   Find A when $b = 4\,\text{cm}$ and $h = 10\,\text{cm}$ **b**   Find b when $A = 30\,\text{cm}^2$ and $h = 5\,\text{cm}$

3 Solve these equations.

     a $24 = 3x + 6$ **b** $24 = \frac{x}{6}$ **c** $24 = \frac{x + 3}{6}$ **d** $24 = \frac{x}{6} + 3$

4   **a**   The square of a number is 9   **b**   The square root of a number is 9
            What is the number?   What is the number?

A **formula** is a rule connecting variables, written using mathematical symbols.

For example, $A = \pi r^2$ is the formula for finding the area of a circle of radius r. Because the formula is written to help you to find A, you can say that A is the **subject** of the formula.

Example 1

Rearrange each formula to make y the subject.

a $A = y + 2$ **b** $B = \frac{y}{5}$ **c** $C = 4y - 3$ **d** $D = \frac{y}{t} + x$

Method

Solution	Commentary
a $A = y + 2$ $-2 \Big(\quad \Big) -2$ $A - 2 = y$ $y = A - 2$	Rearrange the formula in the same way as you would solve an equation. It is more common to write a formula with the subject first.
b $B = \frac{y}{5}$ $\times 5 \Big(\quad \Big) \times 5$ $5B = y$ $y = 5B$	The term involving y is $\frac{y}{5}$. To make y the subject, multiply both sides of the formula by 5 Now you have a formula with y instead of $\frac{y}{5}$
c $C = 4y - 3$ $+3 \Big(\quad \Big) +3$ $C + 3 = 4y$ $\div 4 \Big(\quad \Big) \div 4$ $\frac{C + 3}{4} = y$ $y = \frac{C + 3}{4}$	Start by adding 3 to both sides of the formula, just like you would when solving an equation to find y. Now divide both sides of the formula by 4
	Remember that if two things are equal then the order in which you state them doesn't matter.

d

$$D = \frac{y}{t} + x$$

$-x \Big(\quad \Big) -x$

$$D - x = \frac{y}{t}$$

$\times t \Big(\quad \Big) \times t$

$$t(D - x) = y$$

$$y = t(D - x)$$

Start by subtracting x from both sides as you are trying to isolate y.

Multiply by t to clear the fraction.

Use brackets to show you are multiplying D by t and x by t.

Example 2

Ed rearranges $a = bc^2$ to make c the subject.

Ed's working is shown. He is incorrect.

Identify his mistake and work out the correct answer.

$$a = bc^2$$

$\div b \Big(\quad \Big) \div b$

$$\frac{a}{b} = c^2$$

$\div 2 \Big(\quad \Big) \div 2$

$$\frac{a}{2b} = c$$

Method

Solution	Commentary
The first step is correct.	
In the second step, Ed has divided by 2 instead of taking the square root.	Explain the error clearly.
$$a = bc^2$$ $\div b \Big(\quad \Big) \div b$ $$\frac{a}{b} = c^2$$ $\sqrt{\ } \Big(\quad \Big) \sqrt{\ }$ $$\sqrt{\frac{a}{b}} = c$$ $$c = \sqrt{\frac{a}{b}}$$	The inverse of squaring is taking the square root, not dividing by 2
	The correct second step is to take the square root of both sides.
	It is usual to write the final answer with the new subject on the left-hand side.

Practice

1 Write down the subject of each formula.

a $P = 4l$ **b** $l = \frac{P}{6}$ **c** $F = ma$ **d** $\frac{F}{a} = m$

e $mF + u = v$ **f** $t = ar^n$ **g** $s = \frac{1}{2}(u + v)t$ **h** $\frac{2s}{t} - v = u$

2 Rearrange each formula to make x the subject.

a $a = x + 5$ **b** $a = x - 5$ **c** $a = 5x$ **d** $a = \frac{x}{5}$

3 Rearrange each formula to make r the subject.

a $t = r + k$ **b** $t = r - k$ **c** $t = rk$ **d** $t = \frac{r}{k}$

4 Rearrange each formula to make w the subject.

a $g = 3w + 8$ **b** $g = 3w + h$ **c** $g = 3w - 8$ **d** $g = 3w - h$

e $g = \frac{w}{3} - h$ **f** $g = \frac{w - h}{3}$ **g** $g = h + \frac{w}{3}$ **h** $g = \frac{h + w}{3}$

5 One of these is the correct rearrangement of the formula $c = 3de$ to make d the subject.
Which is it?

$$d = \frac{3e}{c} \qquad d = \frac{c}{e} - 3 \qquad d = \frac{c}{3} - e \qquad d = \frac{c}{3e} \qquad d = \frac{c}{e/3}$$

6 Rearrange each formula to make f the subject.

  a $H = 5fg$ **b** $H = rfg$ **c** $H = 2r + 3f$ **d** $H = fr + gk$

7 The circumference of a circle is given by the formula $C = \pi d$

  a Express d in terms of C.

The area of a triangle is given by the formula $A = \frac{1}{2}bh$

  b Rearrange the formula to make b the subject.

8 Rearrange each formula to make a the subject.

  a $P = a^2$ **b** $P = a^2 + w$ **c** $P = wa^2$ **d** $P = \dfrac{a^2 + w}{5}$

  e $P = \sqrt{a}$ **f** $P = \sqrt{a} - w$ **g** $P = \dfrac{\sqrt{a}}{w}$ **h** $P = w\sqrt{a}$

What do you think?

1 Filipo and Bev are rearranging $g = h - 3p$ to make p the subject.

  a Here are Filipo's workings. Explain his mistake.

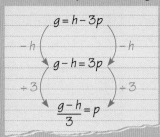

  b Here is the start of Bev's workings. Copy and complete her workings.

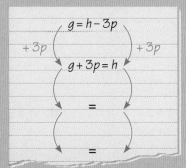

2 Rearrange each formula to make v the subject.

  a $k = 2 - v$ **b** $k = m - v$ **c** $k = 2m - v$ **d** $k = m - 2v$

Consolidate – do you need more?

1 Rearrange each formula to make u the subject.

   a $d = 3u$ **b** $d = \dfrac{u}{3}$ **c** $d = u + 3$ **d** $d = u - 3$

   e $d = u + e$ **f** $d = u - e$ **g** $d = ue$ **h** $d = \dfrac{u}{e}$

2 Rearrange each formula to make t the subject.

   a $p = 5t + k$ **b** $p = 5t - k$ **c** $p = \dfrac{t - k}{5}$

   d $p = \dfrac{t}{5} - k$ **e** $p = k + \dfrac{t}{5}$ **f** $p = \dfrac{k + t}{5}$

3 Rearrange each formula to make y the subject.

   a $a = 3yx$ **b** $a = ryx$ **c** $a = 2x + 3y$ **d** $a = yr - xw$

4 Here are the equations of some straight lines.

   Rearrange them to make y the subject.

   a $y + 6x = 4$ **b** $x = y - 4$ **c** $2y + 6x = 4$ **d** $x = 2y + 4$

5 The area of a circle is given by the formula $A = \pi r^2$

   Rearrange the formula to make r the subject.

Stretch – can you deepen your learning?

1 The area of a trapezium is given by the formula $A = \dfrac{1}{2}(a + b)h$

   a   Write h in terms of A, a and b.

   b   Write a in terms of A, b and h.

2 The formula for converting a temperature in degrees Celsius (C) to degrees Fahrenheit (F) is given by $F = \dfrac{9}{5}C + 32$

   Rearrange this formula to make C the subject.

1 Match a card on the left to the card on the right that shows the correct description. **[1 mark]**

| $4x + 3y + y$ | | Formula |

| $2y + 3 = 17$ | | Expression |

| $\dfrac{2x + y}{m} = w$ | | Equation |

2 Here is a formula to find the area of a triangle: $A = \dfrac{b \times h}{2}$

Work out the area of a triangle with a base of 6 cm and a height of 11 cm.
Give units with your answer. **[3 marks]**

3 Rearrange $t - m = c$ to make t the subject. **[1 mark]**

4 A festival charges £12 for a child ticket and £30 for an adult ticket.

Write a formula for the total cost, £T, of c child tickets and
a adult tickets. **[2 marks]**

5 The volume of a sphere can be found using the formula $V = \dfrac{4}{3}\pi r^3$

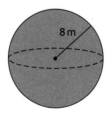

8 m

Use the formula to work out the volume of the sphere. Give your
answer to 2 decimal places. **[3 marks]**

6 When you are h feet above sea level, you can see d miles to the horizon,

where $d = \sqrt{\dfrac{3h}{2}}$

(a) How many miles can you see when you are 6 feet above sea level? **[2 marks]**

(b) How many feet above sea level would you be if you could see
for 9 miles? **[3 marks]**

11 Inequalities

In this block, we will cover...

11.1 Understanding inequalities

Example 1

Write an integer that satisfies each inequality.

a $x > 6$ **b** $x \geqslant 7$ **c** $x < 7$

Method

Solution	Commentary
a 7	Or any whole number greater th⋯
b 7	Or any whole number greater th⋯
c 6	Or any whole number less than 7
d 4	Or any whole number less than o⋯

11.2 Solving inequalities

Practice (A)

1. Solve the inequalities.

   a $g + 2 > 10$ **b** $2a \leqslant 10$

2. Solve the inequalities.

   a $4x + 3 > 27$ **b** $5a - 5 \leqslant 20$

   d $5 + 6f \leqslant 17$ **e** $31 < 2b - 7$

3. **a** Work out the smallest integer value th⋯

   i $6q + 3 > 15$ **ii** $5 + 3p \geqslant 2⋯$

   b Work out the greatest integer value th⋯

11.3 Inequalities on number lines

Consolidate – do you need more⋯

1. Show each inequality on a number line.

   a $x > -2$ **b** $x \geqslant 3$

2. **a** Show these dual inequalities on a num⋯

   i $-3 \leqslant x \leqslant 5$ **ii** $-5 < x \leqslant 3$

   b Write down all the possible integer sol⋯

   i $-3 \leqslant x \leqslant 5$ **ii** $-5 < x \leqslant 3$

3. Write down the inequalities shown on the⋯

Are you ready?

1 Put each set of numbers in order, starting with the lowest.

    a 8, 3, –2   **b** –3, 5, –6   **c** 7, –2, –7, 2

2 Choose a number to complete each statement correctly.

    a $4 > \boxed{}$ **b** $4 < \boxed{}$ **c** $-4 > \boxed{}$

3 Identify the integers in the list below.

    5   6.5   4   6   4.1   41   65

An **inequality** is a comparison between two quantities that are not equal to each other.

You will have seen the signs > (greater than) and < (less than) before. There are two other signs that are used when working with inequalities.

$\geqslant$ means 'greater than or equal to'. If you are told $a \geqslant 5$, then a could have the value 5 or any value greater than 5, such as 5.1, 6, 150, and so on.

In the same way, $\leqslant$ means 'less than or equal to'.

Example 1

Write an integer that satisfies each inequality.

a $x > 6$ **b** $x \geqslant 7$ **c** $x < 7$ **d** $x \leqslant 8$

Method

Solution	Commentary
a 7	Or any whole number greater than 6
b 7	Or any whole number greater than or equal to 7
c 6	Or any whole number less than 7
d 4	Or any whole number less than or equal to 8

Example 2

List all the integer values of n that satisfy each linear inequality.

a $-3 < n \leqslant 4$ **b** $6 \leqslant 2n < 9$

Method

Solution	Commentary
a –2, –1, 0, 1, 2, 3, 4	List all the integers that are greater than –3 and less than or equal to 4
	You include 4 because the inequality includes the $\leqslant$ sign.

b $\quad 6 \leqslant 2n < 9$ $\div 2 \downarrow \quad \downarrow \quad \downarrow \div 2$ $3 \leqslant n < 4.5$	Divide all terms by 2 to isolate n.
n can be 3 or 4	List the integers. You include 3 because the inequality includes the $\leqslant$ sign.

Practice

1   **a**   Write these in words.

     i $x > 6$ **ii** $y < 5$ **iii** $a \leqslant -3$

   b   Write these using symbols.

     i x is less than 2   **ii** a is greater than or equal to 5   **iii**   8 is greater than y

2   Which of these is the same as $b > 4$?

$$4 < b \qquad 4 > b \qquad b \geqslant 4 \qquad b \leqslant 4$$

3   **a**   Write an integer that satisfies each inequality.

     i $e > 5$ **ii** $f < -3$ **iii** $4 \geqslant g$

   b   Write a number that is **not** an integer that satisfies each inequality.

     i $h < 8$ **ii** $k \leqslant -6$ **iii** $-3 > m$

4   List all the integer values of n that satisfy each inequality.

   a $-3 < n \leqslant 1$ **b** $-5 < n < 3$ **c** $-4 \leqslant n \leqslant 2$

   d $-4 \leqslant 2n < 8$ **e** $-3 \leqslant 3n \leqslant 6$ **f** $-4 < 4n \leqslant 10$

Consolidate – do you need more?

1   Write a value of c that satisfies each inequality.

   a $c > 10$ **b** $10 > c$ **c** $-10 > c$ **d** $c > -10$

2   Which of these inequalities does $y = 4$ satisfy?

$$y > 3 \qquad 3 < y \qquad 4 \leqslant y \qquad 4 > y \qquad y \leqslant 4 \qquad 5 > y$$

3 List all the integer values of m that satisfy each inequality.

    a $-4 \leqslant m < 2$ **b** $-1 < m \leqslant 4$ **c** $-3 \leqslant m \leqslant 3$

    d $-3 < 3m \leqslant 9$ **e** $-15 \leqslant 5m \leqslant 10$ **f** $-7 \leqslant 2m < 8$

4 A number p can take the possible integer values $-2, -1, 0, 1, 2, 3$

    Which of these inequalities could be true?

    $\boxed{p \geqslant -2}$ $\boxed{-2 \leqslant p \leqslant 3}$ $\boxed{p < 4}$ $\boxed{p \leqslant 3}$ $\boxed{-3 < p < 4}$

Stretch – can you deepen your learning?

1   **a**   Work out the smallest integer value that satisfies each inequality.

       i $y > \pi$ **ii** $y \geqslant 2\pi$ **iii** $y > \dfrac{\pi}{2}$

    b   Work out the greatest integer value that satisfies each inequality.

       i $y < \pi$ **ii** $y < 2\pi$ **iii** $y \leqslant \dfrac{\pi}{2}$

2 Explain why $w > 3$ and $w \geqslant 4$ are **not** the same inequality.

3 If $24 \leqslant x < 37$, work out the possible values of x if:

    a x is a multiple of 8   **b** x is a square number

    c x is prime   **d** x is a factor of 120

4 Amina says, "If $x > y$, then $x^2 > y^2$."

    Use examples and counterexamples to show that the statement is only sometimes true.

Are you ready? (A)

1 Solve the equations.

   a $2y - 3 = 15$ **b** $6 + 2y = 10$ **c** $4(2y + 1) = 12$

   d $\dfrac{y}{3} = 5$ **e** $5 = \dfrac{y}{3} + 1$ **f** $\dfrac{y + 1}{3} = 5$

2 Write three integers that satisfy each inequality.

   a $x < 3$ **b** $x \geqslant 7$ **c** $-2 < x < 5$

3 Choose a number to complete each statement correctly.

   a $3 > \boxed{}$ **b** $3 < \boxed{}$ **c** $-3 > \boxed{}$ **d** $\boxed{} \leqslant -3$

You can solve inequalities using similar approaches to solving equations. Solutions to inequalities can take a range of numbers that can be integers, negative numbers, fractions or decimals.

Example 1

Work out the sets of values for which:

a $m + 3 > 8$ **b** $n - 5 \leqslant 8.2$ **c** $11 \geqslant 2p$ **d** $\dfrac{q}{3} + 5 < 4$

Method

Solution	Commentary
a $m + 3 > 8$ $-3 \quad\quad -3$ $m > 5$	By subtracting 3 from both sides of the inequality, you see it is true for any value of m that is greater than 5
b $n - 5 \leqslant 8.2$ $+5 \quad\quad +5$ $n \leqslant 13.2$	By adding 5 to both sides of the inequality, you see it is true for any value of n that is less than or equal to 13.2
c $11 \geqslant 2p$ $\div 2 \quad\quad \div 2$ $5.5 \geqslant p$ or $\quad p \leqslant 5.5$	By dividing both sides of the inequality by 2, you see it is true when $5.5 \geqslant p$. This is the same as saying $p \leqslant 5.5$
d $\dfrac{q}{3} + 5 < 4$ $-5 \quad\quad -5$ $\dfrac{q}{3} < -1$ $\times 3 \quad\quad \times 3$ $q < -3$	This is a two-step inequality. The inequality is true for values of q less than -3

Example 2

Solve the inequalities.

a $3(a - 6) > 15$ **b** $30 \leqslant 7(b + 2)$

Method

Solution	Commentary
a $\quad 3(a - 6) > 15$ $\div 3 \quad\quad\quad a - 6 > 5 \quad\quad \div 3$ $+ 6 \quad\quad\quad a > 11 \quad\quad + 6$	As 3 is a factor of 15, you can divide both sides of the inequality by 3 Then add 6 to each side.
b $\quad 30 \leqslant 7(b + 2)$ $\quad\quad 30 \leqslant 7b + 14$ $- 14 \quad\quad\quad\quad\quad\quad - 14$ $\quad\quad 16 \leqslant 7b$ $\div 7 \quad\quad\quad\quad\quad\quad \div 7$ $\quad\quad \dfrac{16}{7} \leqslant b$	As 7 is not a factor of 30, it is easiest to expand the brackets. Then subtract 14 from each side. Then divide both sides by 7 You could also write this solution as $b \geqslant \dfrac{16}{7}$

Practice (A)

1 Solve the inequalities.

a $g + 2 > 10$ **b** $2a \leqslant 10$ **c** $b - 2 < 10$ **d** $\dfrac{h}{2} \geqslant 10$

2 Solve the inequalities.

a $4x + 3 > 27$ **b** $5a - 5 \leqslant 20$ **c** $3g + 15 > 33$

d $5 + 6f \leqslant 17$ **e** $31 < 2b - 7$ **f** $79 \geqslant 19 + 10k$

3 **a** Work out the smallest integer value that satisfies each inequality.

    i $6q + 3 > 15$ **ii** $5 + 3p \geqslant 23$ **iii** $10 \leqslant 2k + 1$

  b Work out the greatest integer value that satisfies each inequality.

    i $3w - 8 \leqslant -2$ **ii** $-17 > 5b - 2$ **iii** $2a + 4 \leqslant 9$

4 Solve the inequalities.

a $\dfrac{x}{2} + 9 \leqslant 15$ **b** $\dfrac{x + 9}{2} \leqslant 15$ **c** $\dfrac{m}{4} - 3 > 18$ **d** $\dfrac{m - 3}{4} > 18$

5 **a** Explain what your first step would be for solving each of these inequalities.

    i $4(x + 2) > 20$ **ii** $2(r - 5) \leqslant 11$ **iii** $20 \geqslant 5(w - 8)$

    iv $4(t + 3) \leqslant 9$ **v** $35 < 7(8 + y)$ **vi** $5(p - 3) \geqslant -15$

  b Solve the inequalities in part **a**.

  c Using your answers to part **b**, work out the smallest or greatest integer value that satisfies each inequality.

6 The area of the rectangle is greater than $40 \, \text{cm}^2$.

  a Work out the range of values that y can take.

  b What is the least possible integer value of y?

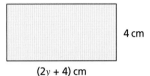

4 cm

$(2y + 4)$ cm

What do you think? 💭

1 **a** Abdullah is trying to solve the inequality $3 - x > 1$

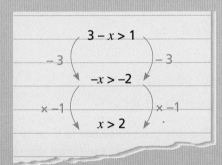

$3 - x > 1$

-3 -3

$-x > -2$

$\times -1$ $\times -1$

$x > 2$

  Show, by substituting any number greater than 2 into the inequality, that Abdullah is incorrect.

  b Copy and complete this working to solve the inequality $3 - x > 1$

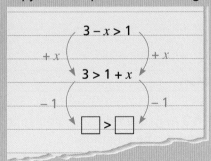

$3 - x > 1$

$+ x$ $+ x$

$3 > 1 + x$

$- 1$ $- 1$

$\square > \square$

2 Solve the inequalities.

  a $9 - y > 4$ **b** $10 - 3x < 1$ **c** $4 \leqslant 8 - 2x$

Are you ready? (B)

1 Solve these equations.

a $3y - 7 = 2y + 4$ **b** $11x - 4 = 5x + 2$ **c** $2p + 44 = 12p - 2$

2 Solve the equations.

a $25 - 4a = 2a + 1$ **b** $7b + 9 = 56 - 3b$ **c** $8 - 3f = f - 2$

Example 1

Work out the range of values of y for which $2y + 7 \leqslant 5y - 2$

Method

Solution	Commentary
$2y + 7 \leqslant 5y - 2$ $-2y \qquad\qquad -2y$ $7 \leqslant 3y - 2$	Start by subtracting $2y$ from both sides as this will leave unknowns on one side only.
$+2 \qquad\qquad +2$ $9 \leqslant 3y$	Next, add 2 to each side.
$\div 3 \qquad\qquad \div 3$ $3 \leqslant y$	Then divide both sides by 3
	This gives the solution set for y.
$y \geqslant 3$	You can write '3 is less than or equal to y' or 'y is greater than or equal to 3' as these mean the same thing. It is more usual to start your solution set with the unknown.

Example 2

Solve the inequality $10 - 3p < 5p + 6$

Method

Solution	Commentary
$10 - 3p < 5p + 6$ $+3p \qquad\qquad +3p$ $10 < 8p + 6$	As there is a negative term in p, add $3p$ to both sides to simplify the equation.
$-6 \qquad\qquad -6$ $4 < 8p$	Now you have an 'ordinary' two-step inequality.
$\div 8 \qquad\qquad \div 8$ $\frac{1}{2} < p$ $p > \frac{1}{2}$	$4 \div 8$ is $\frac{4}{8}$ and this is equivalent to $\frac{1}{2}$

Practice (B)

1 Copy and complete the first step of working to solve these inequalities.

a

$7y > 4y + 6$

$-4y \qquad -4y$

$3y > \square$

b

$5g + 3 \geqslant 9g$

$-5g \qquad -5g$

$3 \geqslant \square$

c

$8w + 4 \leqslant 3w + 19$

$-3w \qquad -3w$

$\square + 4 \leqslant \square$

d

$4m + 21 < 11m - 14$

$-4m \qquad -4m$

$\square < \square - 14$

2 **a** Explain what your first step would be for solving each of these inequalities.

   i $5y + 3 < 2y + 15$ **ii** $5y + 3 < y + 15$

   iii $5y + 3 < 3y + 15$ **iv** $5y + 3 < 4y + 15$

  b Solve the inequalities in part **a**.

3 Solve these inequalities.

  a $5y + 4 > 3y + 20$ **b** $7w + 11 \leqslant 3w + 17$ **c** $7p + 16 \geqslant 13p + 4$

  d $r + 33 < 4r + 3$ **e** $11y + 12 > 42 + 6y$ **f** $3p + 45 \leqslant 4p$

4 **a** What is the same and what is different about these inequalities?

   i $2x + 18 < 6x + 2$ **ii** $2x + 2 < 6x - 18$ **iii** $2x - 2 < 6x - 18$

  b Solve the inequalities.

5 **a** Solve these inequalities.

   i $11y - 8 < 4y + 6$ **ii** $7u - 3 \geqslant 2u + 12$

   iii $7p + 13 \leqslant 8p - 5$ **iv** $7b - 9 \leqslant 3b - 5$

   v $6k - 12 \geqslant 3k - 3$ **vi** $5t + 6 \leqslant 3t - 3$

  b Work out the smallest or greatest integer value that satisfies each inequality in part **a**.

6 Solve these inequalities.

  a $19 - x < 4x + 9$ **b** $4g + 9 > 30 - 3g$

  c $6 - 3y \leqslant 2y - 9$ **d** $22 - 4a \geqslant 3a + 8$

  e $28 - 4n < 7 + n$ **f** $20 - 4g \geqslant 16 - 2g$

Consolidate – do you need more?

1 Solve the inequalities.

  a $b - 4 > 8$ **b** $9 < c + 2$ **c** $4d \leqslant 24$ **d** $\dfrac{e}{7} \geqslant 7$

  e $4f - 3 > 17$ **f** $16 \geqslant 5g + 6$ **g** $\dfrac{h}{4} + 3 > 5$ **h** $18 \leqslant \dfrac{k}{2} - 7$

2 Solve the inequalities.

    a $5(w - 6) < 15$ **b** $4(3 + x) \leqslant 24$ **c** $2(y + 1) \geqslant 11$

    d $3(a + 2) \leqslant 3$ **e** $3(g + 8) > 9$ **f** $4(15 + m) < 32$

3 Work out the smallest integer value of g that satisfies each of these inequalities.

    a $7g + 5 > 3g + 8$ **b** $8g - 8 \geqslant 5g - 2$

    c $7g - 5 < 11g - 21$ **d** $3g + 8 \geqslant g - 11$

4 Solve these inequalities.

    a $3b - 5 > 11 - b$ **b** $15 - 2g \geqslant g + 9$ **c** $5 + 4x > 10 - 6x$

    d $22 - 4y \geqslant 12 - 8y$ **e** $30 - 3p > 14 - 2p$ **f** $4 - 3e \leqslant 3 - 4e$

5 The perimeter of the rectangle is less than 75 cm.

    a   Form an inequality to show this information.

    b   Work out the greatest possible integer value for y.

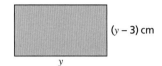

$(y - 3)$ cm

y

Stretch – can you deepen your learning?

1 Flo thinks of a number.

5 less than double her number is greater than 25

Form and solve an inequality to work out the range of possible values for her number.

2 The length of a rectangle is 5 cm greater than its width.

The perimeter of the rectangle is less than 50 cm.

    a   Form and solve an inequality to work out the range of possible values for the width of the rectangle.

    b   How would your answer change if, instead, the perimeter of the rectangle was no more than 50 cm?

3 The angle $(3w - 20)°$ is acute.

Work out the range of values that w can take.

4 Solve these inequalities.

    a $2(x + 3) < 3(x - 1)$ **b** $5(y - 3) \geqslant 3(y + 8)$ **c** $4(2w - 1) \leqslant 3(5 + 5w)$

5 Work out all the integer values of n that satisfy each inequality.

    a $12 \leqslant 5n + 2 < 27$ **b** $2 < 6 + 4n \leqslant 22$

Are you ready?

1 List all the integer values of n that satisfy each inequality.

    a $1 < n \leqslant 5$ **b** $-4 < n < 3$ **c** $-3 \leqslant n \leqslant 2$

2 Solve the inequalities.

    a $3k + 3 > 27$ **b** $4n - 6 \leqslant 22$ **c** $7h + 8 > 43$

3 Solve the inequalities.

    a $5(p + 2) > 20$ **b** $2(r - 4) \leqslant 15$ **c** $12 \geqslant 4(e - 7)$

Inequalities can be represented on number lines.

An inequality compares values. Unlike equations, inequalities do not have a single solution but have **solution sets**. These solution sets can be represented on **number lines**.

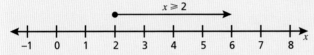

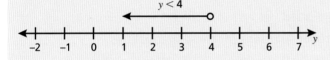

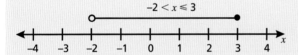

A filled circle shows that a value is included in the solution set and an open circle shows that a value is **not** included in the solution set.

Example 1

Show the solution to $2x - 3 \geqslant 7$ on a number line.

Method

Solution	Commentary
$2x - 3 \geqslant 7$ $+3 \qquad +3$ $2x \geqslant 10$ $\div 2 \qquad \div 2$ $x \geqslant 5$	Solve the inequality.
$x \geqslant 5$ number line from -2 to 9 with filled circle at 5 and arrow to the right	Show the solution set on a number line. 5 is included in the solution set, so the circle is filled.

Example 2

a Describe the inequality shown on the number line.

b List all the integer solutions.

Method

Solution	Commentary
a $-3 < x \leqslant 5$	Notice that although 5 is included in the solution set, -3 is not.
b The integer solutions are $-2, -1, 0, 1, 2, 3, 4, 5$	The integer solutions are the whole numbers included in the range of the inequality. 0 is an integer and is included in this set.

Practice

1 Show each inequality on a number line.

    a $x < 3$ **b** $x \geqslant 5$ **c** $x > -3$ **d** $x \leqslant -5$

2 **a** Show these dual inequalities on a number line.

    i $-3 < x \leqslant 4$ **ii** $-4 \leqslant x < 3$ **iii** $-4 \leqslant x \leqslant -2$

    b Write down all the possible integer solutions for these inequalities.

    i $-3 < x \leqslant 4$ **ii** $-4 \leqslant x < 3$ **iii** $-4 \leqslant x \leqslant -2$

3 Write down the inequalities shown on each number line.

a

b

c

d

4 Solve each inequality and show the solutions on a number line.

    a $4x > 12$ **b** $\dfrac{x}{4} \geqslant 12$ **c** $x + 12 < 4$ **d** $12 \leqslant x - 4$

5 Solve each inequality and show the solutions on a number line.

    a $4x + 1 > 5$ **b** $5 + x < 4$ **c** $5x + 4 \leqslant -1$

    d $-7 \geqslant 3x - 13$ **e** $6 + 2x \leqslant 11$ **f** $5x + 13 > 12$

What do you think?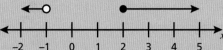

1 Is it possible to write a single linear inequality to represent this solution?

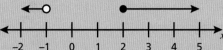

Explain your answer.

Consolidate – do you need more?

1 Show each inequality on a number line.

a $x > -2$ **b** $x \geqslant 3$ **c** $x < 2$ **d** $x \leqslant -3$

2 a Show these dual inequalities on a number line.

i $-3 \leqslant x \leqslant 5$ **ii** $-5 < x \leqslant 3$ **iii** $-5 < x < -1$

b Write down all the possible integer solutions for each inequality.

i $-3 \leqslant x \leqslant 5$ **ii** $-5 < x \leqslant 3$ **iii** $-5 < x < -1$

3 Write down the inequalities shown on these number lines.

a

b

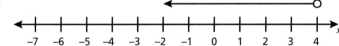

c

d

4 Solve each inequality and show the solutions on a number line.

a $5y < 15$ **b** $\dfrac{y}{5} \leqslant 15$ **c** $x - 15 > 5$ **d** $5 \geqslant y + 15$

5 Solve each inequality and show the solutions on a number line.

a $2x + 1 > 9$ **b** $5 + y < 2$ **c** $3a + 7 \leqslant -2$

d $-3 \geqslant 4b - 11$ **e** $3(y + 1) \leqslant 6$ **f** $5(x + 3) > 12$

Stretch – can you deepen your learning?

1 Jackson states the inequality shown is $-3 \leqslant x \leqslant 3$

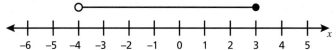

a Explain why Jackson is incorrect.

b Write the inequality shown by the number line.

2 Which of the following inequalities could the number line represent?

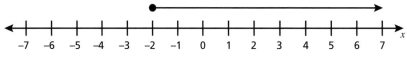

$2x + 10 \geqslant 6$ $4x + 5 > 1$ $6 \leqslant 3x + 12$ $5x + 3 \geqslant -7$

3 Show, on a number line, the range of values of y that satisfy both of the inequalities

$4y + 8 \leqslant 32$ and $2 < 2y + 8$

Inequalities: exam practice

1 Which inequality represents the statement 'x is greater than 4'?

   A $x < 4$ **B** $x \leqslant 4$ **C** $x = 4$ **D** $x > 4$ **[1 mark]**

1–3

2 Write $y \leqslant 2$ in words.   **[1 mark]**

3 $3 \leqslant n < 7$

   n is an integer.

   Write down all the possible values of n.   **[2 marks]**

4 Solve:

   (a) $5x > 10$ **[1 mark]**

   (b) $y + 3 < 1$ **[1 mark]**

5 (a) Show the inequality $y \geqslant -2$ on a copy of the number line below.   **[1 mark]**

                   $-4 \; -3 \; -2 \; -1 \; \; 0 \; \; 1 \; \; 2 \; \; 3 \; \; 4 \; \; 5 \;\;^{y}$

3–5

   (b) An inequality in x is shown on the number line below.

        Write down the inequality.   **[2 marks]**

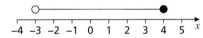

                   $-4 \; -3 \; -2 \; -1 \; \; 0 \; \; 1 \; \; 2 \; \; 3 \; \; 4 \; \; 5 \;\;^{x}$

   (c) Solve the inequality $7m - 3 < 32$ **[2 marks]**

6 t is an integer such that $6t + 4 > 19$

   Write down the smallest value of t.   **[2 marks]**

7 (a) n is an integer such that $-4 \leqslant 2n \leqslant 6$

        Write down the possible values of n.   **[2 marks]**

   (b) Show your answer to part **(a)** on a copy of the number line below.   **[2 marks]**

                   $-4 \; -3 \; -2 \; -1 \; \; 0 \; \; 1 \; \; 2 \; \; 3 \; \; 4 \; \; 5 \;\;^{n}$

In this block, we will cover...

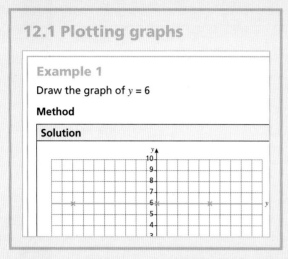

12.1 Plotting graphs

Example 1

Draw the graph of $y = 6$

Method

Solution

12.2 Interpreting straight line equations

Practice

1. State whether each line has a positive or n

   a   b

2. a Draw each of these lines on the same c

   $y = x$ $y = 2x$

   b Explain what happens to a line when y

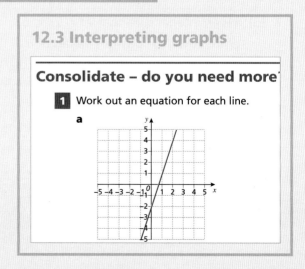

12.3 Interpreting graphs

Consolidate – do you need more

1. Work out an equation for each line.

   a

Are you ready?

1 Here are five pairs of coordinates: (4, 3) (4, 0) (4, –1) (4, 9) (4, –7)

    a What do all the pairs of coordinates have in common?

    b Copy and complete this sentence to describe the coordinates.

       In each pair of coordinates, the _____ value is equal to _____.

2 Write down three pairs of coordinates in which the y value is equal to –3

3 Substitute $x = -1$ into these expressions and evaluate them.

    a $4x + 3$ **b** $3x - 4$ **c** $-3x + 4$ **d** $3 - 4x$

4 Draw a coordinate grid with x-axes and y-axes that go from –5 to 5 and plot these points.

    a (1, 3)   **b** (–1, 3)   **c** (1, –3)   **d** (–1, –3)

The word 'linear' implies 'straight line', so all linear graphs will form a straight line. Knowing this fact helps you to check if you have drawn your graphs correctly.

To plot a linear graph, you can:

- complete a table of values, working out the values of y for some values of x
- plot a point for each pair of coordinates (x, y)
- join the points with a straight line.

If your points do not all lie on the same straight line, you know you have made a calculation error.

Example 1

Draw the graph of $y = 6$

Method

Solution	Commentary
	The equation of the line is 'y is equal to 6' so the y value is equal to 6 in the pair of coordinates for all points on the line. Examples are (–8, 6), (0, 6) and (5, 6).

Example 2

a Complete the table of values for $y = 3x - 4$

x	-2	-1	0	1	2
y					

b Draw the graph of $y = 3x - 4$

Method

Solution	Commentary
a $\times 3$ -4 <table><tr><td>x</td><td>-2</td><td>-1</td><td>0</td><td>1</td><td>2</td></tr><tr><td></td><td>-6</td><td>-3</td><td>0</td><td>3</td><td>6</td></tr><tr><td>y</td><td>-10</td><td>-7</td><td>-4</td><td>-1</td><td>2</td></tr></table> $\times 3$ -4	You can use this function machine to help you to work out the values in the table. Input Output $x \longrightarrow \boxed{\times 3} \longrightarrow \boxed{-4} \longrightarrow y$
b *(graph of $y = 3x - 4$ plotted on coordinate grid from -10 to 10)*	At each point on the line $y = 3x - 4$, the y value is equal to 4 less than 3 times the x value. Note the straight line extends infinitely in both directions, beyond the points given by the table of values.

Practice

1   **a**   Draw a coordinate grid with x-axes and y-axes that go from -10 to 10

     b   **i**   Write down three pairs of coordinates in which the y value is equal to 4

         ii   Plot your coordinates and join them to draw the line $y = 4$

         iii   Choose another point on the line and write down its coordinates.

             How does this confirm that your line has the equation $y = 4$?

     c   On your coordinate grid, draw and label each of these straight lines.

         i $y = 1$ **ii** $y = -3$ **iii** $y = -7$

2 **a** Draw a coordinate grid with x-axes and y-axes that go from -10 to 10

   b **i** Write down three pairs of coordinates in which the x value is equal to 4

     ii Plot your coordinates and join them to draw the line $x = 4$

   c On your coordinate grid, draw and label each of these straight lines.

     i $x = 1$ **ii** $x = -3$ **iii** $x = -7$

   d What is the same and what is different about your answers to part **c**?

   e **i** Write down three pairs of coordinates in which the y value is equal to the x value.

     ii Plot your coordinates and join them to draw the line $y = x$

3 **a** Copy and complete the table of values for $y = 2x + 1$

x	-2	-1	0	1	2
y					

   b Draw the graph of $y = 2x + 1$ for values of x from -2 to 2

4 Here are the equations of three straight lines:

    i $y = x - 4$ **ii** $y = 4x - 1$ **iii** $y = 4(x - 1)$

For each line:

   a complete a table of values using values of x from -2 to 2

   b draw the graph for values of x from -2 to 2

Draw all three graphs on the same coordinate grid. Remember to label each line with its equation.

5 Here are the equations of two straight lines:

    $y = 3x - 2$ $y = 2 - 3x$

   a What is the same and what is different about the lines?

   b For each line:

     i complete a table of values using values of x from -3 to 3

     ii draw the graph for values of x from -3 to 3

6 Here are the equations of four straight lines:

    i $y = -5 + 2x$ **ii** $y = 2(x + 5)$ **iii** $y = 5 - 2x$ **iv** $y = \frac{1}{2}x + 5$

For each line:

   a complete a table of values using values of x from -2 to 2

   b draw the graph for values of x from -2 to 2

What do you think?

1 Junaid has completed a table of values for $y = 3x + 4$

x	−1	0	1	2
y	1	4	7	10

He says, "I only need to find the first value for y, then add 3 each time, as the coefficient of x is 3"

a Is Junaid correct?

b Why won't his method work for the table of values below?

x	0	1	5	10
y				

Consolidate – do you need more?

1 Write down the equation of each line.

a
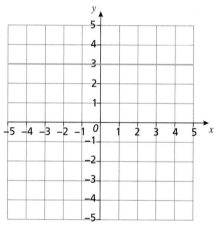

b
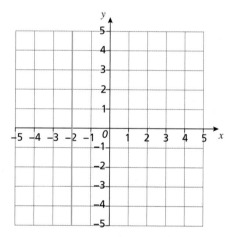

c
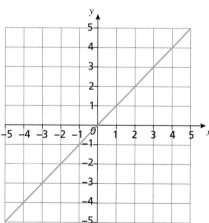

d
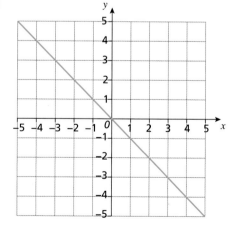

2 **a** Copy and complete each table of values for the given equations.

    i $y = 5x + 2$ **ii** $y = 2x - 5$

x	−2	−1	0	1	2
y					

x	−2	−1	0	1	2
y					

  b Draw a coordinate grid with values of x from −2 to 2

    Plot the graph of each equation on your grid.

3 Here are the equations of four straight lines:

    i $y = -4 + 3x$ **ii** $y = 3(x + 4)$ **iii** $y = 4 - 3x$ **iv** $y = \dfrac{6x + 1}{2}$

For each line:

  a complete a table of values using values of x from −3 to 3

  b draw the graph for values of x from −3 to 3

Stretch – can you deepen your learning?

1 Which of these pairs of values lie on the line $y = 8 - 7x$?

    ((1, 1))   ((13, 3))   ((3, −13))   ((13, −3))

2 Here are the equations of four straight lines:

    i $x + y = 6$ **ii** $x - y = 6$ **iii** $2x + y = 6$ **iv** $3x - 2y = 6$

For each line:

  a complete a table of values using values of x from −2 to 2

  b draw the graph for values of x from −2 to 2

3 Identify the graphs which could show the line $y = 3x + 1$

A 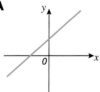   **B** 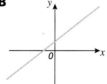   **C** 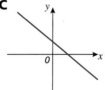   **D**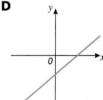

Are you ready?

1 Copy and complete the table of values for $y = 3x - 2$

x	-2	-1	0	1	2
y					

2 Draw a coordinate grid with x- and y-axes that go from -8 to 8

Plot the graph of each equation on your grid.

a $y = 3x$ **b** $y = x + 3$ **c** $y = 2x - 3$

The **general equation of a straight line** can be expressed in the form $y = mx + c$, where m is the gradient and $(0, c)$ is the point at which the line **intercepts** the y-axis.

The gradient of a line is the measure of the steepness of a straight line.

The gradient of a line can be either positive or negative and does not need to be a whole number.

The higher the gradient, the steeper the line. For example, the graph of $y = 5x + 1$ will be steeper than the graph of $y = 2x + 1$

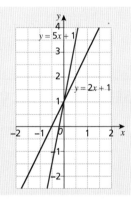

Example 1

a Draw each of these lines on the same coordinate grid.

   i $y = 2x + 1$ **ii** $y = 2x - 4$ **iii** $y = -x + 1$

b Which two lines are parallel?

c Which two lines have the same y-intercept? Why does this happen?

Method

Solution	Commentary
a	Remember to label each line when you have drawn it.
(graph showing lines $y = -x + 1$, $y = 2x + 1$, $y = 2x - 4$ on a grid from -10 to 10)	You can use a table to help you to work out the values of some points on each line.

b $y = 2x + 1$ and $y = 2x - 4$ are parallel because they have the same gradient.	This happens because the equations of these lines have the same coefficient of x. The number before x is 2 in both equations.
c $y = 2x + 1$ and $y = -x + 1$ have the same y-intercept. This is because they cut the y-axis at the same point, (0, 1).	This happens because the equations of these lines have the same constant term (in both equations it is +1), so when $x = 0$ they have the same value.

Example 2

A straight line has equation $y = 5x - 8$

Write down:

a the gradient of the line   **b** the coordinates of the y-intercept.

Method

Solution	Commentary
a 5	The gradient of the straight line is given by the coefficient of x in the equation.
b (0, −8)	The y-intercept of a straight line is given by the constant in the equation. You are asked for the coordinates of the y-intercept, so you need to give the value of x as well as the value of y.

Practice

1 State whether each line has a positive or negative gradient.

a   **b**   **c**   **d**

2 a Draw each of these lines on the same coordinate grid.

$y = x$ $y = 2x$ $y = 3x$ $y = 5x$

b Explain what happens to a line when you increase its gradient.

3 a Draw each of these lines on the same coordinate grid.

$y = 3x$ $y = 3x + 1$ $y = 3x - 2$ $y = 3x + 5$

b Explain how you can tell all lines are parallel from:

i the graphs   **ii** the equations.

c Write down the equation of another line that is parallel to each line above.

4 **a** Draw each of these lines on the same coordinate grid.

$y = 2x + 4$ $y = x + 4$ $y = 4 + 4x$ $y = 4 - x$

b Explain how you can tell all these lines have the same y-intercept from:

i the graphs   **ii** the equations.

c Write down the equation of another line that has the same y-intercept as each line above.

5 Here are the equations of eight straight lines.

a $y = 4x + 1$ **b** $y = 5x - 3$ **c** $y = -3x + 2$ **d** $y = -2x - 5$

e $y = 8x$ **f** $y = 6$ **g** $y = \frac{1}{3}x + 9$ **h** $y = \frac{x}{4} - 7$

For each line, state:

i the gradient   **ii** the coordinates of the y-intercept.

6 Write down the equations of three lines that:

a are parallel to $y = 7x - 12$

b intersect the y-axis at the same point as the line $y = 7x - 12$

c are parallel to $y = 10 - 3x$

d intersect the y-axis at the same point as the line $y = 10 - 3x$

7 Write the equations of the lines with these gradients and y-intercepts.

a gradient 3 and y-intercept (0, 4)   **b** gradient -2 and y-intercept (0, 0)

c gradient 1 and y-intercept (0, -7)   **d** gradient 0.5 and y-intercept (0, -1)

8 Here are the equations of four straight lines.

a $y = 6 + 3x$ **b** $y = 8 - 2x$ **c** $y = 0.5 - 4x$ **d** $y = \frac{3}{4} - \frac{x}{5}$

For each line, state:

i the gradient   **ii** the coordinates of the y-intercept.

What do you think?

1 Some lines are shown in the diagram.

a Which line is **not** parallel to the others? Explain how you know.

b The equation of line D is $y = \frac{1}{2}x - 3$

Suggest equations for the other lines.

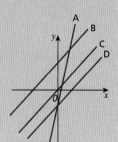

Consolidate – do you need more?

1 Write down the gradients of each of these lines.

    a $y = 9x - 4$ **b** $y = x + 12$ **c** $y = 8 - 5x$ **d** $y = \frac{1}{5}x - 11$

2 Write down the coordinates of the y-intercept of each of these lines.

    a $y = 3x + 14$ **b** $y = 7 - 4x$ **c** $y = 13x - 18$ **d** $y = x + \frac{1}{2}$

3 Write down the equations of three lines that:

    a intersect the y-axis at (0, 5)

    b are parallel to the line $y = 8x - 1$

    c are steeper than the line $y = 3x$

4 Write the equations of the lines with these gradients and y-intercepts.

    a gradient 5 and y-intercept (0, 1)   **b** gradient 12 and y-intercept (0, −5)

    c gradient −8 and y-intercept (0, 0)   **d** gradient −1 and y-intercept $\left(0, \frac{1}{3}\right)$

Stretch – can you deepen your learning?

1 A straight line is parallel to $y = 5x + 8$ and intersects the y-axis at (0, −3).

Write down the equation of the line.

2 Lines l_1 and l_2 are parallel. The equation of l_1 is $y = 9x - 21$

l_2 passes through the point (0, 12).

Work out an equation for l_2

3 Here are the equations of four straight lines.

    a $y = \dfrac{x + 8}{2}$ **b** $y = \dfrac{6x + 5}{3}$ **c** $y = \dfrac{10x - 3}{5}$ **d** $y = \dfrac{7 - 2x}{4}$

For each line, state:

    i the gradient   **ii** the coordinates of the y-intercept.

Are you ready? (A)

1 Here are the equations of some straight lines:

a $y = 3x + 5$ **b** $y = 4x - 1$ **c** $y = -2x + 3$ **d** $y = -5x$

For each line, state:

    i the gradient   **ii** the coordinates of the y-intercept.

2 The gradient of each line is given.

For each graph:

    i write down the coordinates of the y-intercept

    ii work out the equation of the line.

a Gradient = 3

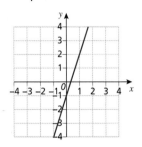

b Gradient = −2

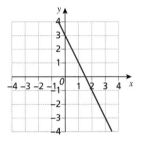

The equation of a straight line can be written in the form $y = mx + c$

The **gradient** of a straight line measures the steepness of the line and can be calculated if two points on the line are known.

The gradient of a line, m, can be found by calculating $\dfrac{\text{change in } y \text{ values}}{\text{change in } x \text{ values}}$

Line 1	Line 2	Line 3	Line 4
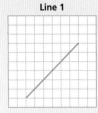	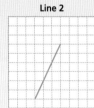	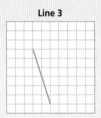	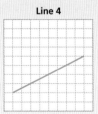

Calculating the gradient:	Calculating the gradient:	Calculating the gradient:	Calculating the gradient:

	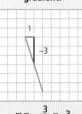	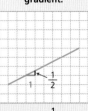	
$m = \dfrac{1}{1} = 1$	$m = \dfrac{2}{1} = 2$	$m = -\dfrac{3}{1} = -3$	$m = \dfrac{1}{2}$

Example 1

The diagram shows a straight line.

a Write down the coordinates of the y-intercept.

b Calculate the gradient.

c Work out the equation of the line.

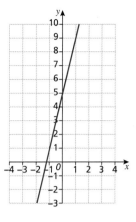

Method

Solution	Commentary
a (0, 5)	The line intersects the y-axis at (0, 5).
b 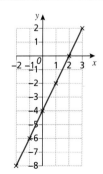 $$\text{gradient} = \frac{\text{change in } y}{\text{change in } x}$$ $$= \frac{4}{1} = 4$$ The gradient is 4	The gradient is how far up or down the line travels for every one unit across. For each 1 square across, the line goes up 4 squares, so the gradient of the line is 4 Use this formula to calculate the gradient.
c $y = 4x + 5$	You can write the equation in the form $y = mx + c$, where m is the gradient and c is where the line cuts the y-axis.

Example 2

Use the graph to solve the equation $2x - 4 = -6$

Method

Solution	Commentary
![graph] $x = -1$	Substitute $2x - 4$ with y to get $y = -6$
	Plot the line $y = -6$
	The solution is at the point where the two lines intersect, therefore $x = -1$

Practice (A)

1 Work out the gradient of each line.

a 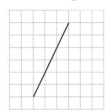   **b** 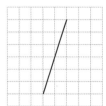   **c** 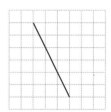   **d**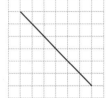

2 For each graph:

   i write down the coordinates of the y-intercept

   ii calculate the gradient

   iii work out the equation of the line.

a 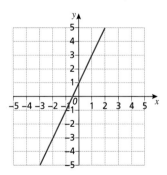   **b**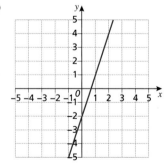

c 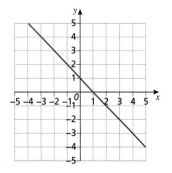   **d**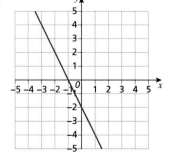

3 Work out an equation for each line.

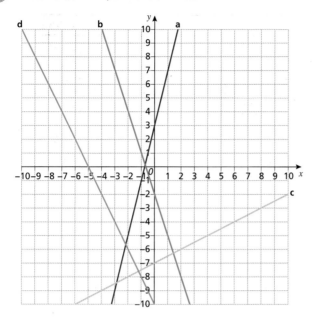

4 The graph shows the cost of a taxi journey for a given number of miles.

   a Use the graph to work out the cost of:

   i a 4-mile journey   **ii** a 1-mile journey.

   b Work out the cost per extra mile travelled.

   c Interpret the gradient and the y-intercept of the line in the context of the question.

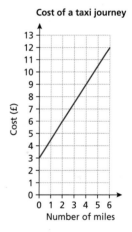

Cost of a taxi journey

5 This is the graph of $y = 3x - 2$

Use the graph to work out the solution to:

   a $3x - 2 = 1$

   b $3x - 2 = -8$

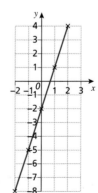

6 Work out the equation of each line.

Work out the scale on each axis first.

a

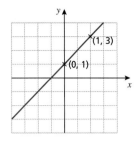

b

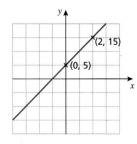

c

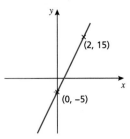

Are you ready? (B)

1 A car travels at an average speed of 45 km/h for 4 hours.

Work out the total distance travelled.

2 Rhys travels 60 miles at an average speed of 24 mph.

How long does his journey take?

Example

Marta drove from Leeds to Liverpool to deliver a parcel, and then she returned home. The graph shows her journey.

a Marta stopped at a service station on the journey from Leeds to Liverpool. For how long did she stop?

b How far is the service station from:

   i Leeds   **ii** Liverpool?

c What was Marta's speed on the return journey from Liverpool to Leeds?

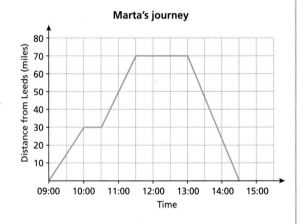

Method

Solution	Commentary
a 30 minutes	Marta stopped after travelling 30 miles.
	The graph is horizontal between 10:00 and 10:30, which means that the car was not moving.
b i 30 miles **ii** 40 miles	The furthest point from Leeds is 70 miles, so the distance from the service station is 70 − 30 = 40 miles
c 70 ÷ 1.5 = 47 mph	The distance travelled was 70 miles and took 1.5 hours (from 13:00 to 14:30).
	Use the formula speed = $\dfrac{\text{distance}}{\text{time}}$

Practice (B)

1 The distance–time graph shows information about Beca's journey.

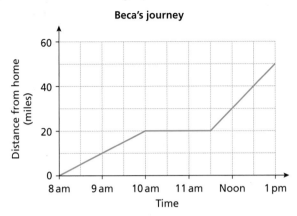

a How far does Beca travel between:

   i 8 am and 10 am   **ii** 11:30 am and 1 pm?

b How long does it take to travel:

   i 10 miles from the start   **ii** 50 miles from the start?

c Beca made one stop on the journey.

   For how long did she stop?

d What was Beca's speed between 8 am and 10 am?

e How long did Beca spend driving altogether?

f Work out her average speed between 8 am and 1 pm.

2 Abdullah sets off from home at 1 pm. He stops to refuel and then continues his journey before returning home later in the day.

The distance–time graph shows information about his journey.

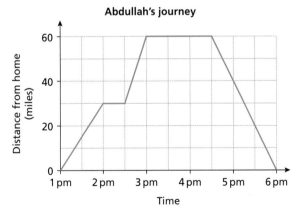

a How many miles did Abdullah travel before stopping?

b For how long did Abdullah stop in total?

c At what time did Abdullah set off on his journey home?

d What was Abdullah's speed:

    i during the first section of his journey

    ii between 2:30 pm and 3:00 pm

    iii on the return journey home?

e Work out Abdullah's average speed for the time he was actually driving.

3 Marta goes on a bike ride.

Here is a description of her journey:

    She left home at 10 am.

    She cycled at a speed of 20 miles per hour for 90 minutes.

    She had a break for 1 hour.

    She then cycled back and returned home at 2:30 pm.

a Represent this information on a distance–time graph.

b Work out:

    i the total distance Marta cycled   **ii** the speed on her return journey home.

Consolidate – do you need more?

1 Work out an equation for each line.

a

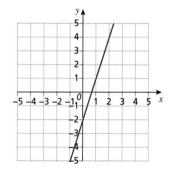

b

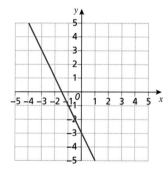

c

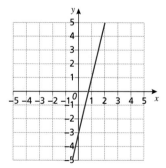

d

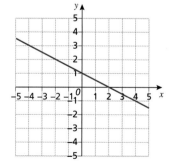

2 The graph shows the cost of delivering a parcel for a given number of miles.

    a How much is the fixed charge?

    b Use the graph to work out the cost of:

       i a 1-mile journey   **ii** a 10-mile journey.

    c Work out the cost per extra mile travelled.

    d Interpret the gradient and the y-intercept of the line in the context of the question.

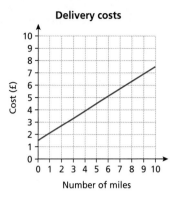

Delivery costs

3 The distance–time graph represents Jakub's walk.

    a How far does Jakub walk in total?

    b Jakub made one stop on the walk.

       For how long did he stop?

    c What was his speed for the second part of the walk?

    d How long did Jakub spend walking altogether?

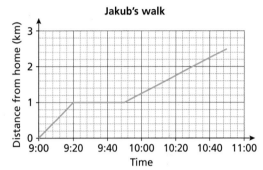

Jakub's walk

Stretch – can you deepen your learning?

1 Work out an equation for each line.

    a

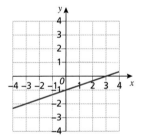

    b

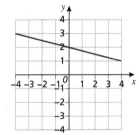

    c

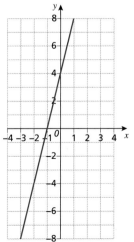

    d

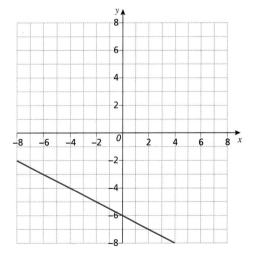

2 The coordinates of point A are (12, −16).

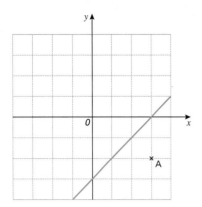

Work out the equation of the line shown.

3 Work out an equation for each line.

a

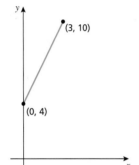

(3, 10)

(0, 4)

b

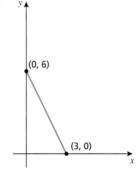

(0, 6)

(3, 0)

4 The cost of hiring a carpet cleaner is £45 plus £10 an hour.

Rhys is going to draw a graph to represent this information.

a Explain why Rhys's graph will be a straight line.

b **i** Work out an equation for Rhys's straight line.

ii Interpret the gradient and the *y*-intercept of the line in the context of the question.

Linear graphs: exam practice

---

1 (a) Plot the points (2, 4) and (2, –3) on a copy of the grid below.
Join your points with a straight line. **[2 marks]**

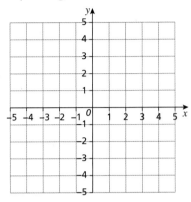

(b) What is the equation of the line you have drawn? **[1 mark]**

(c) On the same grid, draw the line $y = x$. **[1 mark]**

(d) Write down a pair of coordinates that would lie on the line $y = -4$ **[1 mark]**

1–3

---

2 (a) Complete a copy of the table of values for $y = 3x + 4$

x	–2	–1	0	1	2
y					

[2 marks]

(b) Draw the graph of $y = 3x + 4$ on a copy of the grid below.

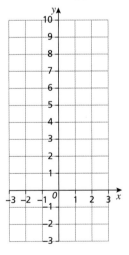

[2 marks]

3–5

---

3 The equation of a line is $y = 7x + 3$

(a) What is the gradient of the line? **[1 mark]**

(b) What is the y-intercept of the line? **[1 mark]**

(c) Write down an equation of a line that would be parallel to $y = 7x + 3$ **[1 mark]**

---

4 Write down three pairs of coordinates that would lie on the line $x + y = 3$ **[2 marks]**

13 Non-linear functions

In this block, we will cover...

13.1 Quadratic graphs

Example

a Draw the graph of $y = x^2 + 2x - 3$ for values of

b Use the graph to estimate the values of x whe

Method

Solution

a

x	−4	−3	−2	−1	0	1	2
$y = x^2 + 2x - 3$	5	0	−3	−4	−3	0	5

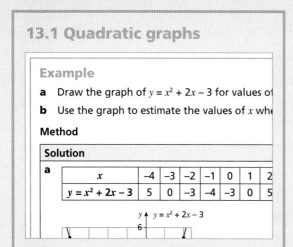

13.2 Cubic and reciprocal graphs

Practice

1 **a** Copy and complete the table of values

x	−2	−1	0	1	
y					

b Draw the graph of $y = x^3 - 4$ for values

2 **a** Copy and complete the table of values

x	−2	−1	0	1	
y					

Are you ready?

1 Work out the value of each expression when $x = 3$

    a x^2 **b** $2x$ **c** $3x - 4$ **d** $x^2 - 6$ **e** $x^2 + 2x$ **f** $x^2 - 2x$

2 Work out the value of each expression when $x = -3$

    a x^2 **b** $2x$ **c** $3x - 4$ **d** $x^2 - 6$ **e** $x^2 + 2x$ **f** $x^2 - 2x$

3 Copy and complete the table of values for the equation $y = 2x - 3$

x	−3	−2	−1	0	1	2	3
y							

A **quadratic** is an expression where the highest power of the variable is 2. Graphs with equations of the form $y = ax^2 + bx + c$, where a, b and c are constants and $a \neq 0$, are quadratic curves. They are always shaped like a 'U' (or an upside down 'U') and have a line of symmetry. These curves are called **parabolas**.

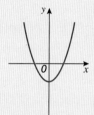 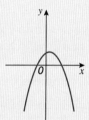

Example

a Draw the graph of $y = x^2 + 2x - 3$ for values of x from −4 to 2

b Use the graph to estimate the values of x when $y = 3$

Method

Solution	Commentary
a 	x	−4	−3	−2	−1	0	1	2	 	$y = x^2 + 2x - 3$	5	0	−3	−4	−3	0	5	 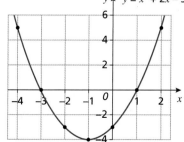	Work out the value of y for each value of x in the given range and write them in a table of values. Be careful with negative values. For example, when $x = -2$, $y = (-2)^2 + 2 \times -2 - 3 = 4 - 4 - 3 = -3$ Plot the points shown in your table of values (−4, 5), (−3, 0) and so on, and join them with a smooth curve. Do not worry if it does not look perfect first time – this takes some practice.

b

$y = x^2 + 2x - 3$

$x = -3.6$ and $x = 1.6$

Add the line $y = 3$ to your graph.

Draw vertical lines from where the line and curve meet down to the x-axis.

Read the values where the vertical lines meet the x-axis.

The values of x when $y = 0$ are called the **roots** of the quadratic equation.

Practice

1. **a** Copy and complete this table of values for $y = x^2$

x	–3	–2	–1	0	1	2	3
y							

   b Draw axes numbered from –3 to 3 in the x direction and from –6 to 10 in the y direction.

   Use the values from your table to plot the graph of $y = x^2$ on your pair of axes. Join the points with a smooth curve.

2. **a** Copy and complete this table of values for $y = x^2 - 3$

x	–3	–2	–1	0	1	2	3
y							

   b Use the values from your table to plot the graph of $y = x^2 - 3$ on the same set of axes you used in question 1. Join the points with a smooth curve.

   c Compare your graphs. What is the same and what is different?

   d Discuss what you think the following graphs will look like.

   $y = x^2 + 3$ $y = x^2 - 5$ $y = x^2 + 5$

   e Check your answers by drawing the three graphs.

3. Emily is drawing the graph of $y = x^2 - 5x$ using this table of values.

x	–1	0	1	2	3	4	5	6
y	6	0	–4	–6	–6	–4	0	6

   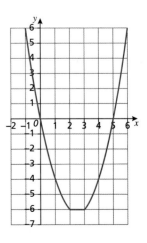

   She notices the x values 2 and 3 have the same y value and thinks that the graph will look as shown.

   Emily is incorrect.

   a At what value of x will the lowest point of the graph be?

   What will the y value be at this point?

   b Draw the graph of $y = x^2 - 5x$ for values of x from –1 to 6

4 **a** Copy and complete this table of values for $y = x^2 + 3x$

x	−5	−4	−3	−2	−1	0	1	2
y								

b Draw axes numbered from −6 to 3 in the x direction and from −3 to 11 in the y direction.

Use the values from your table to plot the graph of $y = x^2 + 3x$ on your pair of axes. Join the points with a smooth curve.

5 Here is the graph of $y = x^2 - 2x - 3$

Use the graph to:

a work out the value of y when $x = -1$

b work out the value of y when $x = 0$

c work out the value of y when $x = 2$

d work out the value of x when $y = -4$

e estimate the value of y when $x = 2.5$

f estimate the values of x when $y = -1$

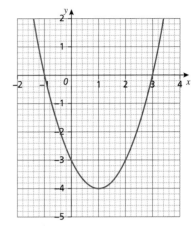

6 **a** Copy and complete this table of values for $y = x^2 - 3x - 2$

x	−3	−2	−1	0	1	2	3
y							

b Draw axes numbered from −4 to 4 in the x direction and from −6 to 18 in the y direction.

Use the values from your table to plot the graph of $y = x^2 - 3x - 2$ on your pair of axes. Join the points with a smooth curve.

c Use the graph to:

i estimate the value of y when $x = 0.5$ **ii** estimate the value of x when $y = 3$

What do you think? 💡

1 Filipo says, "When you square a negative number, you get a positive number, so the graph of $y = -x^2$ will be the same as the graph $y = x^2$."

Copy and complete the table of values and plot the graph of $y = -x^2$ to show that Filipo is incorrect.

x	−3	−2	−1	0	1	2	3
y							

2 Sort the cards under the correct headings in the table.

| $y = 4x + 1$ | $y = 4x^2$ | $y = x - 4$ | $y = x^2 + 4x + 1$ |

| $y = 4 + x$ | $y = 4 - x^2$ | $y = -4 - x$ | $y = (x + 4)^2$ |

Linear graphs	Quadratic graphs

Consolidate – do you need more?

1 a Copy and complete this table of values for $y = x^2 - 4$

x	–3	–2	–1	0	1	2	3
y							

b Draw axes numbered from –3 to 3 in the x direction and from –5 to 6 in the y direction.

Use the values from your table to plot the graph of $y = x^2 - 4$ on your pair of axes. Join the points with a smooth curve.

2 a Copy and complete this table of values for $y = x^2 + 5x$

x	–6	–5	–4	–3	–2	–1	0	1
y								

b Draw axes numbered from –6 to 1 in the x direction and from –7 to 7 in the y direction.

Use the values from your table to plot the graph of $y = x^2 + 5x$ on your pair of axes. Join the points with a smooth curve.

3 a Copy and complete this table of values for $y = x^2 - 5x + 3$

x	–1	0	1	2	3	4	5	6
y								

b Draw axes numbered from –1 to 6 in the x direction and from –4 to 10 in the y direction.

Use the values from your table to plot the graph of $y = x^2 - 5x + 3$ on your pair of axes. Join the points with a smooth curve.

c Use the graph to:

i estimate the value of y when $x = 2.5$ **ii** estimate the value of x when $y = 1$

Stretch – can you deepen your learning?

1   **a**   Copy and complete this table of values for $y = 2x^2 + 3$

x	–3	–2	–1	0	1	2	3
y							

   b   Draw the graph of $y = 2x^2 + 3$ for values of x from –3 to 3

2   Draw the graph of:

   a $y = 2x^2 + x - 10$ for values of x from –3 to 3

   b $y = \frac{1}{2}x^2 + 6$ for values of x from –4 to 4

3   Match each equation to its graph.

$y = x^2 - 7$

$y = x^2 - 7x$

$y = -x^2 + 7$

$y = x^2 + 7x - 7$

1

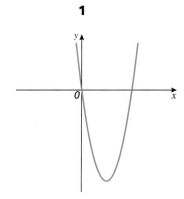

2

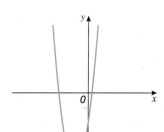

3

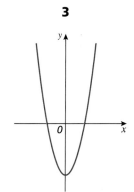

4

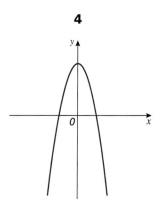

Are you ready?

1 Given $x = 2$, work out the value of:

    a x^3 **b** $x^3 - 4$ **c** $x^3 + 3x$ **d** $x^3 + 3x - 5$

2 Given $x = -2$, work out the value of:

    a x^3 **b** $x^3 - 4$ **c** $x^3 + 3x$ **d** $x^3 + 3x - 5$

3 Evaluate $\dfrac{6}{x}$ when:

    a $x = 3$ **b** $x = 2$ **c** $x = -6$ **d** $x = 0.5$

You have already met quadratic functions, in which the greatest power of x is x^2.

Cubic functions are functions where x^3 is the highest power of x.

Here are some examples of cubic graphs:

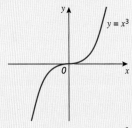
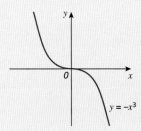

Reciprocal functions are of the form $y = \dfrac{k}{x}$

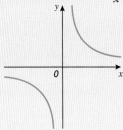
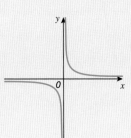

Reciprocal graphs can also look like this:

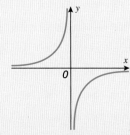

Note that the graph doesn't touch either of the axes; it just gets closer and closer to them. Such lines are called **asymptotes**.

Example 1

a Complete the table of values for $y = x^3 + 1$

x	−2	−1	0	1	2
y		0	1		9

b Draw the graph of $y = x^3 + 1$

Method

Solution	Commentary
a $\quad x = -2$ $y = (-2)^3 + 1$ $\quad = -8 + 1 = -7$	You can work out the remaining values using substitution.
$x = 1$ $y = (1)^3 + 1$ $\quad = 1 + 1 = 2$	
<table><tr><td>x</td><td>−2</td><td>−1</td><td>0</td><td>1</td><td>2</td></tr><tr><td>y</td><td>−7</td><td>0</td><td>1</td><td>2</td><td>9</td></tr></table>	
b	Plot the points and join with a smooth curve.

Example 2

a Complete the table of values and then draw the graph of $y = \dfrac{3}{x}$ for values of x from −4 to 4

x	−4	−3	−2	−1	−0.5	0.5	1	2	3	4
$y = \dfrac{3}{x}$										

b Describe the shape of the graph.

c Why is $x = 0$ not included in the table of values?

Method

Solution	Commentary
a 	Remember, $\frac{3}{x}$ means $3 \div x$. Work out the value of y for each value of x in the given range and write them in the table of values. Join through the points in the first quadrant with a smooth curve. Join through the points in the third quadrant with a smooth curve.

x	−4	−3	−2	−1	−0.5	0.5	1	2	3	4
$y = \dfrac{3}{x}$	−0.75	−1	−1.5	−3	−6	6	3	1.5	1	0.75

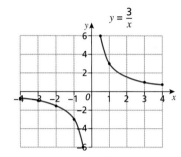

b The graph is in two sections. Both sections get closer and closer to the axes for very small and very large values of x, but they never touch them.	You can check these observations by using graphing software to draw the graph.
c It is impossible to work out $3 \div 0$	You cannot divide a number by 0

Practice

1 a Copy and complete the table of values for $y = x^3 - 4$

x	−2	−1	0	1	2
y					

b Draw the graph of $y = x^3 - 4$ for values of x from −2 to 2

2 a Copy and complete the table of values for $y = x^3 + 2x$

x	−2	−1	0	1	2
y					

b Draw the graph of $y = x^3 + 2x$ for values of x from −2 to 2

3 a Copy and complete the table of values for $y = \dfrac{4}{x}$

x	−4	−2	−1	−0.5	0.5	1	2	4
y								

b Draw the graph of $y = \dfrac{4}{x}$ for values of x from −4 to 4

4 **a** Copy and complete the table of values for $y = 2x^3$

x	-2	-1	0	1	2
y					

b Draw the graph of $y = 2x^3$ for values of x from -2 to 2

5 **a** Copy and complete the table of values for $y = \dfrac{2}{x}$ for values of x from -4 to 4

x	-4	-2	-1	-0.5	0.5	1	2	4
y								

b Draw the graph of $y = \dfrac{2}{x}$ for values of x from -4 to 4

c Work out the value of x when $y = 1.5$

Consolidate – do you need more?

1 **a** Copy and complete the table of values for $y = x^3 + x + 1$

x	-2	-1	0	1	2
y					

b Draw the graph of $y = x^3 + x + 1$ for values of x from -2 to 2

2 **a** Copy and complete the table of values for $y = \dfrac{5}{x}$

x	-5	-2	-1	-0.5	0.5	1	2	5
y								

b Draw the graph of $y = \dfrac{5}{x}$ for values of x from -5 to 5

3 Match each graph to the correct equation.

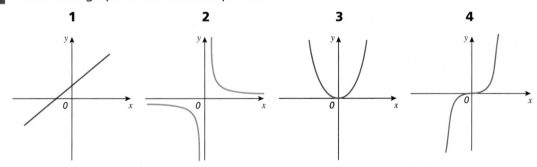

| **1** | **2** | **3** | **4** |

$y = x^2$ $y = \dfrac{2}{x}$ $y = 2x^3$ $y = x + 2$

273

Stretch – can you deepen your learning?

1 **a** Copy and complete the table of values for $y = -x^3$

x	−2	−1	0	1	2
y					

b Draw the graph of $y = -x^3$ for values of x from −2 to 2

2 **a** Copy and complete the table of values for $y = \dfrac{3}{x}$

x	−3	−2	−1	−0.5	0.5	1	2	3
y								

b Draw the graph of $y = \dfrac{3}{x}$

c Give the value of x for which $y = 2$

d On the same axes, draw the line $y = 2x$

e Give the coordinates of the points of intersection of the two graphs.

3 Match each graph to the correct equation.

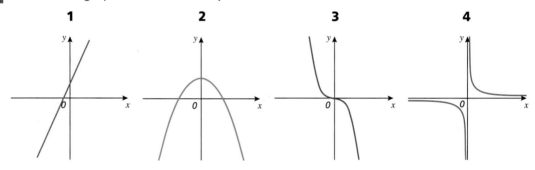

| **1** | **2** | **3** | **4** |

$$y = \frac{1}{x}$$

$$y = -x^3$$

$$y = 2x + 4$$

$$y = -x^2 + 6$$

Non-linear functions: exam practice

---

1 Which of these graphs are linear? **[1 mark]**

1–3

    A   **B**   **C**   **D**

---

2 (a) Complete a copy of the table of values for $y = x^2$ **[2 marks]**

3–5

x	−3	−2	−1	0	1	2	3
y			1	0		4	

(b) Draw the graph of $y = x^2$ on a copy of this grid.   **[2 marks]**

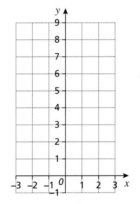

(c) Estimate the value of y when $x = 2.5$ **[1 mark]**

(d) Work out the value of y when $x = 7$ **[1 mark]**

---

3 Complete a copy of the table to match each graph to the correct equation.   **[2 marks]**

    A   **B**   **C**   **D**

Equation	Letter
$y = -x^3 + 3$	
$y = -3x - 1$	
$y = -\dfrac{3}{x}$	
$y = (x - 3)^2$	

In this block, we will cover...

14.1 Quadratic equations

Example 2

Solve the equation $x^2 + 11x = -30$

Method

Solution	Commentary
$x^2 + 11x = -30$ +30 $\qquad$ +30 $x^2 + 11x + 30 = 0$	In order to solve the quadratic exp both sides.
$(x + 6)(x + 5) = 0$	Factorise the qua
Either $x + 6 = 0$ or $x + 5 = 0$	If the product of brackets must eq

14.2 Simultaneous equations

Practice

1. Solve the simultaneous equations by subst

   a $y = x + 1$ $\qquad$ **b** $y = x - 3$

   $\quad y + 3x = 21$ $\qquad\quad$ $y + 2x = 9$

2. Solve the simultaneous equations by elimi

   a $3a + b = 23$ $\qquad$ **b** $c + 5d = 39$

   $\quad 2a + b = 17$ $\qquad\quad$ $c + 2d = 18$

3. Solve the simultaneous equations graphica

Are you ready?

1 Factorise fully:

    a $4y + 12$ **b** $g^2 - 7g$ **c** $5t^2 - 10t$

2 Solve:

    a $x + 4 = 0$ **b** $x - 5 = 0$

3 Write down a pair of numbers that have:

    a a sum of 5 and a product of 4

    b a sum of 13 and a product of 40

    c a sum of 2 and a product of −15

Quadratic equations have a single variable with a greatest power of 2

Examples of quadratic equations are:

$$x^2 + 3x + 2 = 0 \qquad p^2 - 9 = 27 \qquad m^2 - 2m = 3$$

Many quadratic equations can be factorised, and this helps to solve them.

For example, $x^2 + 3x + 2 = 0$ can be written as $(x + 2)(x + 1) = 0$

> How to factorise a quadratic expression is covered in Chapter 8.3

$(x + 2)(x + 1) = 0$ can be solved quite simply using reasoning.

If the product of two quantities equals zero, one of them must be zero.

For example, $6 \times 0 = 0$ or $0 \times 97 = 0$

So in $(x + 2)(x + 1) = 0$, either $x + 2 = 0$ or $x + 1 = 0$

Then there are two solutions to the quadratic equation: $x = -2$ and $x = -1$

Example 1

Solve the equation $x^2 - 5x + 4 = 0$

Method

Solution	Commentary
$x^2 - 5x + 4 = 0$	
$(x - 4)(x - 1) = 0$	Factorise the quadratic expression.
Either $x - 4 = 0$ or $x - 1 = 0$	If the product of the two brackets is zero, one of the brackets must equal zero.
So $x = 4$ or $x = 1$	Solve the two linear equations formed to find the values of x.

Example 2

Solve the equation $x^2 + 11x = -30$

Method

Solution	Commentary
$x^2 + 11x = -30$ $+30$ $\quad$ $+30$ $x^2 + 11x + 30 = 0$	In order to solve a quadratic, it must first be rearranged so the quadratic expression is equal to zero. So add 30 to both sides.
$(x + 6)(x + 5) = 0$	Factorise the quadratic expression.
Either $x + 6 = 0$ or $x + 5 = 0$	If the product of the two brackets is zero, one of the brackets must equal zero.
$x = -6$ or $x = -5$	Solve the two linear equations formed to find the values of x.

Practice

1. Solve these equations.

   a $(x - 3)(x + 1) = 0$ $\qquad$ **b** $(x - 4)(x - 9) = 0$ $\qquad$ **c** $(y + 5)(y - 8) = 0$

   d $(y + 7)(y + 4) = 0$ $\qquad$ **e** $w(w - 2) = 0$ $\qquad$ **f** $w(w + 7) = 0$

2. Solve these equations by factorising.

   a $x^2 + 5x + 6 = 0$ $\quad$ **b** $y^2 + 8y + 15 = 0$ $\quad$ **c** $g^2 + 11g + 28 = 0$ $\quad$ **d** $a^2 + 8a + 12 = 0$

3. Solve these equations by factorising.

   a $m^2 + 3m - 10 = 0$ $\quad$ **b** $b^2 + 2b - 35 = 0$ $\quad$ **c** $h^2 + 2h - 3 = 0$ $\quad$ **d** $w^2 + 5w - 24 = 0$

4. Solve these equations by factorising.

   a $k^2 - 4k - 12 = 0$ $\quad$ **b** $u^2 - 3u - 4 = 0$ $\quad$ **c** $q^2 - 3q - 40 = 0$ $\quad$ **d** $f^2 - f - 30 = 0$

5. Tiff is solving $x^2 - 10x + 24 = 0$

   She says $-4 + -6 = -10$ and $-4 \times -6 = 24$, so the solutions are -4 and -6

   Explain why Tiff is wrong and correctly solve $x^2 - 10x + 24 = 0$

6. Solve these equations by factorising.

   a $x^2 - 11x + 18 = 0$ $\qquad$ **b** $y^2 - 14y + 24 = 0$ $\qquad$ **c** $u^2 - 11u + 24 = 0$

7. Rearrange and solve these quadratic equations.

   a $x^2 + 2x = 15$ $\quad$ **b** $d^2 - 6d = 27$ $\quad$ **c** $h^2 - 10h = -16$ $\quad$ **d** $y^2 + 9y = -20$

8. Solve the equations by factorising.

   a $x^2 - 25 = 0$ $\qquad$ **b** $y^2 - 64 = 0$ $\qquad$ **c** $w^2 - 121 = 0$

   What do you notice about your answers?

Consolidate – do you need more?

1 Solve the equations by factorising.

    a $x^2 + 8x + 7 = 0$ **b** $q^2 + 13q + 36 = 0$ **c** $y^2 + 6y - 16 = 0$

    d $w^2 + 7w - 30 = 0$ **e** $p^2 - 10p - 24 = 0$ **f** $h^2 - 14h + 45 = 0$

2 Solve the equations by factorising.

    a $x^2 - 1 = 0$ **b** $r^2 - 81 = 0$ **c** $g^2 - 169 = 0$

3 Rearrange and solve these quadratic equations.

    a $e^2 - 5e = 36$ **b** $k^2 + 15k = -36$ **c** $m^2 - 13m = -40$

Stretch – can you deepen your learning?

1 Show algebraically that $x^2 + 8x + 16 = 0$ has only one solution.

2 Jackson is trying to solve the equation $y^2 + 6y = 16$

Here are his workings:

$$y^2 + 6y = 16$$
$$y(y + 6) = 16$$
$$y = 16 \text{ or } y + 6 = 16$$
$$\text{So } y = 16 \text{ or } y = 10$$

    a   Explain the mistake Jackson has made.

    b   Work out the correct solutions to the equation.

3 The area of the rectangle is $36\,cm^2$.

Work out the dimensions of the rectangle.

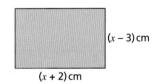

$(x - 3)\,cm$

$(x + 2)\,cm$

Are you ready?

1 Solve the equations.

    a $6x = 24$ **b** $-3x = 24$ **c** $4x = -24$ **d** $-8x = -24$

2 Simplify: **a** $3a - a$ **b** $3a + -2a$ **c** $3a - -2a$

3 Here is an expression: $3x + y$

    Every term in the expression is multiplied by 2

    Write the new expression.

Simultaneous equations are two or more algebraic equations that share variables, e.g. x and y. They are called simultaneous equations because they are both true at the same time.

For example, below are two pairs of simultaneous equations:

$2x + 4y = 10$ $3h + 2j = 14$

$5x + 4y = 18$ $h + 2j = 10$

Each of these pairs of simultaneous equations contains two variables, so that there are two equations and two unknowns.

Each of these equations on its own could have infinite possible solutions. However, when there are at least as many equations as variables, you may be able to solve them using the methods shown in this chapter. Example 1 shows a method known as **substitution**.

Example 1

Solve the simultaneous equations $y = x + 2$

                                      $y + 3x = 14$

Method

Solution	Commentary
$x + 2 + 3x = 14$	Substitute $x + 2$ for y in the second equation.
$4x + 2 = 14$	Simplify the equation by collecting like terms.
$-2 \left(\begin{matrix} 4x + 2 = 14 \\ 4x = 12 \end{matrix} \right) -2$	Subtract 2 from both sides.
$\div 4 \left(x = 3 \right) \div 4$	Divide by 4
$y = 3 + 2$ so $y = 5$	Substitute $x = 3$ into the first equation to find the value of y.
So $x = 3$, $y = 5$	Check the result in the second equation. $5 + 3 \times 3 = 14$ as required.
	The solution to a pair of simultaneous equations represents the intersection of two lines; in this example, they cross at $(3, 5)$.

This is the **substitution** method because you substitute one variable in place of another.

Example 2

Solve the simultaneous equations $h + j = 12$

$$5h + 2j = 39$$

Method

Solution	Commentary
$h + j = 12$ (A) $5h + 2j = 39$ (B)	Label the equations to help make your workings clear.
$2h + 2j = 24$ (C)	Multiply equation A by 2 to create equation C so the coefficients of one of the variables (j) are the same in the two equations.
$3h = 15$ (B) – (C) $h = 5$	Subtract equation C from equation B to eliminate the variable j, as $2j - 2j = 0$. Solve the resulting equation to find h.
$5 + j = 12$ $j = 7$	Substitute $h = 5$ into equation A to find the value of j.

Example 3

Solve the simultaneous equations $12x - 5y = -22$

$$-8x + 4y = 16$$

Method

Solution	Commentary
$12x - 5y = -22$ (A) $-8x + 4y = 16$ (B) (A) × 2 = (C) (B) × 3 = (D) $+\ \dfrac{24x - 10y = -44\ \text{(C)}}{}\ +$ $-24x + 12y = 48$ (D) $\overline{2y = 4}$ $y = 2$	Multiply equation A by 2 and equation B by 3 to generate coefficients of x you can eliminate. Label the new equations C and D. Add equations C and D to eliminate the x variable (as $-24x + 24x = 0$) Solve for y.
$-8x + 4(2) = 16$ $-8x + 8 = 16$ $-8\ \diagdown\qquad\diagup\ -8$ $-8x = 8$ $\div -8\ \diagdown\qquad\diagup\ \div -8$ $x = -1$	Substitute $y = 2$ into equation B to find the value of x. You can substitute $y = 2$ into any of the equations.

Examples 2 and 3 show the **elimination** method because you adjust the equations to eliminate one of the two variables.

Example 4

By drawing graphs, solve the simultaneous equations $\quad y = 3x - 5$

$$2x + y = 5$$

Method

Solution	Commentary
For $y = 3x - 5$ $x = 0, y = 0 - 5 = -5$ $x = 1, y = 3 - 5 = -2$ $x = 2, y = 6 - 5 = 1$	Start by working out some coordinates for each line. You can use a table of values.
For $2x + y = 5$ $x = 0, y = 5$ $y = 0, 2x = 5$ so $x = 2.5$	
   So $x = 2$ and $y = 1$	Note that the graphs must be drawn accurately, not sketches, so that the solution can be read accurately. Draw both graphs on the same set of axes. Since you want both equations to be true simultaneously, you need to find the point that lies on both lines, i.e. where the graphs cross. You can check your answer by substituting the values of x and y into the equations.

Practice

1. Solve the simultaneous equations by substitution.

   a $y = x + 1$

   $y + 3x = 21$

   b $y = x - 3$

   $y + 2x = 9$

   c $y = x + 1$

   $2y + 3x = 17$

   d $y = x - 2$

   $3y + 2x = 34$

2. Solve the simultaneous equations by elimination.

   a $3a + b = 23$

   $2a + b = 17$

   b $c + 5d = 39$

   $c + 2d = 18$

   c $4e + 6f = 42$

   $4e + 3f = 27$

   d $5g + 6h = 31$

   $2g + 6h = 16$

3. Solve the simultaneous equations graphically.

   a $y = 2x + 2$

   $y + x = -4$

   b $y = x$

   $2y = x + 4$

   c $y = 3x + 1$

   $x + y = 7$

④ Here is a pair of simultaneous equations.

$10w + 2x = 18$

$10w + 6x = 14$

Seb has started solving the equations. Here is his working:

a Explain the mistake he has made.

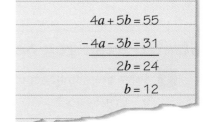

$4x = 4$ so $x = 1$

b Solve the simultaneous equations.

$10w + 2x = 18$

$10w + 6x = 14$

c Solve the simultaneous equations.

i $4k + 3n = 41$	**ii** $8p + 8t = 80$	**iii** $4m + 5q = 7$	**iv** $5r + 7w = 33$
$4k + 5n = 47$	$11p + 8t = 92$	$4m + 3q = 1$	$2r + 7w = 9$

⑤ Solve these simultaneous equations.

a $3x + 3y = 21$	**b** $7m + 2p = 60$	**c** $3w + 9k = 78$	**d** $12h + 4t = 88$
$6x + 2y = 30$	$5m + 8p = 56$	$2w + 3k = 37$	$6h + 5t = 56$

⑥ Zach is solving the simultaneous equations

$4a + 5b = 55$

$4a - 3b = 31$

Zach subtracts the second equation from the first equation to eliminate the coefficients of a.

Shown here is his working to find the value of b.

a Explain the mistake Zach has made when finding the value of b.

b Solve the simultaneous equations.

$4a + 5b = 55$

$- 4a - 3b = 31$

$2b = 24$

$b = 12$

⑦ Solve the simultaneous equations.

a $3x + 5y = 35$	**b** $5f - 2g = 34$	**c** $g + 2h = 17$
$3x - 2y = 7$	$7f + 2g = 62$	$7g - 4h = 11$
d $2m - k = 9$	**e** $3w + 2p = 20$	**f** $3v + 2n = 9$
$5m - k = 21$	$2w - p = 18$	$4v - 5n = 35$

⑧ Which pairs of simultaneous equations have the solution $x = 3$ and $y = -1$?

A

$4x + 2y = 10$

$8x + 3y = 21$

B

$x + y = 2$

$x - y = 4$

C

$x + 5y = -2$

$3x + 2y = 7$

D

$3x + y = 10$

$6x - y = 17$

E

$5x - y = 16$

$x - 2y = 5$

F

$x - 4y = 1$

$3x - 2y = -5$

Consolidate – do you need more?

1 Solve the simultaneous equations by substitution.

a $y = x - 2$
$y + 4x = 18$

b $y = x + 4$
$y + 3x = 16$

c $y = x + 1$
$2y + 3x = 7$

d $y = x - 3$
$3y + 2x = 6$

2 Solve the simultaneous equations by elimination.

a $5a + b = 17$
$2a + b = 8$

b $2e + 7f = 16$
$2e + 4f = 4$

c $5k + 3n = 11$
$5k + 5n = 15$

d $5r + 8w = 9$
$4r + 8w = 12$

3 Solve these simultaneous equations.

a $5m + 4p = 51$
$3m + 8p = 53$

b $10h + 4t = 22$
$5h + 3t = 14$

c $2x + 5y = 28$
$2x - 3y = 12$

d $5f - 5g = 15$
$8f + 5g = -2$

e $5g + 3h = 18$
$6g - 6h = 60$

f $2n - 3k = 11$
$3n - 3k = 21$

4 Make as many pairs of simultaneous equations as you can using these three cards. Solve your pairs.

$a + 2y = 13$ $a - y = 1$ $2a + 3y = 21$

Stretch – can you deepen your learning?

1 Solve the simultaneous equations.

a $9x + 5y = 35$
$4x - 2y = 24$

b $4x - 3y = -10$
$6x + 5y = 23$

c $3x - 2y = 11$
$2x - y = 8$

2 Two cups of tea and five cups of coffee cost £13

Four cups of tea and three cups of coffee cost £12

a How much does a cup of tea cost?

b How much does a cup of coffee cost?

3 Use simultaneous equations to work out the value of each shape in the puzzle.

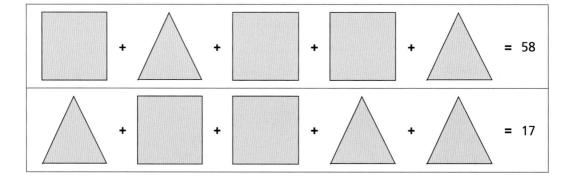

More complex equations: exam practice

1 Solve $x + 6 = 0$ [1 mark]

1–3

2 Write down a pair of numbers that have a sum of 10 and a product of 21 [1 mark]

3–5

3 (a) Factorise $x^2 - 3x$ [1 mark]

(b) Factorise $x^2 - 25$ [2 marks]

(c) Factorise $y^2 - 10y + 16$ [2 marks]

(d) Solve $y^2 - 10y + 16 = 0$ [2 marks]

4 A straight line has the equation $y = 2x + 1$

(a) If $x = 5$, what is the value of y? [1 mark]

(b) Show that $x = 11$ and $y = 23$ is one solution to the equation. [2 marks]

5 Two sandwiches and three drinks cost £5

This can be written as $2s + 3d = 5$

A drink costs 80p.

What is the price of a sandwich? [3 marks]

6 Solve the simultaneous equations

$$4x + 2y = 16$$
$$x + 2y = 1$$

[4 marks]

7 Solve the simultaneous equations

$$2x + 5y = 31$$
$$x + 2y = 13$$

[4 marks]

15 Sequences

In this block, we will cover...

15.1 Generating sequences

Example 2

The nth term of a sequence is given by the rule 6.

a   Work out the eighth term of the sequence.

b   Work out the first term of the sequence that

c   Describe the sequence in words.

Method

Solution	Com
a When $n = 8$, $6n + 3 = 6 \times 8 + 3$ $= 51$	The $n = 8$
b $6n + 3 > 100$	Set

15.2 Linear sequences

Practice

1   Work out the rule for the nth term of each

   a   5, 9, 13, 17 …   b   8, 13, 18, 2

   d   3, 7, 11, 15 …   e   4, 11, 18, 2

2   Here are two sequences:

   4, 8, 12, 16 …   0, 4, 8, 12, 16 …

   Faith says these sequences will have the sa

   Work out the nth term of the two sequenc

3   The nth term of a sequence is $4n + 3$

15.3 Other sequences

Consolidate – do you need more

1   Write the next three terms of these Fibona

   a   2, 7, 9 …   b   20, 21, 41 …

2   All of the following sequences are geomet

   For each sequence:

      i   work out the term-to-term rule

   a   20, 80, 320, 1280 …   b

   c   300, 1200, 4800, 19 200 …   d

Are you ready?

1 Work out the next two terms of each sequence.

    a 1, 7, 13, 19, 25 …   **b** 50, 45, 40, 35, 30 …

    c 18, 25, 32, 39, 46 …   **d** 20, 10, 0, −10, −20 …

2 Here are some patterns of squares:

Pattern 1   **Pattern 2**   **Pattern 3**

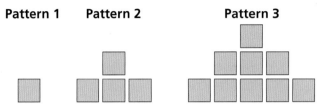

    a Write down the number of squares in each pattern.

    b Draw Pattern 4

A **sequence** is a list of items in a given order which usually follows a rule. Each member of the sequence is called a **term**.

The fourth term in the sequence 6, 9, 12, 15, 18 … is 15

In a **linear** sequence (also known as an **arithmetic** sequence), the terms increase or decrease by the same amount. This is known as the **common difference**.	A **geometric** sequence has a **constant ratio** between terms.

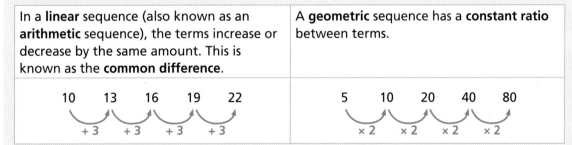

Here are the first few terms of some other common sequences:

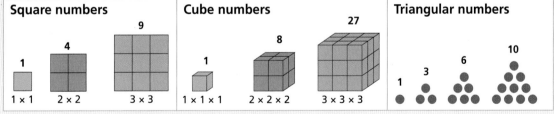

Sequences can be represented using algebraic rules, such as:

$$3n + 1 \qquad 4n - 2 \qquad 10 - 6n$$

These rules tell you the value of the nth term of a sequence; this is the term that is in the nth position.

Rules of the form $an + b$ produce linear sequences.

The next example refers to a **term-to-term rule**. You should be familiar with term-to-term rules from Key Stage 3.

Example 1

a Three sequences all have the first term 10

Work out the fourth term of each sequence if the term-to-term rule is:

i add 7 to the previous term

ii subtract 7 from the previous term

iii double the previous term.

b Are the sequences **linear**, **geometric** or **neither**?

Method

Solution	Commentary
a i 10, 17, 24, 31 The fourth term is 31	Start with 10 each time, and apply the **term-to-term rule** until you get to the fourth term.
ii 10, 3, −4, −11 The fourth term is −11	
iii 10, 20, 40, 80 The fourth term is 80	
b i Linear	There is a constant difference between the terms, so the sequence is linear.
ii Linear	**Linear sequences** can be descending as well as ascending.
iii Geometric	A **geometric sequence** has a constant multiplier between the terms.

Example 2

The nth term of a sequence is given by the rule $6n + 3$

a Work out the eighth term of the sequence.

b Work out the first term of the sequence that is greater than 100

c Describe the sequence in words.

Method

Solution	Commentary
a When $n = 8$, $6n + 3 = 6 \times 8 + 3$ 　　　　　　　　$= 51$	The eighth term means $n = 8$, so substitute $n = 8$ into the expression for the nth term.
b $\qquad -3 \Big(\dfrac{6n + 3 > 100}{6n > 97} \Big) -3$ $\qquad \div 6 \Big(\dfrac{}{n > 16.166...} \Big) \div 6$ The 17th term will be greater than 100	Set up an inequality. Solve this the way you learned in Block 11 n must be an integer.
When $n = 17$, $6n + 3 = 6 \times 17 + 3 = 105$	Substitute into the expression for the sequence to find the required term and check that it is greater than 100

c When $n = 1$,	Find the first two terms by substituting $n = 1$
$6n + 3 = 6 \times 1 + 3 = 9$	and $n = 2$
When $n = 2$,	
$6n + 3 = 6 \times 2 + 3 = 15$	
The difference between the first two terms is $15 - 9 = 6$	You know that the sequence is linear as $6n + 3$ is of the form $an + b$.
The sequence starts at 9 and goes up 6 every time.	You can therefore describe the sequence by giving the first term and the difference.

Practice

1 The first term in a sequence is 50

Work out the third term if the term-to-term rule is:

a add 5 to the previous term **b** subtract 5 from the previous term

c multiply the previous term by 5 **d** divide the previous term by 5

2 Here are the first three terms of a sequence of sticks:

a Draw the next pattern in the sequence.

b Describe the term-to-term rule for the sequence.

3 a Ed is describing a sequence. He says, "The term-to-term rule is subtract 6 every time."

Explain why this is **not** enough information to describe the sequence fully.

b Fully describe these sequences and work out the fifth term of each sequence.

i 5, 9, 13, 17 … **ii** 40, 90, 140, 190 …

iii 38, 30, 22, 14 … **iv** 10, 20, 40, 80 …

4 a The term-to-term rule for a sequence is 'add 15 to the previous term'.

Work out the first five terms of the sequence if the first term is: **i** 2 **ii** 35

b The term-to-term rule for a sequence is 'multiply the previous term by 3 and subtract 5'.

Work out the first five terms of the sequence if the first term is: **i** 10 **ii** 1

5 a Work out the first five terms of the sequences given by these algebraic rules.

i $2n + 5$ **ii** $5n - 2$ **iii** $6n$ **iv** $1 + 6n$

b Describe the sequences in words.

c Work out the 100th term of each sequence.

6 A sequence is given by the rule $20 - 3n$

a Work out the first five terms of the sequence.

b Is the sequence **linear, geometric** or **neither**?

c Work out the: **i** 50th term of the sequence **ii** 300th term of the sequence.

7   Decide if 80 is a term in each of the sequences given by these rules.

$4n - 7$ $\qquad\qquad$ $5n + 25$ $\qquad\qquad$ $3n + 18$ $\qquad\qquad$ $100 - 2n$

8   **a**   Work out the first term in the sequence given by the rule $9n - 4$ that is greater than 350

 b   Work out the last term in the sequence given by the rule $4n - 90$ that is negative.

Consolidate – do you need more?

1   A sequence has a first term 30

Work out the next four terms of the sequence if the term-to-term rule is:

a   add 7 to the previous term $\qquad\qquad$ **b**   subtract 6 from the previous term

c   double the previous term $\qquad\qquad$ **d**   add 45 to the previous term.

2   The term-to-term rule for a sequence is 'multiply the previous term by 5 and subtract 10'.

Work out the first five terms of the sequence if the first term is: **a**   4   **b**   1

3   The table shows the first two terms of a sequence.

Position	1	2	3	4
Term	3	12		

Work out the next two terms of the sequence if it is: **a**   linear   **b**   geometric.

4   **a**   Work out the first five terms of the sequences given by these algebraic rules.

 i $3n + 5$ $\qquad$ **ii** $5n - 8$ $\qquad$ **iii** $10 - 2n$

 b   Describe the sequences in words.

5   The nth term of a sequence is given by the rule $9n + 11$

 a   Which position in the sequence is the term with value 416?

 b   Work out the first term in the sequence that is greater than 800

 c   Show that 1200 is **not** a term in the sequence.

Stretch – can you deepen your learning?

1   A sequence starts with 64

The next term is found by halving the previous term and adding 6

Work out the first five terms of the sequence.

2   The third term of a sequence is 20

Work out the first five terms of the sequence if the term-to-term rule is:

a   add 8 to the previous term

b   subtract 17 from the previous term

c   multiply the previous term by 10

d   add 4 to the previous term and then multiply by 4

Are you ready?

1 A sequence starts 13, 19, 25, 31, 37 …

    a What is the third term of the sequence?

    b Describe how the sequence changes.

    c Work out the next term of the sequence.

2 Draw the next term in the sequence.

 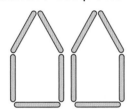 

3 What is the same and what is different about these two sequences?

    1, 3, 5, 7, 9 …   2, 4, 6, 8, 10 …

A **term-to-term** rule explains how one term in a sequence is connected to the next term.

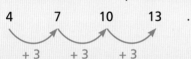

4   7   10   13   …

   + 3   + 3   + 3

The term-to-term rule is 'add three to the previous term'.

A **position-to-term** rule explains how a term in a sequence is connected to its position in the sequence.

The position is usually denoted n, and the rule is also called the nth term rule.

Position (n)	1	2	3	4
Term	4	7	10	13

For a linear sequence, the rule will involve multiplying the position by the common difference, in this case, $3 \times n = 3n$

You can work out the nth term rule by comparing $3n$ with the sequence itself.

n	1	2	3	4
$3n$	3	6	9	12
Term	4	7	10	13

You need to add 1 to each value of $3n$ to get the corresponding term of the sequence.

The rule for the nth term of the sequence is $3n + 1$

Check: 4th term = $3 \times 4 + 1 = 12 + 1 = 13$ ✓

Example

a Work out the rule for the nth term of the sequence 10, 14, 18, 22 …

b Work out the 100th term of the sequence.

c Show that 836 is **not** a term in the sequence.

Method

Solution	Commentary
a 10 14 18 22 $+4$ $+4$ $+4$ The rule will include '$4n$'	Find the common difference. This is the coefficient of n in the rule for the sequence.

n 1   2   3   4
   $4n$ 4   8   12   16
              $+6$ $+6$ $+6$ $+6$
 term 10   14 18 22
	Compare the sequence with $4n$.

The rule is $4n + 6$	As each term in the sequence is 6 greater than $4n$, the rule is $4n + 6$
b When $n = 100$ $4n + 6 = 4 \times 100 + 6$ $= 400 + 6$ $= 406$	Substitute $n = 100$ into the rule you found in part **a**.
c If 836 is in the sequence, then $4n + 6 = 836$ $4n = 830$ $n = 207.5$	Form an equation using your rule. Solve the equation to find the value of n.
The position of a term, n, must be an integer, so 836 is not a term in the sequence.	State your conclusion clearly.

Practice

1 Work out the rule for the nth term of each of these sequences.

    a 5, 9, 13, 17 …   **b** 8, 13, 18, 23 …   **c** 5, 10, 15, 20 …

    d 3, 7, 11, 15 …   **e** 4, 11, 18, 25 …   **f** 11, 12, 13, 14 …

2 Here are two sequences:

    4, 8, 12, 16 …   0, 4, 8, 12, 16 …

    Faith says these sequences will have the same rule as they both increase by 4

    Work out the nth term of the two sequences to show that Faith is incorrect.

3 The nth term of a sequence is $4n + 3$

    Show that the 10th term of the sequence is **not** double the 5th term of the sequence.

4 **a** Work out the 100th term of the sequence 22, 26, 30, 34 …

b Show that 267 is **not** a term of the sequence.

5 Here are two sequences, A and B:

A

6, 8, 10, 12

B

12, 10, 8, 6

a What is the same and what is different about sequence A and sequence B?

b Work out the rule for the nth term of sequence B.

6 Work out the rule for the nth term of these sequences.

a 10, 7, 4, 1 …   **b** 60, 50, 40, 30 …   **c** 24, 19, 14, 9 …

7 **a** Work out the first term in the sequence 37, 46, 55, 64 … that is greater than 900

b Work out the first negative term in the sequence 82, 79, 76, 73 …

Consolidate – do you need more?

1 Work out the rule for the nth term of each of these sequences.

a 6, 9, 12, 15 …   **b** 9, 14, 19, 24 …   **c** 8, 16, 24, 32 …

d 2, 7, 12, 17 …   **e** 6, 7, 8, 9 …   **f** 9, 20, 31, 42 …

2 By working out the first four terms of each sequence, show that the sequences given by the rules $3n - 1$, $3n + 5$ and $3n - 7$ all have a common difference of 3 between consecutive terms.

3 **a** Work out the rule for the nth term of the sequence 4, 11, 18, 25 …

b Work out the 500th term of the sequence.

c Is 1082 a term in the sequence? Explain how you know.

4 Work out the rule for the nth term of these sequences.

a 15, 13, 11, 9 …   **b** 60, 40, 20, 0 …   **c** 89, 80, 71, 62 …

5 The table shows a sequence.

Position	1	2	3	4	5
Term	32	49	66	83	100

a Which position in the sequence is the term with value 270?

b Work out the first term in the sequence that is greater than 1000

c Show that 500 is **not** a term in the sequence.

6 Which term of the sequence 70, 76, 82, 88 … is equal to 178?

Stretch – can you deepen your learning?

1 Here are the first three terms of a sequence of squares.

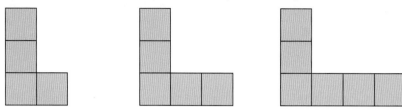

   a Draw the next pattern in the sequence.

   b Work out the rule for the number of squares needed to make the nth shape in the sequence.

   c Explain how your rule is related to the shapes.

2 **a** The term-to-term rule of a sequence is add 8 to the previous term.

     The first term is 9

     Work out the nth term of the sequence.

   b The term-to-term rule of a sequence is subtract 11 from the previous term.

     The first term is 30

     Work out the nth term of the sequence.

   c The term-to-term rule of a sequence is add 15 to the previous term.

     The third term is 10

     Work out the nth term of the sequence.

   d The term-to-term rule of a sequence is subtract 7 from the previous term.

     The third term is −2

     Work out the nth term of the sequence.

Are you ready?

1 A sequence starts   8, 11, 14, 17 …

    a Work out the nth term of the sequence.

    b Work out the 100th term of the sequence.

    c Show that 90 is **not** a term in the sequence.

2 State whether each sequence is linear or not.

    a 5, 10, 15, 20 …   **b** 5, 10, 20, 40 …   **c** 5, 10, 15, 25 …   **d** 5, 0, –5, –10 …

3 Work out the value of these expressions when $n = 4$

    a n^2 **b** $n^2 - 20$ **c** $n(n + 1)$ **d** $\dfrac{1}{n}$

Non-linear sequences do not have common differences.

The rule for a **quadratic sequence** will include an n^2 or an^2 term such as $n^2 + 2$, $n^2 - 3n$ or $2n^2 + 2n - 3$

The square numbers (1, 4, 9, 16 …) form a quadratic sequence given by n^2, and the triangular numbers (1, 3, 6, 10, 15 …) are given by $\frac{1}{2}n(n + 1)$.

In a **Fibonacci sequence**, the next term is found by adding the previous two terms.

For example, 1, 1, 2, 3, 5, 8, 13 … or 3, 4, 7, 11, 18 …

Geometric sequences are sequences where each term is found by multiplying the previous term by a **constant multiplier**. For example, 2, 4, 8, 16, 32 … (multiply by 2 each time) or 81, 27, 9, 3, 1, $\frac{1}{3}$ … (multiply by $\frac{1}{3}$ each time).

Example 1

Work out the next three terms of these Fibonacci sequences.

a 1, 1, 2, 3, 5 …   **b** 5, 7, 12, 19 …

Method

Solution	Commentary
a The next term is 3 + 5 = 8	1 1 2 3 5
Then 5 + 8 = 13	1 + 1 = 2
Then 8 + 13 = 21	1 + 2 = 3 Each term is the sum of
The next three terms are 8, 13 and 21	2 + 3 = 5 the previous two terms.
b The next term is 12 + 19 = 31	
Then 19 + 31 = 50	
Then 31 + 50 = 81	
The next three terms are 31, 50 and 81	

Example 2

The second and the third terms of a geometric sequence are 40 and 80

Work out the first and the fourth terms of the sequence.

Method

Solution	Commentary
$80 \div 40 = 2$ So the term-to-term rule is 'multiply the previous term by 2'. The first term will be $40 \div 2 = 20$ The fourth term will be $80 \times 2 = 160$	First find the term-to-term rule.

Example 3

a Work out the first three terms of the sequences given by these rules.

   i $n^2 + 7$ **ii** $n(n + 1)$ **iii** $10 - 2n^2$

b State whether each of the sequences in part **a** is **linear**, **geometric** or **neither**.

Method

Solution	Commentary

a i When $n = 1$, $n^2 + 7 = 1^2 + 7$
                      $= 1 + 7$
                      $= 8$
	Substitute $n = 1$, 2 and 3 in turn into the rule and evaluate the expression.

When $n = 2$, $n^2 + 7 = 2^2 + 7$
                      $= 4 + 7$
                      $= 11$	

When $n = 3$, $n^2 + 7 = 3^2 + 7$
                      $= 9 + 7$
                      $= 16$	

ii When $n = 1$, $n(n + 1) = 1(1 + 1)$
                      $= 1 \times 2$
                      $= 2$
	Substitute $n = 1$, 2 and 3 in turn into the rule and evaluate the expression. Remember you work out what is in the brackets first and then multiply.

When $n = 2$, $n(n + 1) = 2(2 + 1)$
                      $= 2 \times 3$
                      $= 6$	

When $n = 3$, $n(n + 1) = 3(3 + 1)$
                      $= 3 \times 4$
                      $= 12$	

iii When $n = 1$, $10 - 2n^2 = 10 - 2 \times 1^2$ $= 10 - 2 \times 1$ $= 10 - 2$ $= 8$	Remember: $2n^2$ means $2 \times n^2$, not $2 \times n$ and then square.
When $n = 2$, $10 - 2n^2 = 10 - 2 \times 2^2$ $= 10 - 2 \times 4$ $= 10 - 8$ $= 2$	
When $n = 3$, $10 - 2n^2 = 10 - 2 \times 3^2$ $= 10 - 2 \times 9$ $= 10 - 18$ $= -8$	
b 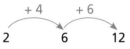 The differences between terms change, so the sequences are not linear.	Work out the differences between successive terms of each sequence to see if these are linear. State your conclusion, giving a reason.
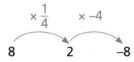 The multipliers between terms change so the sequences aren't geometric either.	Work out the multipliers between successive terms of each sequence, to see if the multipliers are all the same. $8 \times ? = 11$ so $? = \dfrac{11}{8}$ Similarly for $11 \times ? = 16$ You can use decimals to help with the first one. $\dfrac{11}{8} = 1.375$, $\dfrac{16}{11} = 1.4545\ldots$ State your conclusion, giving a reason.

Practice

1. The next term in a Fibonacci sequence is found by adding the previous two terms together.

   Write the next three terms of these Fibonacci sequences.

   a 3, 5, 8 … **b** 10, 12, 22 … **c** –1, 2, 1 … **d** –2, 0, –2 …

2. The first two terms of a sequence are 5 and 15

   Work out the next three terms if the sequence is:

   a geometric **b** Fibonacci.

3 All these sequences are geometric.

For each sequence:

    i work out the term-to-term rule   **ii** work out the next term.

a 20, 60, 180, 540 …   **b** 256, 64, 16, 4 …

c 200, 1000, 5000, 25 000 …   **d** 60, 6, 0.6, 0.06 …

4 The second term of a geometric sequence is 400 and the third term is 1600

Work out the first term of the sequence.

5 **a** Work out the first five terms of the sequences given by each of these rules.

    i $n^2 - 1$ **ii** $n(n + 5)$ **iii** $20 - 3n^2$

b State whether each of the sequences in part **a** is **linear**, **geometric** or **neither**.

6 Work out the 10th term of each sequence given by each of these rules.

a $n^2 - 5$ **b** n^3 **c** $(n + 3)^2$

d $n^2 + 6n$ **e** $\frac{1}{2}n^2$ **f** $n(n + 1)$

7 The diagram shows the first four triangular numbers.

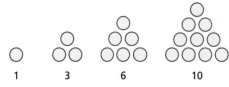

1   3   6   10

a Draw the next triangular number.

The nth triangular number is given by the formula $\frac{1}{2}n(n + 1)$.

b Work out the 100th triangular number.

Consolidate – do you need more?

1 Write the next three terms of these Fibonacci sequences.

a 2, 7, 9 …   **b** 20, 21, 41 …   **c** −2, 3, 1 …   **d** −3, 0, −3 …

2 All of the following sequences are geometric.

For each sequence:

    i work out the term-to-term rule   **ii** work out the next term.

a 20, 80, 320, 1280 …   **b** 162, 54, 18, 6 …

c 300, 1200, 4800, 19 200 …   **d** 90, 9, 0.9, 0.09 …

3 The first and second terms of a sequence are 50 and 100

Work out the next term if the sequence is:

a geometric   **b** Fibonacci.

4 Work out the first five terms of the sequences given by these rules.

    a $n^2 + 7$ **b** $n^2 - n$ **c** $4(n - 1)$ **d** $n(n + 3)$

5 The diagram shows the first three rectangular numbers.

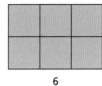
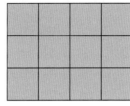

2   6   12

    a Draw the next rectangular number.

The nth rectangular number is given by the formula $n(n + 1)$.

    b Work out the 99th rectangular number.

Stretch – can you deepen your learning?

1 Work out the first five terms of the sequences given by these rules.

    a $\dfrac{1}{n}$ **b** $\dfrac{n}{n + 1}$

2 The first two terms of a Fibonacci sequence are x and $3x$.

Write expressions for the next three terms.

3 Use the rule $-2n(n - 4)$ to copy and complete the table of results.

Position (n)	1	2	3	4	5
Term		8	6		

4 Work out the missing terms in these Fibonacci sequences.

    a 4, _____, _____, 5000

    b 2, _____, _____, _____, 2593

Sequences: exam practice

1 Here is a sequence of circles:

Term 1   Term 2   Term 3

  (a) Draw the fourth term in the sequence. **[1 mark]**

  (b) How many circles would be in the fifth term of the sequence? **[1 mark]**

2 A linear sequence is plotted on a graph.

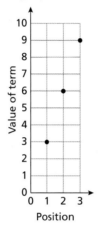

  (a) Write down the first five terms of the sequence. **[2 marks]**

  (b) What does the sequence represent? **[1 mark]**

3 The Fibonacci sequence is such that each term is the sum of the previous two terms.

Write down the next **two** terms of this Fibonacci sequence.

1, 1, 2, 3, _____, _____ **[2 marks]**

4 The first five terms of an arithmetic sequence are 2, 9, 16, 23, 30

  (a) Write down the eighth term of the sequence. **[1 mark]**

  (b) Work out, in terms of n, an expression for the nth term of the sequence. **[2 marks]**

5 Work out the first three terms of the sequence with the rule $(n + 1)^2 - n^2$ **[3 marks]**

Algebra: exam practice

1 Match each expression to its correct description. **[2 marks]**

$y - 4$	y divided by 4
$4 - y$	y multiplied by 4
$4y$	4 less than y
$\dfrac{y}{4}$	4 divided by y
$\dfrac{4}{y}$	y less than 4

2 Here are the first four terms of an arithmetic sequence:

$$-1, 5, 11, 17$$

(a) Work out the next term in the sequence. **[1 mark]**

(b) Work out the 10th term in the sequence. **[1 mark]**

Here are the first four terms of a different sequence:

$$400, 200, 100, 50$$

(c) Write down the next **two** terms of the sequence. **[2 marks]**

(d) Explain how you found your answer. **[1 mark]**

3 $y = 4n - 3d$

$n = 1$

$d = 5$

Work out the value of y. **[2 marks]**

4 Solve:

(a) $m - 19 = 20$ **[1 mark]**

(b) $8 + 3t = 14$ **[2 marks]**

(c) $\frac{1}{2}x = 10$ **[1 mark]**

5 Multiply out $w(w + 12)$ **[2 marks]**

1–3

6 (a) Solve $6(2y + 4) < 60$ **[3 marks]**

(b) Show your answer to part **(a)** on a copy of the number line.   **[2 marks]**

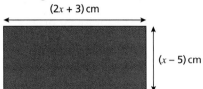

7 The perimeter of this rectangle is 44 cm.

$(2x + 3)$ cm

$(x - 5)$ cm

Work out the value of x.   **[4 marks]**

8 Aisha makes dolls.

She uses the following formula to work out the cost, £C, of making n dolls.

$$C = 5n + 432$$

Aisha makes 160 dolls.

She sells them for £8.50 each.

How much profit does she make?   **[3 marks]**

9 A rectangle is shown.

The length of the rectangle is x cm.

The width of the rectangle is $(x - 4)$ cm.

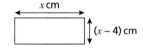

x cm

$(x - 4)$ cm

Two of the rectangles are placed side by side to make this shape.

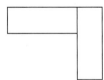

The perimeter of the shape is 58 cm.

Work out the value of x.   **[5 marks]**

10 Solve the equation $x^2 + 9x + 20 = 0$ **[3 marks]**

16 Ratio

In this block, we will cover...

16.1 Ratio notation

Example

Marta has 5 pens and 2 pencils.

a Show this information as a bar model.

b Write down the ratio of pens to pencils Mart

c Write down the ratio of pencils to pens Mart

Method

Solution	Commentary
a Pens Pencils	Use one section of th each pencil.

16.2 Finding parts and wholes

Practice (A)

1 In a mixture of nuts, the ratio of the mass
There are 40 grams of almonds.
How many grams of cashews are there?

2 A recipe requires a ratio of milk to water c
There are 500 millilitres of milk.
How much water is needed?

3 The ratio of lemons to oranges in a fruit ba
There are 15 lemons.

Are you ready? (A)

1 Here are some counters:

    a What fraction of the counters are black?

    b What fraction of the counters are pink?

2 Copy and complete the sentence to describe the fruit.

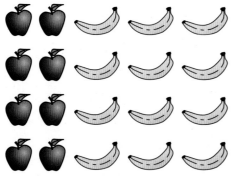

For every _____ apples, there are _____ bananas.

When working with fractions, you are thinking about parts in comparison to the whole.

When working with ratios, you are thinking about parts in comparison to other parts.

Whilst there are similarities between ratios and fractions, the way you represent them is different.

$\frac{2}{3}$ is a fraction meaning 2 out of 3 parts...   ...whereas 2 : 3, which you read as '2 to 3', is a ratio comparing two parts

It is important that you don't confuse the way the two are written.

Ratios are often used in cookery recipes, in construction when mixing cement, and in art and design when drawing scale diagrams.

Example

Marta has 5 pens and 2 pencils.

a Show this information as a bar model.

b Write down the ratio of pens to pencils Marta has.

c Write down the ratio of pencils to pens Marta has.

Method

Solution	Commentary
a Pens Pencils	Use one section of the bar model to represent each pen and each pencil.
b 5:2	There are 5 pens and 2 pencils so the ratio is '5 to 2', which you write using a colon as '5 : 2'.
c 2:5	This is the same information, but with the order changed. There are 2 pencils and 5 pens. It is very important to write the numbers in the correct order when using ratios to represent information.

Practice (A)

1 Kim has 7 boxes and 4 crates.

     a Draw a bar model to show this information.

     b Write the ratio of boxes to crates.

     c Write the ratio of crates to boxes.

2 A shop sells 5 magazines for every 8 newspapers.

     a Draw a bar model to show this information.

     b Write the ratio of magazines to newspapers sold.

     c Write the ratio of newspapers to magazines sold.

3 On a school trip there is 1 adult for every 8 children.

     a Draw a bar model to show this information.

     b Write the ratio of adults to children.

     c Write the ratio of children to adults.

4 The bar model shows the number of pens and pencils in a pot.

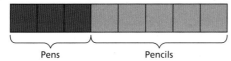

     a Write the ratio of pens to pencils.

     b Write the ratio of pencils to pens.

Are you ready? (B)

1 Write the ratio of cakes to buttons.

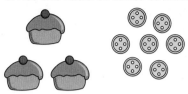

2 Write the ratio of books to pens.

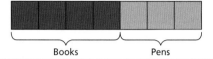

Books   Pens

Example 1

A necklace consists of red beads and white beads.

Are the statements **true** or **false**?

a For every two red beads, there is one white bead.

b The ratio of red beads to white beads is 1 : 2

Method

Solution	Commentary
a True	In each section of three beads, you can see that for every two red beads there is one white bead.
b False	The numbers in the ratio have been written the wrong way round. The ratio of red beads to white beads is 2 : 1

Example 2

Write the ratio of black to white counters in its simplest form.

Method

Solution	Commentary
12 : 9 ÷ 3 ( ) ÷ 3 4 : 3 4 : 3	There are 12 black counters and 9 white counters in total. This means that the ratio of black to white counters is 12 : 9 12 and 9 are both divisible by 3 12 ÷ 3 = 4 and 9 ÷ 3 = 3 This means that the ratio of black to white counters is 4 : 3 4 and 3 don't have any common factors other than 1 This means that the ratio 4 : 3 is in its **simplest** form.

Practice (B)

1 Here are some counters:

a Copy and complete the sentences.

  i For every 9 red counters, there are ___ yellow counters.

  ii For every 3 red counters, there are ___ yellow counters.

b Copy and complete the ratios of red to yellow counters.

  i 9 : ☐   **ii** 3 : ☐

c Write the ratio of yellow to red counters in two different ways.

2 Write the ratio of red to white beads in its simplest form.

3 Write the ratio of circles to squares in its simplest form.

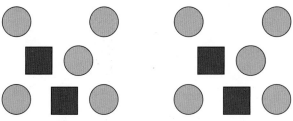

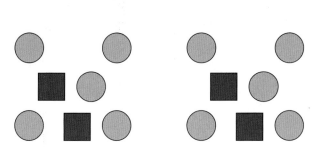

4 Write the ratio of stars to triangles in its simplest form.

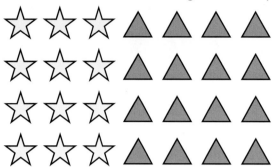

5 Write the ratio of buttons to pencils in its simplest form.

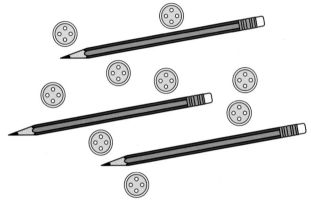

6 Write the ratios in their simplest form.

a 9 : 6	**b** 12 : 15	**c** 8 : 6	**d** 10 : 2	**e** 4 : 12
f 20 : 5	**g** 5 : 15	**h** 22 : 33	**i** 18 : 15	**j** 24 : 36

What do you think? 💡

1 A box contains apples and pears.

The ratio of apples to pears is 3 : 2

How many apples and pears could be in the box? Give three different answers.

Consolidate – do you need more?

1 In a class, for every 4 boys there are 5 girls.

a Draw a bar model to show this information.

b Write the ratio of boys to girls.

c Write the ratio of girls to boys.

2 In a field, for every 3 sheep there are 2 horses.

    a Draw a bar model to show this information.

    b Write the ratio of sheep to horses.

    c Write the ratio of horses to sheep.

3 Write the ratio of apples to strawberries in its simplest form.

4 Write the ratio of blue to white counters in its simplest form.

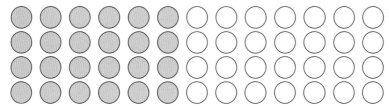

5 Write the ratio of yellow to white buttons in its simplest form.

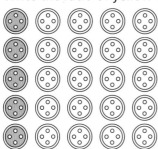

6 Write the ratios in their simplest form.

    a 6 : 3   **b** 12 : 8   **c** 16 : 20   **d** 7 : 35   **e** 24 : 40   **f** 45 : 15

Stretch – can you deepen your learning?

1 For every 2 books in a box, there are 7 bookmarks.

    How many books and bookmarks could be in the box? Give three different answers.

2 For every 3 blue crayons in a tub, there are 5 yellow crayons.

    There are more than 20 but fewer than 40 crayons in total.

    How many of each colour could there be?

3 For every three £2 coins in a jar, there are four £1 coins.

    There are nine £2 coins in the jar.

    How much money is there in total?

Are you ready? (A)

1 Work out 18 × 9

2 Work out 135 ÷ 9

3 The ratio of apples to bananas in a fruit bowl is 3 : 5

Are there more apples or bananas in the fruit bowl?

4 Write 40 : 2000 in its simplest form.

Ratios are used to express the relationship between quantities. When given a part or the whole in a given ratio, you can use this to work out other quantities.

For example, if the ratio of apples to bananas is 3 : 5, and you know there are 16 apples and bananas in total, you can work out how many apples and how many bananas there are.

The ratio '3 to 5' has 8 parts in total. Sharing the 16 between the 8 parts, each part represents 2 pieces of fruit. This means there are 6 apples and 10 bananas.

Always double check that your parts add up to give the whole. If they don't, you must have made a mistake.

Example 1

The ratio of blue pens to black pens in a box is 1 : 4. There are 5 blue pens.

a How many black pens are there?   **b** How many pens are there in total?

Method

Solution	Commentary
a Blue : Black 1 : 4 5 : 20 There are 20 black pens.	Drawing a bar model helps to visualise the problem. Blue | 5 | Black | 5 | 5 | 5 | 5 | ? 4 × 5 = 20
b There are 25 pens in total.	Blue | 5 | Black | 5 | 5 | 5 | 5 | 20 ? 5 × 5 = 25

Example 2

The ratio of red pens to green pens in a box is 3 : 4. There are 12 red pens.

a   How many green pens are there?   **b**   How many pens are there in total?

Method

Solution	Commentary

a   Red : Green
     $\downarrow$ $\downarrow$
    3 : 4
 	
  12 : 16
 	
There are 16 green pens.
	
	$12 \div 3 = 4$
	$4 \times 4 = 16$

b There are 28 pens in total.	 $7 \times 4 = 28$

Practice (A)

1 In a mixture of nuts, the ratio of the mass of almonds to cashews is 1 : 3

There are 40 grams of almonds.

How many grams of cashews are there?

2 A recipe requires a ratio of milk to water of 1 : 4

There are 500 millilitres of milk.

How much water is needed?

3 The ratio of lemons to oranges in a fruit basket is 1 : 2

There are 15 lemons.

a   How many oranges are there?

b   What is the total number of lemons and oranges in the fruit basket?

4 In a music playlist, the ratio of pop songs to rock songs is 1 : 5

There are 20 pop songs.

a   How many rock songs are there?

b   What is the total number of songs in the playlist?

5 The ratio of red to yellow flowers in a garden is 1 : 8

There are 9 red flowers.

a   How many yellow flowers are there?

b   What is the total number of red and yellow flowers in the garden?

6 The ratio of dogs to cats in a pet shop is 3 : 7

There are 21 cats.

a How many dogs are there?

b What is the total number of dogs and cats in the pet shop?

7 A mixture of paint is made by mixing blue paint and green paint in the ratio 2 : 5

There are 16 litres of blue paint.

a How much green paint is there?

b What is the total quantity of paint in the mixture?

8 Students in a class either play basketball or football.

The ratio of students who play football to those who play basketball is 5 : 4

There are 36 students who play basketball.

a How many students play football?

b What is the total number of students who play sports?

9 In a bakery, the ratio of chocolate cakes to vanilla cakes is 4 : 9

There are 36 chocolate cakes.

a How many vanilla cakes are there?

b What is the total number of chocolate and vanilla cakes?

10 In a bag, the ratio of red marbles to blue marbles is 3 : 2

There are 40 blue marbles.

a How many red marbles are there?

b What is the total number of marbles in the bag?

11 The ratio of apples to bananas is 2 : 5

There are 18 apples.

How many more bananas than apples are there?

12 In a school, the ratio of boys to girls is 3 : 5

There are 42 boys.

How many more girls than boys are there in the school?   Draw a pair of bar diagrams.

13 A recipe uses flour and sugar in the ratio 2 : 1

There are 250 grams of flour.

How much sugar is there?

14 In a box, the ratio of red marbles to green marbles is 1 : 3

There are 9 red marbles.

How many more green marbles than red marbles are there?

15 Share £300 in the ratio 2 : 3

Are you ready? (B)

1 The ratio of dogs to cats in a shelter is 1 : 2

There are 18 dogs.

How many cats are there?

2 In a box, the ratio of books to pens is 3 : 5

There are 15 pens.

How many more pens than books are there?

Example 1

In a bag, the ratio of blue counters to yellow counters is 3 : 5. There are 240 counters in total.

a How many blue counters are there?

b How many more yellow counters than blue counters are there?

Method

Solution	Commentary
a 240 ÷ 8 = 30 3 × 30 = 90 90 blue counters	This time the number given doesn't represent a part; it represents the whole. There are 8 parts in total, so you need to divide 240 by 8 to find the value of each part. 240 ÷ 8 = 30 3 × 30 = 90
b 150 − 90 = 60 60 more yellow counters	There are two ways to do this. Either subtract the number of blue counters from the number of yellow counters. Or, given that you know each part is 30, you can multiply the difference in the number of parts (2) by the value of each part. 150 − 90 = 60

Example 2

In a tub, the ratio of pens to pencils is 2 : 5

There are 60 more pencils than pens.

How many pens and pencils are there in total?

Method

Solution	Commentary
$60 \div 3 = 20$	The 60 represents the difference between the parts, which is 3
$7 \times 20 = 140$	To find the value of each part, divide 60 by 3
140	Then to find the whole, multiply this by the total number of parts, which is 7

$60 \div 3 = 20$

$7 \times 20 = 140$

Practice (B)

1 In a bag, the ratio of red marbles to blue marbles is 3 : 5

There are 40 marbles in total.

How many marbles are blue?

2 A recipe for cake requires a ratio of flour to sugar of 2 : 1

If you need 300 grams of flour, how much sugar is required?

3 In a class of 48 students, the ratio of boys to girls is 3 : 5

    a How many boys are there?

    b What is the difference in the number of boys and girls?

4 The ratio of apples to oranges in a fruit basket is 2 : 3

There are 25 fruits in total.

    a How many are apples?   **b** How many are oranges?

5 A mixture of nuts contains almonds and cashews in the ratio of 1 : 4

There are 45 nuts in total.

    a How many are almonds?   **b** How many are cashews?

6 The ratio of lemons to limes in a basket is 2 : 3

If there are 10 more limes than lemons, what is the number of fruit in the basket?

7 The ratio of teachers to students in a school is 1 : 30

There are 20 teachers.

What is the total number of teachers and students in the school?

8 In a box of 28 balls, the ratio of red balls to blue balls is 3 : 4

   a How many balls are red?   **b** How many balls are blue?

9 In a garden, the ratio of roses to tulips is 4 : 5

There are 15 more tulips than roses.

   a How many roses are there?   **b** How many tulips are there?

10 A company has employees in a ratio of men to women of 3 : 5

There are 20 more women than men.

   a How many men are there?   **b** How many women are there?

11 In a basket of 35 fruits, the ratio of pineapples to mangoes is 2 : 5

   a How many are pineapples?   **b** How many are mangoes?

12 16 scoops of ice cream are served at a party in the ratio of chocolate to vanilla of 1 : 3

   a How many are chocolate scoops?   **b** How many are vanilla scoops?

13 In a work of art, the ratio of yellow paint to blue paint used is 3 : 7

The difference in the amount used is 40 units.

   a How much yellow paint is used?   **b** How much blue paint is used?

14 The ratio of triangles to squares in a pattern is 2 : 5

The pattern has only triangles and squares.

If there are 63 shapes in total, how many more squares than triangles are there?

15 Max and Jo share £250 in the ratio 4 : 1

Max gets £150 more than Jo.

How much does Jo get?

What do you think? 💭

1 In a box of 100 chocolates, the ratio of milk chocolates to white chocolates to dark chocolates is 5 : 3 : 2

How many more milk chocolates than white chocolates are there?

Consolidate – do you need more?

1 In a bookstore, the ratio of adventure books to mystery books is 1 : 5

If there are 15 adventure books, how many mystery books are there?

2 A recipe requires sugar to flour in the ratio of 1 : 3

If you need 200 grams of sugar, how much flour is required?

3 The ratio of rabbits to guinea pigs in a pet shop is 1 : 4

There are 20 guinea pigs.

How many rabbits are there?

4 A fruit salad has a ratio of bananas to oranges of 1 : 6

There are 21 bananas.

How many oranges are there?

5 In a charity event, the ratio of adults to children is 3 : 8

If there are 55 attendees in total, how many are adults and how many are children?

6 In a basket of 63 fruits, the ratio of apples to pears is 2 : 5

    a How many are apples?   **b** How many are pears?

7 The ratio of cars to bikes in a parking lot is 1 : 3

If there are 12 more bikes than cars, how many vehicles are there in total?

8 In a class, the ratio of students with glasses to students without glasses is 1 : 8

If there are 28 more students without glasses, how many students are there in total?

9 In a bag of 45 marbles, the ratio of red marbles to blue marbles is 1 : 4

    a How many are red marbles?   **b** How many are blue marbles?

10 In a batch of 63 bakery items, the ratio of cakes to pastries is 2 : 7

How many more pastries than cakes are there?

Stretch – can you deepen your learning?

1 Jack makes purple paint by mixing red paint and blue paint in the ratio 2 : 3

Red paint costs £50 for 10 litres. Blue paint costs £20 for 10 litres.

Jack wants to make 200 litres of purple paint.

How much will he need to spend?

2 The angles in a triangle are in the ratio 8 : 5 : 5

    a Explain why the triangle must be isosceles.

    b Work out the size of the largest angle.

3 The ratio of cakes to cookies in a box is 3 : 7

Half of the cakes are vanilla and the rest are chocolate.

One-quarter of the cookies are double chocolate and the rest are chocolate chip.

There are 600 cakes and cookies in total.

How many of each type of cake and cookie are there?

Ratio: exam practice

1 For every 1 red counter in a bag, there are 5 blue counters.

    (a) Write down the ratio of red counters to blue counters. **[1 mark]**

    (b) Write down the ratio of blue counters to red counters. **[1 mark]**

1–3

2 A chocolate cookie has white chocolate, milk chocolate and dark chocolate chips.

    The ratio of white to milk to dark chocolate chips is 3 : 5 : 2

    (a) Write the ratio of milk chocolate to white chocolate chips. **[1 mark]**

    (b) Write the ratio of dark chocolate to white chocolate chips. **[1 mark]**

    (c) Mo really likes dark chocolate.

        Suggest a ratio he could use that would have more dark chocolate chips. **[1 mark]**

3 Write 15 : 18 in its simplest form. **[1 mark]**

4 Share £40 in the ratio 5 : 3 **[3 marks]**

3–5

5 A piece of wood is 54 cm in length.

    The piece is divided in the ratio 7 : 2

    Work out the length of each part. **[3 marks]**

6 Purple paint is made by mixing red paint and blue paint in the ratio 5 : 4

    How much more red paint than blue paint is in 450 ml of the mixture? **[3 marks]**

7 The angles in a triangle are in the ratio 2 : 1 : 5

    What is the size of the largest angle in the triangle? **[3 marks]**

8 Mo, Aisha and Jack share £400 in the ratio 3 : 1 : 4

    How much more money does Jack get than Aisha? **[3 marks]**

9 $x : y = 4 : 9$

    $x + y = 65$

    (a) Work out the value of x. **[1 mark]**

    (b) Work out the value of y. **[1 mark]**

17 Proportion

In this block, we will cover...

17.1 Fractions and equal ratios

Example 1

The ratio of blue pens to black pens in a box is 1

What fraction of the pens are blue?

Method

Solution	Commentary
1 + 4 = 5 $\frac{1}{5}$	The bar model shows the ratio 1 There is 1 part blue and 4 parts

Blue	Black	Black	Black

There are 5 parts in total, so the

17.2 Direct proportion

Practice (A)

1. A box contains 20 pencils.
   a How many pencils are there in 2 boxes
   b How many pencils are there in 5 boxes
   c How many pencils are there in 10 boxe

2. The mass of a bag of sugar is 4 kg.
   a What is the mass of 2 bags of sugar?
   b What is the mass of 10 bags of sugar?
   c What is the mass of 20 bags of sugar?

17.3 Inverse proportion

Consolidate – do you need more

1. Sarah can decorate 60 cookies in 2 hours.
   a How long will it take Sarah and her fri together?
   b What assumption have you made?

2. It takes 4 machines 10 hours to manufactu
   a How long will it take 8 machines to ma
   b What assumption have you made?

Are you ready?

1 Here are some counters:

    a What fraction of the counters are yellow?

    b What fraction of the counters are red?

    c What is the ratio of yellow to red counters?

Ratios and fractions both represent the relationship between parts and wholes.

However, ratios are mainly used for comparing quantities, such as in recipes, while fractions are used to show parts of a whole, as when cutting a pizza into equal slices.

Both ratios and fractions play important roles in daily life, whether in the kitchen, at school, or in other everyday situations.

Example 1

The ratio of blue pens to black pens in a box is 1 : 4

What fraction of the pens are blue?

Method

Solution	Commentary
1 + 4 = 5 $\frac{1}{5}$	The bar model shows the ratio 1 : 4 There is 1 part blue and 4 parts black. <table><tr><td>Blue</td><td>Black</td><td>Black</td><td>Black</td><td>Black</td></tr></table> There are 5 parts in total, so the fraction of pens that are blue is $\frac{1}{5}$

Example 2

$\frac{2}{7}$ of the counters in a bag are red. The rest are blue.

Write the ratio of red counters to blue counters.

Method

Solution	Commentary
7 − 2 = 5 2 : 5	2 of the 7 parts are red. 5 of the 7 parts are blue. The ratio of red to blue counters is 2 : 5 <table><tr><td>Red</td><td>Red</td><td>Blue</td><td>Blue</td><td>Blue</td><td>Blue</td><td>Blue</td></tr></table>

Practice

1 Here are some counters:

   a What fraction of the counters are red?

   b What fraction of the counters are yellow?

   c Write the ratio of red to yellow counters.

   d Write the ratio of yellow to red counters.

2 Max has some pens and pencils:

   a Write the ratio of pens to pencils.

   b Write the ratio of pencils to pens.

   c What fraction of Max's equipment are pens?

   d What fraction of Max's equipment are pencils?

3 Here is some fruit:

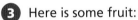

   a What fraction of the fruit are bananas?

   b What fraction of the fruit are apples?

   c Write the ratio of bananas to apples.

   d Write the ratio of apples to bananas.

4 Here are some coins:

   a Write the ratio of 50p coins to £1 coins.

   b Write the ratio of £1 coins to 50p coins.

   c What fraction of the coins are 50p coins?

   d What fraction of the coins are £1 coins?

5 $\frac{2}{5}$ of a pile of counters are blue.

The rest are green.

   a Draw a bar model to represent the counters.

   b Write the ratio of blue to green counters.

6 Kim and Jack share some money in the ratio 5 : 8

   a Draw a bar model to represent this information.

   b What fraction of the money does Kim get?

7 $\frac{4}{9}$ of a pile of books are fiction.

The rest are non-fiction.

Write the ratio of non-fiction books to fiction books.

8 The ratio of males to females in a supermarket is 10 : 11

What fraction of the people in the supermarket are female?

9 On a desk, there are books, notepads and envelopes.

The ratio of books to notepads to envelopes is 3 : 4 : 1

What fraction of the items on the desk are envelopes?

10 In a bag, there are red sweets and yellow sweets.

$\frac{3}{4}$ of the sweets are red.

Write the ratio of red to yellow sweets.

11 Ron has some 50p and 20p coins.

$\frac{2}{3}$ of the coins are 20p coins.

Write the ratio of 50p coins to 20p coins.

12 A bowl contains some apples, oranges and pears.

$\frac{1}{2}$ of the fruit are apples.

There is an equal number of oranges and pears.

Write the ratio of apples to oranges to pears.

Consolidate – do you need more?

1 There are 13 marbles in a bag.

9 marbles are blue and the rest are red.

    a What fraction of the marbles are blue?

    b What fraction of the marbles are red?

    c Write the ratio of blue to red marbles.

    d Write the ratio of red to blue marbles.

2 Sam has 17 items of stationery.

She has 6 pens and the rest are pencils.

    a Write the ratio of pens to pencils.

    b Write the ratio of pencils to pens.

    c What fraction of Sam's items are pens?

    d What fraction of Sam's items are pencils?

3 There are 23 pieces of fruit in a bowl: 15 bananas and the rest are apples.

   a   What fraction of the fruit are bananas?

   b   What fraction of the fruit are apples?

   c   Write the ratio of bananas to apples.

   d   Write the ratio of apples to bananas.

4 Lucy and Ben share some money in the ratio 4 : 7

   a   Draw a bar model to represent this information.

   b   What fraction of the money does Lucy get?

5 The ratio of men to women in a store is 5 : 7

   What fraction of the people in the store are women?

Stretch – can you deepen your learning?

1 The ratio of dogs to cats in a pet shop is 5 : 7

   The pet shop only has dogs and cats.

   If there are 63 cats, what is the ratio of dogs to the total number of pets?

2 The ratio of men to women to children in a cinema is 2 : 3 : 5

   $\frac{1}{5}$ of the children are boys.

   What fraction of the people in the cinema are girls?

3 There are blue marbles, red marbles and green marbles in a box.

   50% of the marbles are blue.

   $\frac{1}{5}$ of the marbles are red.

   The rest are green.

   Write the ratio of blue marbles to red marbles to green marbles.

17.2 Direct proportion

Are you ready? (A)

1 A box contains 5 pens.

How many pens are there in 4 boxes?

2 The ratio of blue marbles to red marbles in a bag is 3 : 7

What fraction of the marbles are blue?

3 Two-thirds of a box of counters are green. The rest are purple.

What is the ratio of green counters to purple counters?

When two quantities are in **direct proportion**, as one quantity increases or decreases, the other quantity increases or decreases at the same rate.

Double number lines can show the multiplicative relationship between quantities in direct proportion, such as exchange rates.

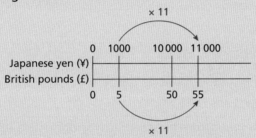

When one of the quantities is multiplied or divided by an amount, the other quantity is also multiplied or divided by the same amount.

Proportion diagrams

These can be used to show different multiplicative relationships and are helpful when problem-solving.

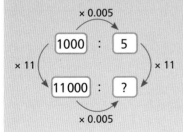

Conversion graphs

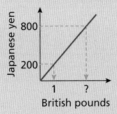

These are useful for converting from one currency to another and for converting between different units of measurement.

Example 1

Eggs are sold in boxes of 6

How many eggs are there in 8 boxes?

Method

Solution	Commentary
6 × 8 = 48	The number of eggs is directly proportional to the number of boxes.
	For every one box, there are 6 eggs. So multiply 6 by 8

Example 2

A set of bags contain marbles. The number of marbles is directly proportional to the number of bags.

Complete the table.

Number of bags	0	1	2	10	100
Number of marbles		15			

Method

Solution	Commentary
0 × 15 = 0	Don't get caught out by the zero in the table – if there are no bags, there are no marbles.
2 × 15 = 30	The number of marbles is directly proportional to the number of bags. This means that there are the same number of marbles in each bag. One bag contains 15 marbles, so to work out two bags multiply 2 by 15
10 × 15 = 150	For 10 bags, multiply 10 by 15
100 × 15 = 1500	For 100 bags, multiply 100 by 15

Number of bags	0	1	2	10	100
Number of marbles	0	15	30	150	1500

Practice (A)

1. A box contains 20 pencils.

   a How many pencils are there in 2 boxes?

   b How many pencils are there in 5 boxes?

   c How many pencils are there in 10 boxes?

2. The mass of a bag of sugar is 4 kg.

   a What is the mass of 2 bags of sugar?

   b What is the mass of 10 bags of sugar?

   c What is the mass of 20 bags of sugar?

3 Cupcakes are sold in boxes of 8

    a How many cupcakes are there in 4 boxes?

    b How many cupcakes are there in 20 boxes?

    c Jack needs 40 cupcakes. How many boxes should he buy?

4 There are 25 desks in each classroom.

    a How many desks are there in 4 classrooms?

    b How many desks are there in 9 classrooms?

    c How many desks are there in 15 classrooms?

5 A bakery bakes 40 loaves of bread daily.

    a What is the total number of loaves baked in 3 days?

    b What is the total number of loaves baked in 8 days?

    c What is the total number of loaves baked in 12 days?

6 A store sells packets of 12 marker pens.

    a How many marker pens are there in 5 packets?

    b How many marker pens are there in 15 packets?

    c If Sarah needs 60 marker pens, how many packets should she buy?

7 The cost of buying sweets is directly proportional to the number of sweets bought.

Copy and complete the table.

Number of sweets	0	1	2	10	100
Cost (pence)		20			

8 The time taken to mow a lawn is directly proportional to the area of the lawn.

Copy and complete the table.

Area (m²)	5	10	20	100	50
Time (hours)			4		

Which value did you find easiest to work out? Which one did you find hardest?

Are you ready? (B)

1 A multi-pack of crisps costs £1.50

How much do 4 multi-packs cost?

2 A packet of sweets costs 35p.

How much do 5 packets cost?

3 The distance travelled by a train is directly proportional to the time spent travelling.

Copy and complete the table.

Time (hours)	1	3	5	10	15
Distance (miles)	30				

Example 1

Pens come in boxes of 6 or boxes of 10

Which box of pens is better value?

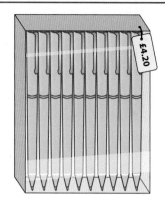

Method

Solution	Commentary
Box of 6: £2.70 ÷ 6 = £0.45	This method works out the cost per pen in each box.
Box of 10: £4.20 ÷ 10 = £0.42	
Box of 10	The cost per pen is cheaper in the box of 10 so this is better value.

Example 2

A 150 ml bottle of shampoo costs £1.50

A 200 ml bottle of the same brand of shampoo costs £3

Which is better value?

Method A

Solution	Commentary
£3 buys one bottle of 200 ml or two bottles of 150 ml (a total of 300 ml). The 150 ml bottle is better value.	You can buy two bottles of 150 ml for the same price as one bottle of 200 ml and you get more, so this bottle is better value.

Method B

Solution	Commentary
For the 150 ml bottle: £1.50 ÷ 150 ml = 150p ÷ 150 ml = 1p/ml So 1 ml of shampoo in this bottle costs 1p.	Alternatively, work out the cost per millilitre for each bottle.
For the 200 ml bottle: £3 ÷ 200 = 300p ÷ 200 ml = 1.5p/ml So 1 ml of shampoo in this bottle costs 1.5p.	
The 150 ml bottle is better value.	The 150 ml bottle is better value because it costs less per millilitre.

Practice (B)

1 Two shops sell the same type of cupcake.

Shop A: 5 cupcakes are sold for £1.80

Shop B: 3 cupcakes are sold for £1.05

At which shop are the cupcakes better value?

2 Handwash comes in two different sized bottles.

A 500 ml bottle costs £2.50

A 300 ml bottle costs £1.20

Which bottle is better value?

3 Two shops sell the same pencils.

Shop A: 6 pencils cost £2.50

Shop B: 4 pencils cost £1.80

At which shop are they cheaper per pencil?

4 Two boxes of the same type of chocolates can be bought.

Box X: Contains 24 chocolates for £8

Box Y: Contains 15 chocolates for £6

Which box offers a better price per chocolate?

5 Two Internet providers have different subscription plans.

Provider A: 100 Mbps cost £30 per month

Provider B: 120 Mbps cost £40 per month

Which provider offers a better value in terms of cost per Mbps?

6 Two cinemas have different pricing schemes.

Cinema A: Tickets are £10 each, but a group of 4 can get tickets for £35 in total.

Cinema B: Tickets are £8 each, but a group of 5 can get tickets for £36 in total.

Which cinema offers the better value for a group of 22 people?

7 There are two taxi services in a town.

X Taxis: Charges £2.50 per mile with a £2 initial fee

Y Taxis: Charges £3 per mile with no initial fee

Which taxi service is more cost-effective for a 6-mile trip?

Are you ready? (C)

1 A box of pencils costs £3.80

How much do 5 boxes cost?

2 A pack of 9 toilet rolls costs £3.51

A pack of 4 of the same toilet roll costs £1.68

Which pack is better value?

If y is directly proportional to x, it can be written as:

$y \propto x$ where $\propto$ means 'directly proportional to'.

This can also be written as:

$y = kx$ where k is the **constant of proportionality**.

The graph of a proportional relationship $y = kx$ is a straight line going through the origin (0, 0).

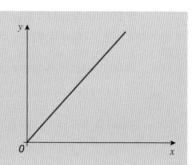

Example

The graph can be used to convert between British pounds and Canadian dollars.

Use the graph to convert:

a £40 into Canadian dollars

b $48 into pounds

c £350 into Canadian dollars.

Conversion graph for British pounds and Canadian dollars

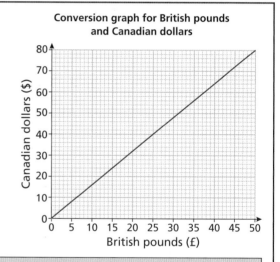

Method

Solution	Commentary
a £40 is equivalent to $64	Find £40 on the British pounds axis and draw a line to the graph. Then draw a horizontal line to the Canadian dollars axis to help you to read off the value.

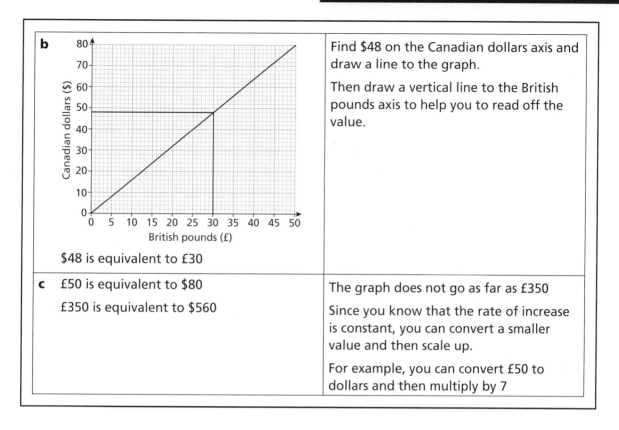

b

$48 is equivalent to £30

| Find $48 on the Canadian dollars axis and draw a line to the graph. |
| Then draw a vertical line to the British pounds axis to help you to read off the value. |

c £50 is equivalent to $80

£350 is equivalent to $560

The graph does not go as far as £350

Since you know that the rate of increase is constant, you can convert a smaller value and then scale up.

For example, you can convert £50 to dollars and then multiply by 7

Practice (C)

1 Here are the ingredients needed to make 400 g of pastry:

| 250 g plain flour | 125 g butter | 2 tbsp milk |

How much of each ingredient is needed to make:

a 800 g of pastry **b** 200 g of pastry **c** 1 kg of pastry?

2 Here are the ingredients needed to make 10 chocolate chip cookies:

| 120 g butter | 150 g sugar | 1 egg |
| 1 tsp vanilla extract | 180 g plain flour | 150 g chocolate chips |

a How much of each ingredient is needed to make 15 chocolate chip cookies?

b Tom has 300 g of sugar and plenty of each of the other ingredients.

Does he have enough ingredients to make 30 chocolate chip cookies? Explain your answer.

3 Which of the graphs represent two quantities that are directly proportional to each other?

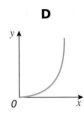

4 The graph shows how to convert between euros (€) and Turkish lira (₺).

Use the graph to convert:

a €15 into Turkish lira

b ₺60 into euros

c €50 into Turkish lira.

d The exchange rate between US dollars and euros is $1 = €0.90

Write the exchange rate for US dollars to Turkish lira.

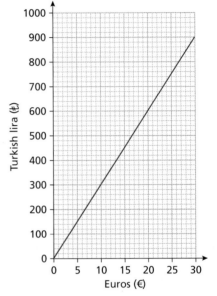

Consolidate – do you need more?

1 A construction site uses 80 bags of cement per week.

What is the total number of bags used in:

a 4 weeks   **b** 7 weeks   **c** 12 weeks?

2 An online game awards 50 points for each level completed.

a How many points are earned by completing 5 levels?

b How many points are earned by completing 15 levels?

c If Maria wants to get 200 points, how many levels does she need to complete?

3 Two mobile data plans are offered by a provider.

Plan A: 10 gigabytes cost £15 per month

Plan B: 6 gigabytes cost £10 per month

Which plan provides better value per gigabyte of data?

4 Two stores sell packs of the same type of battery.

Store X: 12 batteries cost £5.40

Store Y: 8 batteries cost £3.20

In which store are the batteries cheaper per battery?

5 Here are the ingredients needed to make 10 servings of pasta:

> 200 g spaghetti   500 ml pasta sauce

How much spaghetti and pasta sauce are needed to make:

a 20 servings   **b** 30 servings?

6 Here are the ingredients for a fruit salad recipe serving 6 people:

> 3 apples   2 bananas   1 cup of grapes

How many apples, bananas and cups of grapes are needed to make:

a 12 servings   **b** 18 servings?

Stretch – can you deepen your learning?

1 A machine makes 10 000 tools in 20 hours.

a How many tools will it make in 100 hours?

b How long will it take the machine to make 7500 tools?

c What assumptions have you made?

2 A pack of 4 tins of soup costs £3.40

A pack of 9 tins of soup costs £7.20

Ron wants to buy 44 tins of soup. What is the cheapest way to do it?

3 Here is a recipe for banana bread that makes 8 servings:

140 g butter	140 g sugar	2 eggs
140 g self-raising flour	1 tsp baking powder	2 bananas

Kim has 350 g of butter and plenty of the other ingredients.

Jack has 6 eggs and plenty of the other ingredients.

Who can make more banana bread?

How many more servings can they make?

4 Mo has boxes of two different sizes.

He has two boxes that can each hold 500 marbles, and plenty of boxes that can each hold 100 marbles.

Mo has 1756 marbles in total. He wants to use the least number of boxes.

How many of each box will Mo need?

White Rose MATHS

Are you ready?

1 When 4 people share a pizza, they each get 3 slices.

If more people share the same pizza, will they get more or fewer slices?

If two quantities are in **inverse proportion**, when one quantity increases, the other decreases at the same rate.

For example, a factory uses machines to make milk bottles.

It takes one machine 4 working days to produce 1000 bottles.

However, if there are *more* of the same machines, it will take *less* time. If there are two identical machines, the 1000 milk bottles can be produced in half the time, 2 days.

Number of machines	Number of working days
1	4
2	2

$\times 2$ ( ) $\div 2$

If there are three machines, it will take $1\frac{1}{3}$ days to produce 1000 milk bottles.

Number of machines	Number of working days
1	4
3	$1\frac{1}{3}$

$\times 3$ ( ) $\div 3$

This is an example of inverse proportion. When you increase the number of machines, the time taken decreases at the same rate.

Number of machines	Number of days	
1	4	$1 \times 4 = 4$
2	2	$2 \times 2 = 4$
6	$\frac{2}{3}$	$6 \times \frac{2}{3} = 4$

$\times 2$ $\div 2$
$\times 3$ $\div 3$

Notice that the product of the number of machines and the number of days is always 4. This is because 4 days' worth of production time are being shared between the machines.

Recall from Chapter 17.2 that if y is directly proportional to x, this can be written as $y \propto x$ or $y = kx$.

If y is inversely proportional to x, this is the same as

y being directly proportional to $\frac{1}{x}$

Therefore, this can be written as $y \propto \frac{1}{x}$ or $y = \frac{k}{x}$

The graph of $y = \frac{k}{x}$ is a **reciprocal** curve.

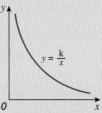

$y = \frac{k}{x}$

$y = \frac{k}{x}$ can be rearranged to $xy = k$, which shows that the product of the two variables is always constant.

Example 1

It takes 8 people 5 hours to clean a building.

a   How long would it take 2 people to clean the same building?

b   What assumption have you made?

Method

Solution	Commentary
a 8 × 5 = 40	If it takes 8 people 5 hours, this means it takes 40 hours of people's time in total because 8 × 5 = 40
40 ÷ 2 = 20 20 hours	If only 2 people were doing the same task, it would take 20 hours because 40 ÷ 2 = 20
b The assumption is that each person works at the same rate.	When you work out this type of problem, you assume that every person works at exactly the same speed or rate. In reality, this isn't likely to be the case.

Example 2

The variables x and y are inversely proportional to each other.

x	10	20	30		
y	24			6	4

a   Complete the table.

b   Write this relationship in the form $xy = k$, where k is a constant to be found.

c   Using each pair of x and y values, plot the graph of $xy = k$.

Method

Solution	Commentary
a <table><tr><td>x</td><td>10</td><td>20</td><td>30</td><td>**40**</td><td>**60**</td></tr><tr><td>y</td><td>24</td><td>**12**</td><td>**8**</td><td>6</td><td>4</td></tr></table> 20 = 10 × 2, so when $x = 20$, $y = 24 ÷ 2 = 12$ 30 = 10 × 3, so when $x = 30$, $y = 24 ÷ 3 = 8$ 6 = 12 ÷ 2, so when $y = 6$, $x = 20 × 2 = 40$ 4 = 24 ÷ 6, so when $y = 4$, $x = 10 × 6 = 60$	Since x and y are inversely proportional, as x decreases at a rate, y increases at the same rate. So if x is halved then y is doubled.
b $xy = 240$	The product of x and y is a constant, in this case 240. This is always true of two variables that are inversely proportional to each other.
c	Draw a coordinate grid, then plot the points and join them with a smooth curve. Notice that the graph approaches both the x- and y-axes but never reaches either.

Practice

1 Ron can paint a fence in 8 hours.

    a How long would it take if Ron and a friend paint the fence together?

    b What assumption have you made?

2 It takes 6 days for 15 machines to complete a job.

    a How many days would it take 30 machines, working at the same rate, to complete the same job?

    b Why do you **not** need to make an assumption when answering part **a**?

3 Using one tap, it takes 9 hours to fill a swimming pool.

    How long would it take to fill the pool using two of the same taps together?

4 18 people can complete a project in 12 days.

    How many days would it take 6 people to complete the same project, assuming they work at the same rate?

5 It takes 4 people 3 days to build a wall.

    How many days would it take 6 people to build the same wall, assuming they work at the same rate?

6 A garden hose can fill a water tank in 12 minutes.

    How long would it take to fill the tank using two of the same hoses together?

7 It takes Jack 3 hours to mow the lawn.

    How long would it take Jack and two friends to mow the lawn if they all work at the same rate?

Consolidate – do you need more?

1 Sarah can decorate 60 cookies in 2 hours.

    a How long will it take Sarah and her friend to decorate the same 60 cookies together?

    b What assumption have you made?

2 It takes 4 machines 10 hours to manufacture a batch of toys.

    a How long will it take 8 machines to manufacture the batch of toys?

    b What assumption have you made?

3 A factory produces 400 units in 5 days.

    a How many days will it take two factories to produce the same 400 units if they work at the same pace?

    b Why do you **not** need to make an assumption in this problem?

4 A water tank can be filled by one tap in 6 hours.

How long would it take two identical taps to fill the tank?

5 12 students can complete a school project in 8 days.

How many days would it take 6 students, working at the same rate, to complete the same project?

6 It takes 5 workers 4 days to assemble a structure.

How many days would it take 8 workers to assemble the same structure if they work at the same rate?

7 A printer can print a book in 20 minutes.

How long will it take two identical printers to print the same book together?

8 It takes Emily 6 hours to clean a church.

How long would it take Emily and two friends to clean the church if they work at the same rate?

Stretch – can you deepen your learning?

1 The table shows how long it takes for different numbers of cleaners to clean the same building. Each cleaner works at the same rate.

Copy and complete the table.

Number of cleaners	1	2	3	4	5
Time taken (minutes)	240				

2 The table shows how long it takes for different numbers of machines to make a batch of toys.

Each machine works at the same rate.

Copy and complete the table.

Number of machines	1	2	5	10	20
Time taken (hours)		15			

3 The area of a rectangle is fixed.

As the length increases, the width decreases.

When the length is 12 cm, the width is 8 cm.

What is the length of the rectangle when the width is 6 cm?

4 Quantity a is inversely proportional to quantity b.

Copy and complete the table.

a	1	2	6	16	24
b	48				

Proportion: exam practice

1 Here are some shapes:

   ■ ■ ■ ▲ ▲ ▲ ▲

   (a) Write down the ratio of squares to triangles.   **[1 mark]**

   (b) Write down the ratio of triangles to squares.   **[1 mark]**

   (c) What fraction of the shapes are triangles?   **[1 mark]**

2 In a box of pens, $\frac{5}{6}$ are black. The rest of the pens are red.

   (a) What fraction of the pens are red?   **[1 mark]**

   (b) Write the ratio of black pens to red pens.   **[1 mark]**

   There are 5 red pens.

   (c) How many black pens are there?   **[2 marks]**

3 A box contains 15 pencils.

   How many pencils are there in 3 boxes?   **[1 mark]**

4 Here is a list of ingredients for making 12 cakes:

Ingredients for 12 cakes	
Butter	180 g
Sugar	180 g
Plain flour	200 g
Baking powder	1 teaspoon
Eggs	2

   (a) Amir is going to make 24 of the cakes.

       Work out how much butter he needs.   **[2 marks]**

   (b) Sophie is going to make 18 of the cakes.

       Work out how much flour she needs.   **[2 marks]**

   (c) Suggest why it might **not** be possible to get exact ingredients
       for 3 cakes.   **[1 mark]**

5 The mass of a piece of wire is directly proportional to its length.

   A piece of wire is 20 cm long and has a mass of 7 grams.

   Another piece of the same wire is 25 cm long.

   Calculate the mass of the 25 cm piece of wire.   **[2 marks]**

6 It takes 2 decorators 3 days to paint a community hall.

   How long would it take 6 decorators to paint the community hall?   **[2 marks]**

1–3

3–5

18 Rates

In this block, we will cover...

18.1 Speed

Example 1

A car travels 90 miles in 2 hours at a constant spe

At what speed is the car travelling?

Method

Solution		Commentary
Distance	Time	The unit for sp
90 miles	2 hours	Work out how
÷ 2	÷ 2	You are told
45 miles	1 hour	work out how
		Remember

18.2 Density

Practice

1. Use density $= \dfrac{\text{mass}}{\text{volume}}$ to work out the den
   Give units with your answers.

   a mass = 45 g, volume = 5 mm³   **b**

   c mass = 500 mg, volume = 2 m³   **d**

2. Use density $= \dfrac{\text{mass}}{\text{volume}}$ to work out the mas
   Give units with your answers.

   a density = 26 g/cm³, volume = 5 cm³   **b**

   c density = 4 kg/m³, volume = 0.5 m³   **d**

Are you ready? (A)

1 How many minutes are there in 3 hours?

2 Write 150 minutes in hours and minutes.

3 How many minutes are there in $5\frac{1}{2}$ hours?

4 A teacher can mark 10 tests in 1 hour.

How many tests can they mark in 2 hours?

Make sure you understand what is meant by **distance**, **time** and **speed**.

What it means	Example
Distance is how far something travels	The distance between London and Manchester is 200 miles.
Time is how long something takes	A car journey from London to Manchester takes approximately 5 hours.
Speed is how fast something is travelling	A car travelling from London to Manchester might be travelling at 65 miles per hour at a particular moment on the motorway.
Average speed is the speed across a whole journey, particularly useful when there is a change of speed.	200 miles in 5 hours means the average speed is $200 \div 5 = 40$ miles per hour.
Constant speed describes when the speed doesn't change.	A car cruising at a steady 60 miles per hour is travelling at a constant speed.

There is a formula that links these together: $\text{Speed} = \dfrac{\text{Distance}}{\text{Time}}$

So, if you know the distance of a journey and how long it took, you can work out the speed.

Or, if you know the speed and the time, you can find the distance travelled.

$\text{Speed} = \dfrac{\text{Distance}}{\text{Time}}$ so $\text{Distance} = \text{Speed} \times \text{Time}$

Some units of speed are:

- mph (miles per hour)
- km/h (kilometres per hour)
- m/s (metres per second).

Example 1

A car travels 90 miles in 2 hours at a constant speed.

At what speed is the car travelling?

Method

Solution	Commentary
Distance Time ÷ 2 ⟨ 90 miles 2 hours ⟩ ÷ 2 45 miles 1 hour 45 miles per hour 45 mph	The unit for speed in this example is miles per hour (mph). Work out how many miles the car travels each hour. You are told the car travels 90 miles in 2 hours so to work out how far it travels in 1 hour, divide by 2 Remember that 45 mph means travelling 45 miles in 60 minutes. This can help when working out how far a car travels in half an hour (30 minutes).

Example 2

A drone flies at a constant speed of 20 m/s for 5 seconds.

How far does it travel?

Method

Solution	Commentary
Distance Time × 5 ⟨ 20 m 1 s ⟩ × 5 100 m 5 s 100 m	m/s means metres per second. So 20 m/s means the drone flies 20 metres in 1 second. To work out how far it travels in 5 seconds, multiply by 5

Practice (A)

1 A car is travelling at a constant speed of 30 mph.

How far will it travel in:

   a 2 hours   **b** 5 hours   **c** half an hour   **d** $3\frac{1}{2}$ hours?

2 A bus is travelling at a constant speed of 40 mph.

How long will it take the bus to travel:

   a 120 miles   **b** 400 miles   **c** 20 miles   **d** 10 miles?

3 A motorbike travels 200 miles at a constant speed.

How fast does the motorbike travel if the journey lasts:

   a 10 hours   **b** 5 hours   **c** 8 hours   **d** 100 hours?

4 A sprinter runs at a constant speed of 10 m/s.

How long will it take the sprinter to run 200 m?

5 A train travels at a constant speed of 50 km/h for 3 hours.

How far does the train travel?

6 A plane flies at a constant speed of 500 mph.

How long will it take the plane to fly 3000 miles?

7 Ron cycles 9 miles in $1\frac{1}{2}$ hours.

What is his average speed?

8 Jack runs 1000 m in 5 minutes.

Eva runs 400 m in $2\frac{1}{2}$ minutes.

Who ran faster?

Are you ready? (B)

1 A car travels 60 miles in 3 hours.

What is the average speed of the car?

2 A ball flies at a constant speed of 40 m/s.

How far will it fly in 20 seconds?

3 A train travels at a constant speed of 80 mph for 5 hours.

How far will the train travel?

Example 1

Use the formula speed $= \dfrac{\text{distance}}{\text{time}}$ to work out the distance travelled by a particle moving at 10 m/s for 30 seconds.

Method

Solution	Commentary
Speed = 10 m/s Time = 30 s	First, write out the values you know.
$10 = \dfrac{?}{30}$? = 10 × 30 ? = 300	Then substitute them into the formula. To work out the value of the unknown, multiply by 30 to get 300
The particle travels 300 m	Now think about the units: the speed was in m/s so the distance must be metres (m).

Example 2

How long does it take to travel 100 km at a constant speed of 2.5 km/s?

Method

Solution	Commentary
Distance = 100 km Speed = 2.5 km/s	First, write down the values you know.
$2.5 = \dfrac{100}{?}$ $2.5 \times ? = 100$ $? = \dfrac{100}{2.5}$	Then substitute them into the formula. Multiply both sides by the ? to start to solve the problem. Then divide by 2.5
? = 40 seconds	The units of speed in this example are km/s so the unit of time is seconds.

Practice (B)

1 Use the formula speed $= \dfrac{\text{distance}}{\text{time}}$ to work out the speed given the following distances and times.

    a   distance = 200 miles, time = 4 hours   **b**   distance = 500 metres, time = 25 seconds

    c   distance = 1000 km, time = 20 hours   **d**   distance = 30 miles, time = $1\frac{1}{2}$ hours

2 Use the formula speed $= \dfrac{\text{distance}}{\text{time}}$ to work out the distance travelled given the following speeds and times.

    a   speed = 30 mph, time = 3 hours   **b**   speed = 10 m/s, time = 27 seconds

    c   speed = 40 km/h, time = $7\frac{1}{2}$ hours   **d**   speed = 20 m/s, time = 1 minute

3 Use the formula speed $= \dfrac{\text{distance}}{\text{time}}$ to work out the time taken to travel given the following speeds and distances.

    a   speed = 20 km/h, distance = 100 km   **b**   speed = 40 mph, distance = 120 miles

    c   speed = 50 m/s, distance = 175 m   **d**   speed = 500 km/h, distance = 250 km

4 A bird flies at a constant speed of 70 km/h for $4\frac{1}{2}$ hours.

    How far does it travel?

5 A helicopter flies 235 miles in $1\frac{1}{2}$ hours.

    What is the average speed of the helicopter?

6 Max cycles 10 km in 30 minutes.

    What is his average speed in km/h?

7 A plane flies 1925 km at a constant speed of 550 km/h.

How long does it take?

8 The speed limit on a road is 30 mph.

A driver travels 8 miles in 15 minutes along the road.

Is the driver exceeding the speed limit? You must show your working.

Consolidate – do you need more?

1 A car is travelling at a constant speed of 40 mph.

How far will it travel in:

a 2 hours   b 3 hours   c 10 hours   d half an hour?

2 A motorbike is travelling at a constant speed of 60 mph.

How long will it take for the motorbike to travel:

a 120 miles   b 240 miles   c 30 miles   d 20 miles?

3 A van travels 100 miles at a constant speed.

How fast was the van travelling if the journey lasted:

a 12 hours   b 5 hours   c 10 hours   d 100 hours?

4 A runner runs at a constant speed of 12 m/s.

How long will it take the sprinter to run 240 m?

5 A bus travels at a constant speed of 60 km/h for $3\frac{1}{2}$ hours.

How far does the bus travel?

6 Use the formula speed $= \dfrac{\text{distance}}{\text{time}}$ to work out the speed given the following distances and times.

a distance = 400 miles, time = 4 hours   b distance = 500 metres, time = 50 seconds

c distance = 500 km, time = 20 hours   d distance = 30 miles, time = 2.5 hours

7 Use the formula speed $= \dfrac{\text{distance}}{\text{time}}$ to work out the distance travelled given the following speeds and times.

a speed = 60 mph, time = 7 hours   b speed = 15 m/s, time = 42 seconds

c speed = 26 km/h, time = 3.5 hours   d speed = 20 km/h, time = half an hour

8 Use the formula speed $= \dfrac{\text{distance}}{\text{time}}$ to work out the time taken to travel given the following speeds and distances.

a speed = 40 km/h, distance = 200 km   b speed = 36 mph, distance = 180 miles

c speed = 25 m/s, distance = 175 m   d speed = 280 km/h, distance = 140 km

Stretch – can you deepen your learning?

1 Two athletes run a 1 km race.

Athlete A runs at a speed of 500 metres per minute.

Athlete B runs at a speed of 15 m/s.

Who will finish first?

2 A spacecraft travels at a constant speed of 8 km/s.

The distance to Mars is approximately 380 000 000 km.

Approximately how long will it take in seconds for the spacecraft to get to Mars?

3 The distance between Alder and Benford is 75 miles.

The distance between Benford and Carton is 40 miles.

Jack drives from Alder to Benford at a constant speed of 50 mph.

He then drives from Benford to Carton at a constant speed of 20 mph.

How long will the journey take him in total?

4 Jo walks 600 m in 20 minutes.

Work out her average speed in km/h.

Are you ready?

1 Which of these units are used to measure mass?

| ml | g | cm | kg | inches |

2 Which of these units are used to measure volume?

| cm | cm² | cm³ | mm² | m³ |

3 Use speed = $\dfrac{\text{distance}}{\text{time}}$ to work out the distance travelled by a bus moving at 30 mph for 2 hours.

4 What is the volume of the cuboid?

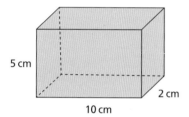

5 cm
2 cm
10 cm

Mass is a measure of how much matter there is in an object.

Volume is a measure of the space inside an object.

Density is a measure of how tightly packed the matter in that object is.

There is a formula that links these together:

Density = Mass ÷ Volume   or   Density = $\dfrac{\text{Mass}}{\text{Volume}}$

This means if you know the mass and volume, you can work out the density.

Alternatively, if you know the density and the volume, you can find the mass.

Density = $\dfrac{\text{Mass}}{\text{Volume}}$ so   Mass = Density × Volume   and   Volume = $\dfrac{\text{Mass}}{\text{Density}}$

So to find mass: Multiply density by volume

And to find volume: Divide mass by density

Some units of density are:

• kg/m³ (kilograms per metre cubed)

• g/cm³ (grams per centimetre cubed).

The units of density are always a 'unit of mass' per 'unit of volume'.

Example 1

A piece of wood has a mass of 300 g and a volume of 10 cm^3.

What is the density of the piece of wood?

Method

Solution	Commentary
Mass = 300 g Volume = 10 cm^3	First, write down the values you know.
Density = $\dfrac{\text{Mass}}{\text{Volume}}$ Density = $\dfrac{300}{10}$ = 30	Then substitute the values into the formula.
Density = 30 g/cm^3	In this example, the unit of mass is g and the unit of volume is cm^3. Therefore, the unit of density is g/cm^3 (grams per centimetre cubed).

Example 2

A box has a volume of 120 cm^3.

Its density is 5 g/cm^3.

What is the mass of the box?

Method

Solution	Commentary
Density = 5 g/cm^3 Volume = 120 cm^3	First, write down the values you know.
$5 = \dfrac{?}{120}$ 5 × 120 = ?	Then substitute the values into the formula.
? = 600 g	You are working out mass and the density in this example is in g/cm^3, so the unit of mass must be grams.

Practice

1 Use density = $\dfrac{\text{mass}}{\text{volume}}$ to work out the density given the following masses and volumes.
Give units with your answers.

   a mass = 45 g, volume = 5 mm^3 **b** mass = 240 g, volume = 30 cm^3

   c mass = 500 mg, volume = 2 m^3 **d** mass = 320 kg, volume = 40 m^3

2 Use density = $\dfrac{\text{mass}}{\text{volume}}$ to work out the mass given the following densities and volumes.
Give units with your answers.

   a density = 26 g/cm^3, volume = 5 cm^3 **b** density = 5 mg/mm^3, volume = 40 mm^3

   c density = 4 kg/m^3, volume = 0.5 m^3 **d** density = 0.8 g/mm^3, volume = 8.5 mm^3

3 Use density $= \dfrac{\text{mass}}{\text{volume}}$ to work out the volume given the following masses and densities.

Give units with your answers.

    a   mass = 500 g, density = 100 g/cm³   **b**   mass = 28 kg, density = 7 kg/m³

    c   mass = 2400 mg, density = 80 mg/mm³   **d**   mass = 6 g, density = 0.5 g/cm³

4 The masses and volumes of some pieces of metal are shown.

Work out the density of each metal.

a

Radium
Mass = 40 g
Volume = 8 cm³

b

Technetium
Mass = 110 g
Volume = 10 cm³

c

Vanadium
Mass = 12 000 kg
Volume = 2 m³

d

Platinum
Mass = 225 g
Volume = 18 cm³

5 The density of brass is 8.5 g/cm³.

The volumes of some pieces of brass are given.

Work out the mass of each piece.

    a   Volume = 26 cm³   **b**   Volume = 17 cm³   **c**   Volume = 0.5 cm³

6 The density of cobalt is 8860 kg/m³.

A block of cobalt has a volume of $3\frac{1}{2}$ m³.

What is the mass of the block?

7 The density of silver is 10.5 g/cm³.

The masses of some pieces of silver are given.

Work out the volume of each piece.

    a   210 g   **b**   84 g   **c**   3360 g   **d**   2.1 kg

8 The density of aluminium is 2600 kg/m³.

An aluminium statue has a mass of 6500 kg.

What is the volume of the statue?

Consolidate – do you need more?

1 Use density $= \dfrac{\text{mass}}{\text{volume}}$ to work out the density given the following masses and volumes.

Give units with your answers.

    a   mass = 72 g, volume = 9 cm³   **b**   mass = 500 mg, volume = 25 cm³

    c   mass = 2.5 kg, volume = 5000 cm³   **d**   mass = 800 g, volume = 1000 cm³

2 Use density = $\dfrac{\text{mass}}{\text{volume}}$ to work out the mass given the following densities and volumes.

Give units with your answers.

a density = 12 g/cm³, volume = 3 cm³   **b** density = 6 kg/m³, volume = 9 m³

c density = 0.5 g/mm³, volume = 40 mm³   **d** density = 18 mg/mm³, volume = 6 mm³

3 Use density = $\dfrac{\text{mass}}{\text{volume}}$ to work out the volume given the following masses and densities.

Give units with your answers.

a mass = 250 g, density = 50 g/cm³   **b** mass = 5.6 kg, density = 0.8 kg/m³

c mass = 1200 mg, density = 100 mg/mm³   **d** mass = 14 g, density = 2 g/cm³

Stretch – can you deepen your learning?

1 The density of this cuboid is 18 g/cm³.

Work out the mass of the cuboid.

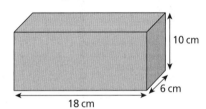

2 The mass of this cube is 72 kg.

Work out the density of the cube.

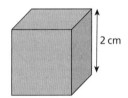

3 This cuboid has a density of 18 g/mm³.
The mass of the cuboid is 432 g.

Work out the length of the unknown side.

4 The table shows the density of some common metals in g/cm³.

Metal	Density (g/cm³)
Bronze	8
Silver	10.5
Gold	19.3
Platinum	21.5
White metal	7.05

A piece of platinum has a volume of 560 cm³.

A piece of silver has a volume of 1.2 m³.

Which is heaver, the piece of platinum or the piece of silver?

Rates: exam practice

1 Ron drives at a constant speed for 1 hour.

He travels a distance of 50 miles.

What is Ron's speed? **[1 mark]**

2 Alex runs at a constant speed of 9 miles per hour.

How many miles does Alex run in 30 minutes? **[1 mark]**

3 The distance–time graph shows Flo's journey from home to the beach, and back.

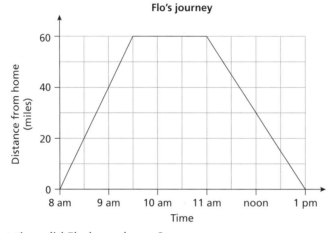

(a) At what time did Flo leave home? **[1 mark]**

(b) How long did Flo spend at the beach? **[1 mark]**

(c) The traffic was heavier on the way home.

Describe how the graph shows this. **[1 mark]**

4 A piece of wood has a mass of 15 g and a volume of 20 cm³.

Work out the density of the wood. **[2 marks]**

5 A cyclist makes a journey from Town A to Town B.

The distance between Town A and Town B is 180 kilometres.

The cyclist travels at an average speed of 25 kilometres per hour.

How long does it take for the cyclist to travel from Town A to Town B? **[2 marks]**

Ratio, proportion and rates of change: exam practice

1 100 g of butter is needed to make 12 shortbread biscuits.

How much butter is needed to make 24 shortbread biscuits? **[1 mark]**

2 Simplify the ratio 16 : 20 : 32 **[2 marks]**

3 Chloe travels at a constant speed of 40 mph for 15 minutes.

How far does Chloe travel? **[1 mark]**

4 Aisha, Max and Amir share £90 in the ratio 5 : 2 : 3

How much money does each person get? **[3 marks]**

5 Two shops, *Shake Shack* and *Milks*, both sell milkshakes.

Shake Shack

5 for £12

Milks

3 for £8.50

At which shop are milkshakes the better value for money?

You must show all your working. **[3 marks]**

6 A force of 50 Newtons acts on an area of 15 cm^2.

The force stays the same but the area increases.

What happens to the pressure?

$$\text{pressure} = \frac{\text{force}}{\text{area}}$$ **[1 mark]**

7 Jack went to France.

He changed £400 into euros (€).

The exchange rate was £1 = €1.17

How many euros did he get? **[2 marks]**

8 Max is saving to buy a computer game that costs £26

He saves 5p, 10p and 50p coins in a jar.

The ratio of 5p to 10p to 50p coins is 2 : 5 : 1

There are 120 coins in the jar.

How much more money does he need to save? **[4 marks]**

9 Forty people take part in a show.

The ratio of children to adults in the show is 3 : 5

$\frac{3}{5}$ of the people are dancers.

The rest of the people are singers.

There are three children who are dancers in the show.

How many more adults are dancers than singers? **[4 marks]**

10 Richard drives his car for 240 miles in three-and-a-half hours.

Calculate his average speed, in miles per hour. Give your answer to the nearest mph. **[3 marks]**

11 300 students are asked how they travel to school.

- 60% say they walk to school.

- The remaining students travel by car or by bus.

- The ratio of students that travel by car to those that travel by bus is 2 : 3

How many more students walk to school than travel by car? **[4 marks]**

12 A florist is making some bunches of flowers for a wedding.
Each bunch contains some carnations, roses and lilies.
Each bunch is identical.

- A bunch contains 40 flowers.

- 60% of the flowers are lilies.

- The ratio of carnations to roses is 3 : 5

The florist only has 130 carnations.

How many bunches of flowers can the florist make? **[4 marks]**

Glossary

Acceleration – the rate at which an object's speed is changing

Algebraic fraction – a fraction whose numerator and/or denominator are algebraic expressions

Annual – covers a period of one year

Arithmetic sequence – an ordered set of numbers that have a common difference between each consecutive term

Axis – a line on a graph that you can read values from

Balance – an amount of money in an account

Base – the number that gets multiplied when using a power/index

Best buy (or **best value**) – the item which is cheapest when equal-sized amounts of different items are compared

Bill – shows how much money is owed for goods or services

Binomial – an expression with two terms

Coefficient – a number in front of a variable, e.g. for $4x$ the coefficient of x is 4

Collect like terms – put like terms in an expression together as a single term

Common denominator – two or more fractions have a common denominator when their denominators are the same

Commutative – when an operation can be performed in any order

Conjecture – a statement that might be true that has not yet been proved

Consecutive – following on, e.g. 14, 15, 16… or n, $n + 1$, $n + 2$…

Constant – not changing

Constant of proportionality – a constant number that is used in direct or inverse proportion

Convert – change from one form to another, e.g. a percentage to a decimal

Coordinate – an ordered pair used to describe the position of a point

Counterexample – an example that disproves a statement

Credit – an amount of money paid into an account

Cube root – the cube root of a number is that number which is multiplied three times to get the original, e.g. the cube root of 64 is 4 because $4^3 = 4 \times 4 \times 4 = 64$

Cubic graph – a graphical representation of a cubic function

Curve – a line on a graph showing how one quantity varies with respect to another

Debit – an amount of money taken out of a bank account

Decimal – a number with digits to the right of the decimal point

Decimal places – the number of digits to the right of the decimal point in a number

Decrease – make something smaller

Decreasing (or descending) sequence – a sequence where every term is smaller than the previous term

Degree of accuracy – how precise a number is

Denominator – the bottom number in a fraction; it shows how many equal parts one whole has been divided into

Density – the ratio of the mass of an object to its volume; the mass per unit volume

Deposit – an amount of money paid into a bank account

Depreciate – reduce or decrease in value

Descending – decreasing in size

Difference – in arithmetic, the result of subtracting a smaller number from a larger number; in sequences, the gap between numbers in a sequence

Digits – the numerals used to form a number

Direct proportion – two quantities are in direct proportion when as one increases or decreases, the other increases or decreases at the same rate

Directed numbers – numbers that can be negative or positive

Divide in a ratio – share a quantity into two or more parts so that the shares are in a given ratio

Dividend – the amount you are dividing

Division – the process of splitting a number into equal parts

Divisor – the number you are dividing by

Double number line – two lines used to represent ratio problems

Equal – having the same value; the sign = is used between numbers and calculations that are equal in value, and the sign ≠ is used when they are not

Equation – a statement with an equals sign, which states that two expressions are equal in value

Equivalent – numbers or expressions that are written differently but are always equal in value

Error interval – the range of values a number could have taken before being rounded

Estimate – give an approximate answer

Evaluate – work out the numerical value of

Even number – a number that is divisible by 2

Exchange rate – the rate at which the currency of one country is exchanged for that of another

Expand – multiply to remove brackets from an expression

Exponential – containing, involving or expressed as a power (exponent); an exponential function is of the form $y = k^x$

Express – write, often in a different form

Expression – a collection of terms involving mathematical operations

Factor – a positive integer that divides exactly into another positive integer

Factor pair – a pair of numbers that multiply together to give a number

Factorise – find the factors you need to multiply to make an expression

Fibonacci sequence – the next term in a Fibonacci sequence is found by adding the previous two terms together

Find – work out the value of

Formula (plural: **formulae**) – a rule connecting variables written with mathematical symbols

Fraction – a number that compares equal parts of a whole

Function – a relationship with an input and an output

General term – an expression that generates the terms of a sequence

Geometric sequence – a sequence is geometric if the value of each successive term is found by multiplying or dividing the previous term by the same number

Gradient – the steepness of a line

Graph – a diagram showing how values change

Graphical – using a graph

Highest common factor (HCF) – the greatest number that is a factor of every one of a set of numbers

Identity – a statement that is true no matter what the values of the variables are

Improper fraction – a fraction in which the numerator is greater than the denominator

Inclusive – including the end points of a list

Increase – make something larger

Increasing (or ascending) sequence – a sequence where every term is greater than the previous term

Index (plural: **indices**) – an index number (or power) tells you how many times to multiply a number by itself

Inequality – a comparison between two quantities that are not equal to each other

Inequality symbol – a symbol comparing values showing which is greater and which is smaller

Integer – a whole number

Intercept – the point at which a graph crosses, or intersects, a coordinate axis

Interest – a percentage fee paid when borrowing money or a percentage earned when you deposit money into a savings account

Inverse – the opposite of a mathematical operation; it reverses the process

Inverse function – the inverse function of a function f is a function that undoes the operation of f

Inverse proportion – if two quantities are in inverse proportion, when one quantity increases, the other decreases at the same rate

Irrational number – a number that cannot be written in the form $\frac{a}{b}$ where a and b are integers

Is equivalent to ≡ – equal to for all values of the variables in an expression

Iteration – repeating a process

Kilo- – one thousand

Like terms – terms whose variables are the same, e.g. $7x$ and $12x$

Linear – forming a straight line

Linear equation – an equation with a simple unknown like a, b, or x, i.e. there are no squared terms, cubed terms, etc.

Linear sequence – a sequence whose terms are increasing or decreasing by a constant difference

Loss – if you buy something and then sell it for a smaller amount; loss = amount paid – amount received

Lower bound – the bottom limit of a rounded number

Lowest common multiple (LCM) – the smallest number that is a multiple of every one of a set of numbers

Maximum point – the point on a graph where a function has its greatest value

Mental strategy – a method that enables you to work out the answer in your head

Midpoint – the point halfway between two others

Minimum point – the point on a graph where a function has its least value

Mixed number – a number presented as an integer and a proper fraction

Multiple – the result of multiplying a number by a positive integer

Multiplier – a number you multiply by

Negative numbers – numbers less than zero

Non-linear – not forming a straight line

Non-linear sequence – a sequence whose terms are not increasing or decreasing by a constant difference

Non-unit fraction – a fraction with a numerator that is not 1

Notation – a symbolic system for the representation of mathematical items and concepts

Number line – a line on which numbers are marked at intervals

Numerator – the top number in a fraction that shows the number of parts

Odd number – a number that when divided by 2 gives a remainder of 1, e.g. 1, 17, 83

Operation – a mathematical process such as addition, subtraction, multiplication or division

Order of magnitude – size of a number in powers of 10

Order of operations – the rules that tell you the order in which to perform each part of a calculation

Origin – the point where the x-axis and y-axis meet

Original value – a value before a change takes place

Parabola – a type of curve that is approximately U-shaped and has a line of symmetry

Parallel – in the same direction; parallel lines have the same gradient

Per annum – every year

Per cent – parts per hundred

Percentage – the number of parts per hundred

Perpendicular – at right angles to

Piece-wise graph – a graph that consists of more than one straight line

Plot – to draw a representation of a function or data on a graph

Point of intersection – the point where two graphs cross each other

Position-to-term rule – the rule that links the position of the term to the value of the term

Power (or exponent) – this is written as a small number to the right and above the base number, indicating how many times to use the number in a multiplication, e.g. the 5 in 2^5

Powers of 10 – the result of multiplying 10 by itself a number of times to give a value such as 10, 100, 1000, 10 000, and so on

Prime factor decomposition – writing numbers as a product of their prime factors

Prime number – a positive integer with exactly two factors, 1 and itself

Priority – a measure of the importance of something

Product – the result of a multiplication

Profit – if you buy something and then sell it for a higher amount; profit = amount received – amount paid

Proof – an argument that shows that a statement is true

Proportion – a part, share, or number considered in relation to a whole

Prove – show that something is always true

Quadratic – of the form $ax^2 + bx + c$

Quadratic formula – the formula that gives the roots of a quadratic equation

Quotient – the result of a division

Ratio – a ratio compares the sizes of two or more values

Rational number – a number that can be written in the form $\frac{a}{b}$ where a and b are integers

Real number – all positive and negative numbers, including decimals and fractions

Reciprocal – the result of dividing 1 by a given number; the product of a number and its reciprocal is always 1

Recurring decimal – a recurring decimal has digits that are in a repeating pattern like 0.3333… or 0.171717…

Reduce – make something smaller

Reflection – a type of geometrical transformation, where an object is flipped to create a mirror image

Remainder – the amount left over after dividing one integer by another

Repeated percentage change – when an amount is changed by one percentage followed by another

Reverse percentage – a problem where you work out the original value

Root – the nth root of a number x is the number that is equal to x when multiplied by itself n times

Round – give an approximate value of a number that is easier to use

Satisfy – make an equation or inequality true

Sector – a part of a circle formed by two radii and a fraction of the circumference

Sequence – a list of items in a given order, usually following a rule

Significant figures – the most important digits in a number that give you an idea of its size

Simplify – rewrite in a simpler form, e.g. rewrite $8 \times h$ as $8h$

Simultaneous – at the same time

Solution – a value you can substitute in place of the unknown in an equation to make it true

Solve – find a value that makes an equation true

Speed – the rate at which an object is moving

Square number – a positive integer that is the result of an integer multiplied by itself

Square root – a square root of a number is a value that, when multiplied by itself, gives the number

Standard form – a number written in the form $A \times 10^n$ where A is at least 1 and less than 10, and n is an integer

Subject – the variable in a formula that is expressed in terms of the other variables

Substitute – replace letters with numerical values

Successive – coming after another term in a sequence

Tangent (to a circle) – a straight line that touches the circumference of a circle at one point only

Term – in algebra, a single number or variable, or a number and variable combined by multiplication or division; in sequences, one of the members of a sequence

Term-to-term rule – a rule that describes how you get from one term of a sequence to the next

Terminating decimal – a decimal fraction with a finite number of digits, e.g. 0.75

Translation – a type of geometrical transformation, where an object is moved left or right and/or up or down

Trial and improvement – a method of finding a solution to a mathematical problem where you make a guess (a trial), see if it works in the problem, and then refine it to get closer to the actual answer (improvement)

Triangular number – a positive integer that is the sum of consecutive positive integers starting from 1

Truncate – remove digits from a decimal number

Turning point – the point at which a quadratic graph changes direction

Unit fraction – a fraction with a numerator of 1

Unknown – a variable (letter) whose value is not yet known

Unlike terms – terms whose variables are not exactly the same, e.g. $7x$ and 12 or $5a$ and $5a^2$

Upper bound – the top limit of a rounded number

Variable – a numerical quantity that might change, often denoted by a letter, e.g. x or t

y-intercept – the point at which a graph crosses or intersects the y-axis

Zero pairs – a pair of numbers whose sum is zero, e.g. +1 and –1 make 0

Answers

Block 1 Integers

Chapter 1.1

Are you ready? (A)
1 a 100   **b** 1000   **c** 10 000
2 $10^2 = 10 \times 10 = 100$

Practice (A)
1 a 10^2 **b** 10^5 **c** 10^3
  d 10^6 **e** 10^7 **f** 10^9
2 a 10 000   **b** 100 000   **c** 100
  d 1 000 000 000
3 10 000 000 10^6 ten thousand 10^3
  one hundred   10
4 a 10^6 **b** 10^6 **c** 10^6
  d 10^9 **e** 10^4 **f** 10^{10}

What do you think? (A)
1 False

Are you ready? (B)
1 a 0.1   **b** 0.01   **c** 0.001
2 a $0.01 > 0.001$ **b** $0.01 < \frac{1}{10}$ **c** $\frac{1}{1000} = 0.001$
3 a 100   **b** 10   **c** 1

Practice (B)
1 a 10^{-1} **b** 10^{-3} **c** 10^{-6} **d** 10^{-8}
2 a 0.0001   **b** 0.000 01   **c** 0.000 000 01
  d 0.000 000 0001
3 10^{-7} 0.000 01 10^{-4} one-hundredth
  one-tenth   0

What do you think? (B)
1 10^{-4} or 10^{-5}

Consolidate
1 a 10^4 **b** 10^8 **c** 10^{-2}
  d 10^{-5} **e** 10^6 **f** 10^{-6}
2 a 1000   **b** 0.001   **c** 0.1
  d 1 000 000   **e** 0.000 000 001
  f 10 000 000 000
3 a 10^5 **b** 10^5 **c** 10^4
  d 10^8 **e** 10^3 **f** 10^5
4 a $10^2 < 10^4$ **b** $10^{-4} < 10^{-2}$ **c** $10^2 > 10^{-4}$
5 10^6 one hundred   1   0.01
  one-thousandth 10^{-6}

Stretch
1 a 10^2 **b** 10^{-3}
2 1000 times greater

Chapter 1.2

Are you ready? (A)
1 a i 30   **ii** 300   **iii** 3000
  b i 400   **ii** 40   **iii** 4
2

Length	Mass	Volume
metres millimetres centimetres	kilograms grams	millilitres litres

Practice (A)
1 a 300 cm   **b** 900 cm   **c** 1700 cm
2 a 5 m   **b** 8 m   **c** 11 m
3 a 5 cm   **b** 12 cm   **c** 50 cm
4 a 3.2 kg   **b** 3.24 kg   **c** 43.5 kg
5 a 1200 ml   **b** 2430 ml   **c** 12 430 ml

What do you think? (A)
1 No, 2.6 kilometres is equal to 2600 metres.

Are you ready? (B)
1

Length	Mass	Volume
miles inches feet	stones pounds	gallons pints

2 a gallons   **b** miles   **c** stones

Practice (B)
1 a 56 pints   **b** 136 pints   **c** 376 pints
2 a 8 gallons   **b** 18 gallons   **c** 37 gallons
3 a 84 pounds   **b** 322 pounds   **c** 434 pounds
4 a 6 pounds   **b** 14 pounds   **c** 33 pounds
5 a 8 inches ≈ **20** centimetres
  b 32 inches ≈ **80** centimetres
  c 246 inches ≈ **615** centimetres
6 a 20 miles ≈ **32** kilometres
  b 22.5 miles ≈ **36** kilometres
  c 27.5 miles ≈ **44** kilometres

What do you think? (B)
1 Any answer from: 4   5   6

Consolidate
1 a 2 m   **b** 8 m   **c** 12 m
2 a 60 mm   **b** 130 mm   **c** 260 mm
3 a 4000 g   **b** 5400 g   **c** 8310 g
4 a 5 l   **b** 6.5 l   **c** 11.63 l
5 a 3 stones   **b** 14 stones   **c** 25 stones
6 a 108 inches   **b** 192 inches   **c** 372 inches
7 a 5 centimetres ≈ **2** inches
  b 7.5 centimetres ≈ **3** inches
  c 25 centimetres ≈ **10** inches
8 a 16 kilometres ≈ **10** miles
  b 32 kilometres ≈ **20** miles
  c 88 kilometres ≈ **55** miles

Stretch
1 Sven is 27 cm taller.
2 25 miles or 40 km

Chapter 1.3

Are you ready? (A)
1 a 213 452   **b** 40 603   **c** 13 205 043
2 a 900   **b** 100 000   **c** 1000
3 a 8521   **b** 90 406   **c** 215 350

Practice (A)
1 a Tens   **b** Thousands   **c** Hundreds
  d Ones   **e** Hundred thousands   **f** Millions
2 a 6342   **b** 60 302   **c** 316 500   **d** 2 006 006
3 a Three thousand, four hundred and fifty-seven
  b Thirty-two thousand, nine hundred and eight
  c Four million, five hundred and sixty-one thousand, three hundred and eighty-two
  d Fourteen million, two thousand, seven hundred and forty

What do you think? (A)
1 Three thousand, six hundred and three

Are you ready? (B)
1 a 6013 and 4564 (6013 is greater)
  b 7398 and 7402 (7402 is greater)
2 a 204   **b** 1302   **c** 897
3 a greater than   **b** greater than   **c** equal to

Practice (B)

1 a 4501 < 6874 **b** 397 < 3970 **c** 11042 > 9856
2 a 21806, 40998, 41687, 41712
b 41712, 41687, 40998, 21806
The order of the numbers is reversed.
3 a 2841, 2840, 2481, 2418, 1840
b 367, 376, 3670, 30670, 36007
4 Nile, Amazon, Yangtze, Mississippi

What do you think? (B)

1 136, 163, 316, 361, 613, 631

Consolidate

1 a 300 **b** 3000 **c** 300000
d 30000 **e** 30 **f** 3000000
2 a 2415 **b** 13107 **c** 534600 **d** 5063011
3 a Two thousand, one hundred and ninety-eight
b One million, four hundred and thirty-two thousand, seven hundred and forty-six
c Twenty-four thousand and eighteen
d Eleven million, thirty thousand, nine hundred and two
4 a 1067 **b** 7610
5 124603, 125002, 142633, 214360, 214630
6 a 7980, 7960, 7691, 7690, 6790
b 9342, 12395, 12935, 100359, 124359
7 Car D, car A, car C, car B

Stretch

1 Any price between £2490000 and £2500000, e.g. £2496310
2 41, 4, 0, −4, −14, −41

Chapter 1.4

Are you ready? (A)

1 a

+	3	6	1
9	12	15	10
7	10	13	8
5	8	11	6

b

+	6	8	4
2	8	10	6
9	15	17	13
7	13	15	11

2 a 130 **b** 13000 **c** 500

Practice (A)

1 a 7936 **b** 121 **c** 636
d 490 **e** 2720 **f** 1502
2 a 52 **b** 46 **c** 208
d 15341 **e** 377 **f** 2676
3 57 marbles
4 183
5 £382
6 1038 units

What do you think? (A)

1 $x = 8$
2 a

$$\begin{array}{r} 2\ 4\ \boxed{8} \\ +\ 6\ \boxed{5}\ 3 \\ \hline \boxed{9}\ 0\ 1 \end{array}$$

b

$$\begin{array}{r} \boxed{3}\ 4\ 2 \\ -\ 1\ \boxed{0}\ 6 \\ \hline 2\ 3\ \boxed{6} \end{array}$$

3 31

Are you ready? (B)

1

×	3	6	9
5	15	30	45
7	21	42	63
4	12	24	36

2 a 120 **b** 120 **c** 1200 **d** 12000
3 a 12 **b** 9 **c** 6
d 18 **e** 2 **f** 3
4 a 1 **b** 2 **c** 2
d 3 **e** 2 **f** 1

Practice (B)

1 a 69 **b** 144 **c** 1472
d 5658 **e** 51952 **f** 36288
2 a 26 **b** 137 **c** 256 **d** 349
3 a 26 **b** 17 **c** 14 **d** 23
4 a 126 r4 or 126.8
b 142 r1 or 142.14 to 2 d.p.
c 105 r3 or 105.23 to 2 d.p.
d 26 r12 or 26.55 to 2 d.p.
5 1888 grams
6 4900 straws
7 17 coaches

What do you think? (B)

1 a 4932 cm² **b** 532 mm²
2

$$\begin{array}{r} 3\ 2\ \boxed{4} \\ \times\ \ \ \boxed{3} \\ \hline \boxed{9}\ \boxed{7}\ \boxed{2} \end{array}$$

Are you ready? (C)

1 0 − 3 and 3 − 8
2 a 6, 5, 4, 3, 2, 1, 0, −1, −2, −3, −4, −5, −6
b 6, 4, 2, 0, −2, −4, −6
c 6, 3, 0, −3, −6
3 a −2 **b** 2

Practice (C)

1 a −5 **b** 11 **c** 3
d −7 **e** 18 **f** 36
2 −3 + 5 = 2 or 5 + −3 = 2
6 − −4 = 10
3

×	−4	−3	−2	−1	0	1	2	3	4
4	−16	−12	−8	−4	0	4	8	12	16
3	−12	−9	−6	−3	0	3	6	9	12
2	−8	−6	−4	−2	0	2	4	6	8
1	−4	−3	−2	−1	0	1	2	3	4
0	0	0	0	0	0	0	0	0	0
−1	4	3	2	1	0	−1	−2	−3	−4
−2	8	6	4	2	0	−2	−4	−6	−8
−3	12	9	6	3	0	−3	−6	−9	−12
−4	16	12	8	4	0	−4	−8	−12	−16

4 a −20 **b** −42 **c** 45
d −138 **e** 1652 **f** −1536
5 a −4 **b** −3 **c** 3 **d** −36

Consolidate

1 a 1075 **b** 1136 **c** 639
d 3403 **e** 4838 **f** 508
2 a −2 **b** 6 **c** 8
d −11 **e** −11 **f** 249

3 a 124   **b** 168   **c** 1118
  d 4864   **e** 44226   **f** 24225
4 a 157   **b** 236   **c** 213
  d 304   **e** 149 r1 or 149.33 to 2 d.p.
  f 142 r3 or 142.25
5 £78
6 a −56   **b** −9   **c** −132
  d 273   **e** −12   **f** 13
7 29 more cars
8 −1°C

Stretch
1 35519
2 36 cm
3 −3 × −7 = 21

Chapter 1.5

Are you ready?
1 57 − 28 = 85   57 = 85 + 28
2 3 × 7 = 21   7 × 3 = **21**   21 ÷ 7 = **3**   21 ÷ 3 = 7
3

×	2	8	5
3	6	24	15
9	18	72	45
4	8	32	20

Practice
1 139 + 267 = 406   267 + 139 = 406
  406 − 139 = 267   406 − 267 = 139
2 a 4060   **b** 26700
3 a 1954   **b** 5172   **c** 3218
  d 517200   **e** 32180
4 a 240   **b** 2400   **c** 60   **d** 4
5 a 67   **b** 140   **c** 93800
  d 938   **e** 67
6 a Not correct   **b** Correct
  c Correct   **d** Not correct

What do you think?
1 No, Flo is incorrect. $b + c = a$

Consolidate
1 8956 − 6085 = 2871   8956 − 2871 = 6085
2 a 3310   **b** 847   **c** 8470   **d** 331000
3 a 320   **b** 320000   **c** 800   **d** 8
4 a 9576   **b** 9576   **c** 63
  d 95760   **e** 1520
5 $z + y = x$ $x = y + z$ $z = x − y$

Stretch
1 Input = 8
2 279
3 £40

Integers: exam practice
1 a 7 and 16   **b** 138
2 a 6 kg   **b** 800 cm
3 −3, −1, 6, 9, 15, 19
4 a 4071   **b** 80000
5 4998
6 £40
7 a 5232   **b** 6540   **c** 5232000
8 a 1°C   **b** 12°C

Block 2 Fractions and decimals

Chapter 2.1

Are you ready? (A)
1 a $\frac{2}{7}$ **b** $\frac{4}{9}$ **c** $\frac{5}{8}$

2 a $\frac{1}{3}$ **b** $\frac{3}{5}$ **c** $\frac{3}{4}$

3 Example answer:

Practice (A)
1 a $\frac{1}{3} = \frac{\mathbf{5}}{15}$ **b** $\frac{1}{6} = \frac{3}{\mathbf{18}}$ **c** $\frac{3}{5} = \frac{\mathbf{9}}{15}$

  d $\frac{2}{3} = \frac{14}{\mathbf{21}}$ **e** $\frac{12}{18} = \frac{\mathbf{2}}{3}$ **f** $\frac{11}{55} = \frac{1}{\mathbf{5}}$

2 a $\frac{2}{6} = \frac{5}{\mathbf{15}}$ **b** $\frac{6}{18} = \frac{\mathbf{7}}{21}$ **c** $\frac{3}{9} = \frac{4}{12}$

3 a $\frac{1}{2}$ **b** $\frac{1}{3}$ **c** $\frac{3}{4}$

4 a 6   **b** 3   **c** 18
  d $\frac{5}{4}$ **e** $\frac{3}{2}$ **f** $\frac{6}{5}$

5 a $\frac{11}{4}$ **b** $\frac{17}{6}$ **c** $\frac{25}{7}$

6 a $3\frac{2}{3}$ **b** $2\frac{1}{4}$ **c** $3\frac{4}{5}$

Are you ready? (B)
1 a $\frac{1}{4}$ **b** $\frac{1}{5}$ **c** $\frac{1}{2}$ **d** $\frac{17}{100}$

2 $\frac{7}{1000}$

3 a 0.5   **b** 0.75   **c** 0.6   **d** 0.125

Practice (B)
1 a 0.1   **b** 0.3   **c** 0.9
  d 0.01   **e** 0.87   **f** 0.21
  g 0.001   **h** 0.179   **i** 0.991
2 a 0.5   **b** 0.4   **c** 0.75   **d** 0.35
  e 0.16   **f** 0.98   **g** 0.255   **h** 0.625

3 a $\frac{1}{10}$ **b** $\frac{3}{10}$ **c** $\frac{1}{100}$

  d $\frac{3}{100}$ **e** $\frac{71}{100}$ **f** $\frac{23}{100}$

  g $\frac{1}{1000}$ **h** $\frac{9}{1000}$ **i** $\frac{107}{1000}$

4 a $\frac{4}{5}$ **b** $\frac{1}{2}$ **c** $\frac{1}{4}$

  d $\frac{6}{25}$ **e** $\frac{19}{20}$ **f** $\frac{1}{8}$

5 $\frac{3}{5}$ $\frac{3}{4}$ $\frac{3}{10}$ $\frac{3}{8}$

  0.3   0.6   0.75   0.375

Consolidate
1 a $\frac{1}{4} = \frac{\mathbf{4}}{16}$ **b** $\frac{1}{8} = \frac{3}{\mathbf{24}}$ **c** $\frac{4}{5} = \frac{\mathbf{20}}{25}$

  d $\frac{2}{5} = \frac{14}{\mathbf{35}}$ **e** $\frac{2}{8} = \frac{3}{12}$ **f** $\frac{3}{9} = \frac{5}{15}$

2 a $\frac{3}{5}$ **b** $\frac{2}{5}$

3 0.5   0.55   0.005   0.05

$\frac{1}{20}$ $\frac{11}{20}$ $\frac{1}{2}$ $\frac{1}{200}$

4 a 0.9   **b** 0.25   **c** 0.5
  d 0.21   **e** 0.65   **f** 0.862

5 a $\frac{7}{10}$ **b** $\frac{3}{10}$ **c** $\frac{7}{100}$
  d $\frac{59}{100}$ **e** $\frac{871}{1000}$ **f** $\frac{13}{1000}$

6 a $\frac{3}{5}$ **b** $\frac{1}{5}$ **c** $\frac{3}{4}$
  d $\frac{1}{25}$ **e** $\frac{11}{20}$ **f** $\frac{21}{40}$

7 a $\frac{11}{3}$ **b** $\frac{21}{8}$ **c** $\frac{67}{6}$

8 a $5\frac{1}{2}$ **b** $2\frac{1}{5}$ **c** $3\frac{1}{2}$

9 $\frac{7}{20}$

Stretch

1 $\frac{3}{5}$

2 $\frac{5}{3}$

3 $1\frac{1}{4}$

Chapter 2.2

Are you ready? (A)

1 a i $\frac{1}{4} > \frac{1}{5}$ **ii** $\frac{2}{5} < \frac{2}{3}$
  b When the numerators are equal, the greater the denominator, the **smaller** the fraction.

2 a i $\frac{2}{9} < \frac{7}{9}$ **ii** $\frac{5}{8} > \frac{3}{8}$
  b When the denominators are equal, the greater the numerator, the **greater** the fraction.

3 a $\frac{2}{3} > \frac{5}{9}$ **b** $\frac{2}{3} < \frac{3}{4}$

Practice (A)

1 a $\frac{1}{3} > \frac{1}{8}$ **b** $\frac{5}{7} > \frac{5}{9}$ **c** $\frac{1}{6} < \frac{5}{6}$
  d $\frac{3}{8} < \frac{3}{5}$ **e** $\frac{3}{4} > \frac{5}{8}$ **f** $\frac{2}{3} > \frac{7}{12}$
  g $\frac{2}{9} < \frac{4}{13}$ **h** $\frac{4}{5} > \frac{3}{4}$ **i** $\frac{7}{8} > \frac{5}{6}$

2 $\frac{5}{9}$ $\frac{11}{18}$ $\frac{2}{3}$ $\frac{8}{9}$

3 $\frac{4}{5}$ $\frac{1}{2}$ $\frac{3}{8}$ $\frac{3}{16}$

4 $2\frac{2}{9}$ $3\frac{7}{11}$ $4\frac{3}{4}$ $4\frac{5}{6}$

5 $\frac{11}{3}$ $\frac{7}{2}$ $\frac{13}{6}$ $\frac{17}{12}$

6 $\frac{31}{8}$ $\frac{17}{4}$ $4\frac{3}{8}$ $4\frac{11}{16}$

7 Bev, Amina, Samira

What do you think? (A)

1 Benji could be correct if his house had an equal or greater value than Ed's house originally, as $\frac{3}{4}$ is a greater increase than $\frac{2}{3}$
Benji could also be incorrect depending on how much each house was worth originally.

2 $\frac{17}{6}$

Are you ready? (B)

1 a 0.4   **b** 126.3
2 a 0.7 > 0.3   **b** 0.43 < 0.46   **c** 0.265 < 0.301

Practice (B)

1 a 0.5 < 0.9   **b** 0.5 > 0.09   **c** 0.64 < 0.67
  d 3.24 > 3.22   **e** 0.978 < 1.2   **f** 1.068 < 1.203
2 0.005   0.08   0.4   0.9   1.1
3 6.76   6.67   6.61   6.3   6.03
4 Rob, Huda, Chloe, Filipo
5 Amina, Beca, Samira, Marta, Emily
6 0.09 $\frac{1}{4}$ 0.8   0.84 $\frac{19}{20}$

What do you think? (B)

1 No, Ed is incorrect because he has not taken into account the place value of the digits.
The correct order is 2.13, 2.1, 2.09

Consolidate

1 a $\frac{3}{4} > \frac{3}{11}$ **b** $\frac{5}{7} < \frac{6}{7}$ **c** 0.03 < 0.07
  d $\frac{2}{3} > \frac{8}{15}$ **e** 0.2 > 0.08   **f** 4.53 > 4.39
  g $\frac{5}{7} > \frac{2}{3}$ **h** 3.015 < 3.021

2 $\frac{7}{12}$ $\frac{3}{4}$ $\frac{19}{24}$ $\frac{5}{6}$

3 $\frac{5}{7}$ $\frac{2}{3}$ $\frac{5}{9}$ $\frac{2}{5}$

4 1.989   2.07   2.1   2.34   2.39

5 $\frac{35}{6}$ $5\frac{7}{12}$ $5\frac{1}{4}$ $\frac{10}{3}$

6 Abdullah, Sven, Jackson

7 Rufus, Patch, Barney, Millie

8 0.13 $\frac{1}{2}$ 0.6   0.64 $\frac{7}{10}$

Stretch

1 7.5 m
2 46.86 cm

3 a $3\frac{3}{10}$ 3.95   4.2 $4\frac{3}{4}$
  b The smallest number, $3\frac{3}{10}$, is closer to 4 than $4\frac{3}{4}$

Chapter 2.3

Are you ready? (A)

1 a $\frac{2}{7} + \frac{3}{7} = \frac{5}{7}$ **b** $\frac{4}{5} - \frac{3}{5} = \frac{1}{5}$

2 a $\frac{4}{5}$ **b** $\frac{7}{9}$ **c** $\frac{8}{11}$
  d $\frac{5}{7}$ **e** $\frac{3}{8}$ **f** $\frac{10}{13}$

3 a $\frac{13}{5}$ **b** $\frac{29}{6}$ **c** $\frac{51}{8}$

4 a $1\frac{2}{5}$ **b** $1\frac{1}{3}$ **c** $2\frac{1}{8}$

5 a 12   **b** 20   **c** 24

Practice (A)

1 a $\frac{8}{9}$ **b** $\frac{2}{3}$ **c** $\frac{7}{12}$
  d $\frac{1}{3}$ **e** $\frac{1}{2}$ **f** $\frac{1}{6}$

2 a i $1\frac{2}{9}$ **ii** $1\frac{5}{16}$ **iii** $1\frac{1}{15}$
  b i $\frac{7}{8}$ **ii** $\frac{2}{3}$ **iii** $\frac{1}{3}$

3 a $\frac{7}{12}$ **b** $\frac{13}{20}$ **c** $\frac{1}{6}$

  d $\frac{24}{35}$ **e** $\frac{5}{24}$ **f** $\frac{13}{18}$

4 a $1\frac{5}{12}$ **b** $1\frac{9}{40}$ **c** $1\frac{1}{36}$

5 a $5\frac{7}{9}$ **b** $3\frac{5}{8}$ **c** $2\frac{2}{3}$

  d $8\frac{1}{10}$ **e** $8\frac{17}{36}$ **f** $\frac{13}{15}$

6 $1\frac{7}{8}$ cans

7 $1\frac{9}{20}$ km

What do you think?

1 $3\frac{1}{4}$ m

2 $\frac{13}{20}$ m

Are you ready? (B)

1 a 32   **b** 27   **c** 30   **d** 42   **e** 72

2 a $\frac{1}{4}+\frac{1}{4}+\frac{1}{4}=3\times\frac{1}{4}$

  b $\frac{1}{8}+\frac{1}{8}+\frac{1}{8}=3\times\frac{1}{8}$

  c $\frac{2}{3}+\frac{2}{3}+\frac{2}{3}+\frac{2}{3}=4\times\frac{2}{3}$

  d $5\times\frac{1}{6}=\frac{1}{6}+\frac{1}{6}+\frac{1}{6}+\frac{1}{6}+\frac{1}{6}$

  e $4\times\frac{3}{4}=\frac{3}{4}+\frac{3}{4}+\frac{3}{4}+\frac{3}{4}$

3 a $\frac{3}{20}$ **b** $\frac{4}{9}$

Practice (B)

1 a $\frac{4}{7}$ **b** $\frac{2}{3}$ **c** $\frac{5}{9}$

  d $\frac{6}{7}$ **e** $\frac{2}{3}$ **f** $\frac{4}{5}$

2 a $1\frac{1}{3}$ **b** $2\frac{2}{3}$ **c** $2\frac{1}{4}$

  d $1\frac{1}{5}$ **e** $1\frac{1}{2}$ **f** $4\frac{1}{2}$

3 a $\frac{1}{6}$ **b** $\frac{1}{12}$ **c** $\frac{1}{25}$ **d** $\frac{1}{28}$

4 a $\frac{6}{35}$ **b** $\frac{25}{72}$ **c** $\frac{3}{10}$ **d** $\frac{1}{2}$

5 a $1\frac{2}{3}$ **b** $2\frac{5}{8}$ **c** $4\frac{1}{12}$

  d $2\frac{4}{5}$ **e** $7\frac{1}{5}$ **f** $31\frac{8}{9}$

6 $2\frac{2}{5}$ tins

7 $30\frac{7}{8}$ m²

Are you ready? (C)

1 a $1\frac{2}{5}$ **b** $2\frac{3}{4}$ **c** $3\frac{1}{3}$

2 a $\frac{7}{3}$ **b** $\frac{27}{4}$ **c** $\frac{24}{7}$

3 a $1\frac{2}{5}$ **b** $3\frac{3}{4}$ **c** $4\frac{1}{2}$ **d** $2\frac{2}{5}$

  e $\frac{1}{24}$ **f** $\frac{5}{8}$ **g** $4\frac{1}{12}$ **h** $3\frac{47}{54}$

4 a $\frac{1}{6}$ **b** 4   **c** $\frac{1}{11}$

  d $\frac{5}{2}$ **e** $\frac{7}{3}$ **f** $\frac{1}{21}$

Practice (C)

1 a i 4   **ii** 12   **iii** 48

  b i 4   **ii** 12   **iii** 48

2 a i 5×3 **ii** 9×6 **iii** 13×5

  b i 15   **ii** 54   **iii** 65

3 a 6   **b** 10   **c** 18

  d 6   **e** 32   **f** 14

4 a $\frac{21}{2}$ **b** $\frac{25}{3}$ **c** $\frac{27}{2}$

  d $\frac{48}{5}$ **e** $\frac{120}{7}$ **f** $\frac{49}{3}$

5 a $\frac{16}{15}$ **b** $\frac{9}{5}$ **c** $\frac{16}{15}$

  d $\frac{27}{25}$ **e** $\frac{9}{14}$ **f** $\frac{9}{22}$

6 a $1\frac{1}{15}$ **b** $3\frac{1}{5}$ **c** $4\frac{2}{7}$

  d $2\frac{5}{8}$ **e** $1\frac{31}{32}$ **f** $2\frac{6}{11}$

7 a $\frac{28}{3}$ or $9\frac{1}{3}$ **b** $\frac{39}{5}$ or $7\frac{4}{5}$ **c** $\frac{40}{7}$ or $5\frac{5}{7}$

  d $\frac{33}{5}$ or $6\frac{3}{5}$ **e** $\frac{32}{35}$ **f** $\frac{145}{24}$ or $6\frac{1}{24}$

8 a $\frac{8}{5}$ or $1\frac{3}{5}$ **b** $\frac{51}{22}$ or $2\frac{7}{22}$ **c** $\frac{10}{7}$ or $1\frac{3}{7}$

  d $\frac{81}{46}$ or $1\frac{35}{46}$ **e** $\frac{67}{20}$ or $3\frac{7}{20}$ **f** $\frac{155}{108}$ or $1\frac{47}{108}$

Consolidate

1 a i $1\frac{1}{4}$ **ii** $1\frac{1}{3}$ **iii** $1\frac{9}{20}$

  b i $\frac{5}{8}$ **ii** $\frac{1}{2}$ **iii** $\frac{1}{4}$

2 a $\frac{13}{14}$ **b** $\frac{5}{12}$ **c** $\frac{5}{36}$ **d** $1\frac{1}{12}$

3 a $6\frac{7}{9}$ **b** $3\frac{3}{8}$ **c** $2\frac{11}{12}$ **d** $10\frac{13}{30}$

4 a $2\frac{1}{4}$ **b** $2\frac{3}{5}$ **c** $1\frac{3}{5}$ **d** $5\frac{1}{4}$

5 a $\frac{8}{15}$ **b** $\frac{5}{24}$ **c** $\frac{6}{35}$ **d** $\frac{49}{90}$

6 a $1\frac{7}{8}$ **b** $2\frac{2}{9}$ **c** $11\frac{3}{8}$

  d $3\frac{3}{10}$ **e** $9\frac{8}{15}$ **f** $18\frac{13}{32}$

7 a 8   **b** 10   **c** 8   **d** 21

8 a $2\frac{1}{4}$ **b** $1\frac{2}{3}$ **c** $3\frac{8}{9}$ **d** $1\frac{1}{5}$

9 a $5\frac{1}{4}$ **b** $6\frac{1}{2}$ **c** $8\frac{1}{16}$ **d** $6\frac{1}{24}$

  e $1\frac{19}{20}$ **f** $5\frac{5}{7}$ **g** $2\frac{5}{56}$ **h** $3\frac{1}{5}$

Stretch

1 $\frac{27}{100}$

2 a $2\frac{22}{25}$ **b** $\frac{25}{96}$

3 9 tins are required, so it will cost £45

Chapter 2.4

Are you ready? (A)

1 a 587   **b** 472   **c** 639
   d 643   **e** 6201   **f** 4406
2 a 351   **b** 218   **c** 337
   d 148   **e** 5439   **f** 267
3 6.4 + 2.5 = 8.9   8.9 − 6.4 = 2.5
   2.5 + 6.4 = 8.9   8.9 − 2.5 = 6.4
4 a 130   **b** 13 000   **c** 600   **d** 60
5 a 0.7   **b** 0.05   **c** 0.8   **d** 0.13

Practice (A)

1 a 85.23   **b** 5.81   **c** 8.22
   d 9.25   **e** 15.01   **f** 1.406
2 a 5.2   **b** 15.31   **c** 3.17
   d 3.03   **e** 4.19   **f** 0.148
3 a 12.03   **b** 2.83   **c** 26.002
   d 360.47   **e** 32.78   **f** 6.787
4 16.1 cm
5 a £3.10   **b** £6.90
6 Building B: 25.06 m
   Building C: 23.22 m
7 $x = 7.15$

What do you think? (A)

1 a

$$
\begin{array}{r}
3 \cdot 8 \boxed{5} \\
+\; 5 \cdot \boxed{1}\, 6 \\
\hline
\boxed{9} \cdot 0 \quad 1
\end{array}
$$

b

$$
\begin{array}{r}
\boxed{4}\; 6 \cdot 3 \\
-\; 2 \quad \boxed{0} \cdot 8 \\
\hline
2 \quad 5 \cdot \boxed{5}
\end{array}
$$

2 3.1

Are you ready? (B)

1 a i 210   **ii** 21   **iii** 2.1   **iv** 20.1
   b i 2.1   **ii** 0.21   **iii** 0.021   **iv** 0.201
2

×	4	8	6
2	8	16	12
7	28	56	42
3	12	24	18

3 a 180   **b** 1800   **c** 18 000
   d 30   **e** 3   **f** 60
4 a i 6   **ii** 6   **iii** 6
   b i 3   **ii** 3   **iii** 3

Practice (B)

1 a 19.2   **b** 19.2   **c** 0.192   **d** 0.192
2 a 15572   **b** 15.572
3 a 6.9   **b** 1.44   **c** 147.2
   d 565.8   **e** 4.08   **f** 3.888
4 a 0.7   **b** 0.07   **c** 0.4   **d** 7
5 a 123   **b i** 1230   **ii** 123   **iii** 12.3
6 a 1260   **b** 23 100   **c** 245
   d 462   **e** 38   **f** 1450
7 530.6 g
8 £389.85

What do you think? (B)

1 a 22.09 cm²   **b** 10.54 mm²
2 28 strips of wallpaper

Consolidate

1 a 10.75   **b** 11.36   **c** 78.66
   d 39.76   **e** 2.38   **f** 4.72
2 a 24.4   **b** 28.2   **c** 11.18
   d 0.456   **e** 44.226   **f** 52.725
3 a 1.53   **b** 0.237   **c** 2.14   **d** 0.306
4 a 5620   **b** 2160   **c** 304
   d 51.3   **e** 143   **f** 206

5 £69.15
6 £282.36
7 £9.05

Stretch

1 45.9 mm
2 a 11.976   **b** 485
3 1.48 cm²

Fractions and decimals: exam practice

1 a 0.5   **b** 30%   **c** $\frac{67}{100}$
2 $\frac{9}{20}$
3 0.8
4

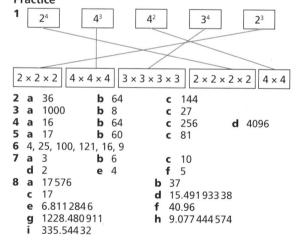

5 $\frac{3}{7}$
6 $\frac{2}{5}$
7 a $\frac{43}{35} = 1\frac{8}{35}$ **b** $\frac{4}{18} = \frac{2}{9}$
8 a $\frac{2}{5}, \frac{1}{2}, \frac{2}{3}, \frac{3}{4}$ **b** $\frac{1}{2}, \frac{3}{5}, 70\%, \frac{8}{10}, 0.82$
9 43.18 cm²

Block 3 Moving on with number

Chapter 3.1

Are you ready?

1 a i 9 cm²   **ii** 25 cm²
   b i 9   **ii** 25
2 a i 27 cubes   **ii** 125 cubes
   b i $3^3 = 3 \times 3 \times 3 = 27$ **ii** $5^3 = 5 \times 5 \times 5 = 125$
3 a 625   **b** 243   **c** 343

Practice

1

2^4 4^3 4^2 3^4 2^3

$2 \times 2 \times 2$ $4 \times 4 \times 4$ $3 \times 3 \times 3 \times 3$ $2 \times 2 \times 2 \times 2$ 4×4

2 a 36   **b** 64   **c** 144
3 a 1000   **b** 8   **c** 27
4 a 16   **b** 64   **c** 256   **d** 4096
5 a 17   **b** 60   **c** 81
6 4, 25, 100, 121, 16, 9
7 a 3   **b** 6   **c** 10
   d 2   **e** 4   **f** 5
8 a 17 576   **b** 37
   c 17   **d** 15.491 933 38
   e 6.811 284 6   **f** 40.96
   g 1228.480 911   **h** 9.077 444 574
   i 335.544 32

What do you think?

1 No, if you cube a number less than 1 the answer will be smaller, e.g. $0.2^3 = 0.008$

Consolidate

1 a 16   **b** 49   **c** 121
   d 1   **e** 81   **f** 169
2 a 1   **b** 125   **c** 64
   d 8   **e** 1000   **f** 729
3 a 36   **b** 216   **c** 1296   **d** 7776
4 a 43   **b** 89   **c** 145

5 8, 64, 1, 27
6 a 12   **b** 7   **c** 1
   d 6   **e** 7   **f** 10
7 a 29 791   **b** 27   **c** 16

Stretch

1 4 and 6
2 100 and 125
3 3^2 2^3 $\sqrt[3]{125}$ $\sqrt{16}$
4 49, 64 and 81

Chapter 3.2

Are you ready?

1 a i 10   **ii** 10
   b i −4   **ii** −4
2 a i 16   **ii** 16
   b i 24   **ii** 24
3 a $3 + 4 \times 2 = 11$ **b** $(3 + 4) \times 2 = 14$
4 a

   b

Practice

1 a 11   **b** 11   **c** 11
   d 19   **e** 5   **f** 5
2 a 36   **b** 36   **c** 36
   d 81   **e** 81   **f** 1
3 a 3   **b** 45   **c** 123
   d 63   **e** 19   **f** 5
4 $25 + 3 \times 4$
5 a 53   **b** 83   **c** 95
   d 7   **e** 63   **f** 149
6 a Rhys has calculated from left to right without taking into account the correct order of operations.
   b 32
7 a 34   **b** 52   **c** 40
   d 20   **e** 67   **f** 22
8 a $(6 + 4) \times 2 + 3 = 23$ **b** $6 + 4 \times 2 + 3 = 17$
   c $(6 + 4) \times (2 + 3) = 50$ **d** $6 + 4 \times (2 + 3) = 26$

What do you think?

1 a Addition and subtraction have equal priority. If a calculation has only additions and subtractions, just work from left to right.
   b 10
2 a $5 + 3^2 \times (4 − 2) = 23$
   b Various possible answers, e.g. $(5 + 3)^2 \times 4 − 2 = 254$

Consolidate

1 a 10   **b** 10   **c** 10   **d** 10   **e** 18
   f 18   **g** 8   **h** 8   **i** 2
2 a 3   **b** 57   **c** 48
   d 3   **e** 33   **f** 15
3 a 67   **b** 26   **c** 190
   d 90   **e** 62   **f** 62
4 a 47   **b** 64   **c** 49
   d 24   **e** 14   **f** 21
5 $(24 − 3) \times 6$
6 a 38   **b** 23   **c** 35   **d** 80
7 a $8 + 12 \div (4 − 2) = 14$ **b** $(8 + 12) \div 4 − 2 = 3$
   c $8 + 12 \div 4 − 2 = 9$ **d** $(8 + 12) \div (4 − 2) = 10$

8 Marta has not taken into account the order of operations.
$950 + 50 \times 7 = 950 + 350 = 1300$
Their pay is £1300

Stretch

1 Various possible answers for each number, e.g.
$2 = (4 − 3) \times 2 \times 1$
And $16 = 4 \times (3 − 1 + 2)$

Chapter 3.3

Are you ready? (A)

1 a 100   **b** 49   **c** 32   **d** 125
2 10^4
3

10^4	10 000
10^3	1000
10^2	100
10^1	10
10^0	1
10^{-1}	0.1
10^{-2}	0.01
10^{-3}	0.001
10^{-4}	0.0001

Practice (A)

1 a 6×10^4 **b** 9×10^5 **c** 4×10^3
   d 5×10^8 **e** 2×10^9 **f** 7×10^6
2 a 6.3×10^5 **b** 1.9×10^4 **c** 3.5×10^3
   d 2.1×10^8 **e** 1.57×10^6 **f** 7.29×10^5
3 a 3×10^{-1} **b** 3×10^{-2} **c** 7×10^{-5}
   d 4×10^{-4} **e** 9×10^{-8} **f** 6×10^{-7}
4 a 3.2×10^{-1} **b** 3.2×10^{-2} **c** 6.4×10^{-4}
   d 2.14×10^{-2} **e** 9.08×10^{-4} **f** 2.067×10^{-3}
5 No, the correct answer is 2.34×10^6
A should always be a number between 1 and 10
6 1.5×10^8 km
7 7×10^{-9} mm³

Are you ready? (B)

1 a 1000   **b** 100 000   **c** 1
2 a 0.001   **b** 0.000 001   **c** 0.000 000 000 1
3 a 3200   **b** 4000   **c** 105 000
   d 0.004   **e** 0.046   **f** 0.000 030 2

Practice (B)

1 a 20 000   **b** 6 000 000   **c** 3000
   d 40 000 000   **e** 9 000 000 000   **f** 50
2 a 24 000   **b** 510 000   **c** 330
   d 7 410 000   **e** 1080   **f** 52 500 000
3 a 0.0002   **b** 0.000 006   **c** 0.009
   d 0.000 03   **e** 0.08   **f** 0.000 000 000 1
4 a 0.000 26   **b** 0.0045   **c** 0.000 001 9
   d 0.000 032 4   **e** 0.0801   **f** 0.000 000 030 56
5 No, the correct answer is 740. Jakub has made the mistake of just adding zeros.
6 21 300 000 m

Are you ready? (C)

1 a 3×10^4 is greater than 9×10^3
As ordinary numbers, they are 30 000 and 9000
30 000 > 9000
You can also compare the indices.
$10^4 > 10^3$

b 4×10^{-2} is greater than 2×10^{-6}
  As ordinary numbers, they are 0.04 and 0.000 002
  0.04 > 0.000 002
  You can also compare the indices.
  $10^{-2} > 10^{-6}$

2 a 3006   **b** 9000
3 a 10^5 **b** 10^6 **c** 10^6
   d 10^1 **e** 10^3 **f** 10^{-1}

Practice (C)

1 a 7×10^3 **b** 9×10^5 **c** 7.7×10^2
   d 4×10^3 **e** 1×10^5 **f** 1.3×10^4
2 a 4.3×10^4 **b** 3.05×10^5 **c** 6.42×10^3
   d 6.7×10^3 **e** 7.94×10^5 **f** 4.6×10^6
3 a 6×10^6 **b** 1.2×10^7 **c** 1.2×10^7
   d 8.4×10^8 **e** 2×10^2 **f** 3.3×10^2
4 a 6 000 000   **b** 12 000 000   **c** 12 000 000
   d 840 000 000   **e** 200   **f** 330

What do you think?

1 a $3 \times 10^3 + 5 \times 10^3 = 8 \times 10^3$
   b $(3.4 \times 10^3) \times (2 \times 10^2) = 6.8 \times 10^5$
   c $(9 \times 10^2) \div (3 \times 10^4) = 3 \times 10^{-2}$

Consolidate

1 a 3×10^4 **b** 1.4×10^4 **c** 4.2×10^3
   d 2.3×10^7 **e** 2.59×10^6 **f** 1.04×10^5
2 a 2×10^{-1} **b** 6×10^{-2} **c** 5×10^{-5}
   d 4.1×10^{-3} **e** 8.9×10^{-7} **f** 6.01×10^{-4}
3 a 40 000   **b** 50 000 000   **c** 3100
   d 624 000   **e** 30 500   **f** 2 220 000
4 a 0.004   **b** 0.000 07   **c** 0.8
   d 0.000 31   **e** 0.0127   **f** 0.000 003 08
5 a 7×10^3 **b** 3×10^6 **c** 9.3×10^4
   d 6×10^4 **e** 4×10^5 **f** 2.7×10^3
6 a 5.002×10^5 **b** 2.005×10^5 **c** 3.83×10^3
   d 8.7×10^4 **e** 5.98×10^4 **f** 3.1×10^3
7 a 8×10^8 **b** 1.5×10^7 **c** 1.5×10^7
   d 6.8×10^6 **e** 2×10^0 **f** 4.3×10^3

Stretch

1 90
2 a Mercury, Earth, Uranus, Saturn   **b** 12 800 000 m
   c 12 800 km   **d** 6.99×10^7 m
3 4.987×10^2 seconds

Moving on with number: exam practice

1 a 16   **b** 6   **c** 2^5
2 19
3 49
4 $2^3 = 2 \times 2 \times 2 = 8$
   Sophie has incorrectly done $2 \times 3 = 6$
5 89
6 a Ron because multiplication should be done first.
   b 25   **c** $(9 + 2) \times (6 - 3) = 33$
7 a 4.57×10^4 **b** 3×10^{-5} **c** 7×10^6
8 10 cm
9 a 410 000   **b** 0.000 006 04
10 3×10^{12}

Block 4 Factors, multiples and primes

Chapter 4.1

Are you ready?

1 a 1, 2, 3, 6, 9, 18
   b 1, 5, 25
   c 1, 2, 3, 4, 6, 8, 12, 16, 24, 48
   d 1, 2, 3, 4, 6, 7, 12, 14, 21, 28, 42, 84
2 a 4, 8, 12, 16, 20   **b** 8, 16, 24, 32, 40
   c 12, 24, 36, 48, 60   **d** 23, 46, 69, 92, 115
3 a 12 and 30   **b** 64 and 128

Practice

1 a Factors of 24: 1, 2, 3, 4, 6, 8, 12, 24
   Factors of 42: 1, 2, 3, 6, 7, 14, 21, 42
   b 1, 2, 3, 6
   c 6
2 a 18   **b** 12   **c** 24
   d 40   **e** 14   **f** 18
3 a Multiples of 3: 3, 6, 9, 12, 15, 18, 21, 24, 27, 30
   Multiples of 4: 4, 8, 12, 16, 20, 24, 28, 32, 36, 40
   b 12
4 a Multiples of 9: 9, 18, 27, 36, 45, 54, 63, 72, 81, 90
   Multiples of 6: 6, 12, 18, 24, 30, 36, 42, 48, 54, 60
   b 18
5 a 15   **b** 24   **c** 60
   d 72   **e** 144   **f** 60
6 24 seconds

What do you think?

1 3:29 pm
2 No, Flo is incorrect. Sometimes the LCM can be found by multiplying the two numbers together e.g. the LCM of 3 and 4 is 12, but sometimes the LCM is not the product of the two numbers e.g. the LCM of 3 and 6 is 6

Consolidate

1 a 14   **b** 15   **c** 21
   d 25   **e** 27   **f** 30
2 a 12   **b** 36   **c** 175
   d 42   **e** 48   **f** 180
3 24 days
4 7 packs of pens and 13 packs of pencils so a total of 91 pens and 91 pencils
5 a 120 minutes   **b** 12:50 pm
6 6 metres

Stretch

1 a 4:54 pm   **b** 4:36 pm   **c** 5:48 pm
2 $a = 15$, $b = 6$, $c = 14$
   $a = 30$, $b = 3$, $c = 7$

Chapter 4.2

Are you ready? (A)

1 2, 3, 5, 7, 11, 13
2 a 1, 2, 3, 4, 6, 12
   b 1, 2, 3, 6, 9, 18
   c 1, 2, 4, 8, 16, 32
   d 1, 2, 3, 4, 6, 7, 12, 14, 21, 28, 42, 84
3 a 12   **b** 45   **c** 42   **d** 49
4 a 2^3 **b** 5^2 **c** 3^4

Practice (A)

1 a $30 = 2 \times 3 \times 5$
   b $28 = 2 \times 2 \times 7$
   c $90 = 2 \times 3 \times 3 \times 5$

2 a   84

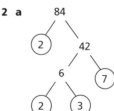

   b $84 = 2 \times 2 \times 3 \times 7$
3 a $48 = 2 \times 2 \times 2 \times 2 \times 3$ **b** $18 = 2 \times 3 \times 3$
   c $25 = 5 \times 5$ **d** $90 = 2 \times 3 \times 3 \times 5$
   e $68 = 2 \times 2 \times 17$ **f** $160 = 2 \times 2 \times 2 \times 2 \times 2 \times 5$
4 a $48 = 2^4 \times 3$ **b** $50 = 2 \times 5^2$
   c $49 = 7^2$ **d** $96 = 2^5 \times 3$
   e $140 = 2^2 \times 5 \times 7$ **f** $248 = 2^3 \times 31$

5 a 12   **b** 45   **c** 42
   d 99   **e** 126   **f** 200
6 a Zach has included 9, which is not a prime number.
   b $2 \times 2 \times 3 \times 3 \times 5$

What do you think? (A)
1 a $48 = 2^4 \times 3$ **b** $240 = 2^4 \times 3 \times 5$
   c $480 = 2^5 \times 3 \times 5$ **d** $120 = 2^3 \times 3 \times 5$

Are you ready? (B)
1 a $2 \times 2 \times 3$ **b** $2 \times 2 \times 3 \times 3$
2 a 12   **b** 36
3

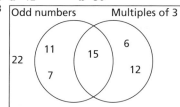

Practice (B)
1 a 12   **b** 72
2 a 30   **b** 220   **c** 10   **d** 660
3 a $18 = 2 \times 3 \times 3$
   b $60 = 2 \times 2 \times 3 \times 5$
   c

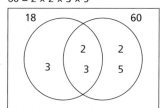

   d 6
   e 180
4 a i $16 = 2 \times 2 \times 2 \times 2, 48 = 2 \times 2 \times 2 \times 2 \times 3$
    ii HCF = 16, LCM = 48
  b i $24 = 2 \times 2 \times 2 \times 3, 60 = 2 \times 2 \times 3 \times 5$
    ii HCF = 12, LCM = 120
  c i $45 = 3 \times 3 \times 5, 120 = 2 \times 2 \times 2 \times 3 \times 5$
    ii HCF = 15, LCM = 360
  d i $40 = 2 \times 2 \times 2 \times 5, 75 = 3 \times 5 \times 5$
    ii HCF = 5, LCM = 600
  e i $110 = 2 \times 5 \times 11, 115 = 5 \times 23$
    ii HCF = 5, LCM = 2530
  f i $96 = 2 \times 2 \times 2 \times 2 \times 2 \times 3, 132 = 2 \times 2 \times 3 \times 11$
    ii HCF = 12, LCM = 1056

What do you think? (B)
1 $y = 2^2 \times 3 \times 7$

Consolidate
1 a i 54   **ii** 126

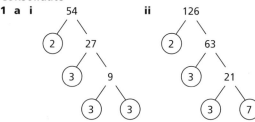

   b $54 = 2 \times 3 \times 3 \times 3$
      $126 = 2 \times 3 \times 3 \times 7$
2 a $24 = 2 \times 2 \times 2 \times 3$
   b $36 = 2 \times 2 \times 3 \times 3$
   c $72 = 2 \times 2 \times 2 \times 3 \times 3$
   d $80 = 2 \times 2 \times 2 \times 2 \times 5$
   e $112 = 2 \times 2 \times 2 \times 2 \times 7$
   f $320 = 2 \times 2 \times 2 \times 2 \times 2 \times 2 \times 5$

3 a $24 = 2^3 \times 3$ **b** $60 = 2^2 \times 3 \times 5$
   c $84 = 2^2 \times 3 \times 7$ **d** $125 = 5^3$
   e $250 = 2 \times 5^3$ **f** $360 = 2^3 \times 3^2 \times 5$
4 a 18   **b** 105   **c** 44
   d 90   **e** 825   **f** 675
5 a 18   **b** 84   **c** 6   **d** 252
6 a $16 = 2 \times 2 \times 2 \times 2$
   b $80 = 2 \times 2 \times 2 \times 2 \times 5$
   c

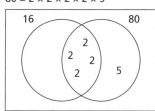

   d 16
   e 80
7 a i $21 = 3 \times 7, 56 = 2 \times 2 \times 2 \times 7$
    ii HCF = 7, LCM = 168
  b i $32 = 2 \times 2 \times 2 \times 2 \times 2, 40 = 2 \times 2 \times 2 \times 5$
    ii HCF = 8, LCM = 160
  c i $55 = 5 \times 11, 100 = 2 \times 2 \times 5 \times 5$
    ii HCF = 5, LCM = 110
  d i $60 = 2 \times 2 \times 3 \times 5, 85 = 5 \times 17$
    ii HCF = 5, LCM = 1020
  e i $120 = 2 \times 2 \times 2 \times 3 \times 5, 125 = 5 \times 5 \times 5$
    ii HCF = 5, LCM = 3000
  f i $86 = 2 \times 43, 146 = 2 \times 73$
    ii HCF = 2, LCM = 6278

Stretch
1 a 6 is because $2 \times 3 = 6$
   b 15 is because $3 \times 5 = 15$
   c 65 is not because 65 cannot be made by
     multiplying any of the prime factors.
   d 45 is because $3 \times 3 \times 5 = 45$
   e 140 is not because 140 cannot be made by
     multiplying any of the prime factors.
2 a is 3 times greater than b because a and b share all
   the same prime factors apart from a has an extra
   3 as a prime factor.
3 a $2^2 \times 5$ **b** $2^4 \times 5$
   c $2^3 \times 5^2$ **d** $2^4 \times 5^2$

Factors, multiples and primes: exam practice
1 a 23   **b** 8 or 32   **c** 15
2 2, 3, 5, 7, 11
3 No, because prime numbers have exactly two factors.
   1 only has one factor, itself.
4 1, 2, 3, 5, 6, 10, 15, 30
5 a 60   **b** 2
6 $3845 \div 5 = 769$, therefore it has more than two factors
   and is not prime.
7 350
8 Three times (including the first one)
9 a $2 \times 2 \times 2 \times 3 \times 5$ or $2^3 \times 3 \times 5$ **b** 5

Block 5 Percentages

Chapter 5.1

Are you ready?
1 a $\frac{1}{2} = \frac{50}{100}$ **b** $\frac{1}{5} = \frac{20}{100}$ **c** $\frac{4}{5} = \frac{80}{100}$ **d** $\frac{21}{25} = \frac{84}{100}$

2 a 0.1   **b** 0.7   **c** 0.01   **d** 0.19

3 a $\frac{9}{10}$ **b** $\frac{1}{5}$ **c** $\frac{13}{20}$ **d** $\frac{9}{50}$

4 a 0.125   **b** 0.625   **c** 0.175   **d** 0.875
5 a 2.3   **b** 3.23   **c** 5.25   **d** 10.6

Practice

1 **a** 37% **b** 12% **c** 2%
   d 20% **e** 50% **f** 90%
2 **a** 33% **b** 9% **c** 40%
   d 20% **e** 54% **f** 75%
3 **a** 50% **b** 25% **c** 12.5%
   d 37.5% **e** 62.5% **f** 87.5%
4 **a** 23% **b** 23.6% **c** 23.6%
   d 50.2% **e** 13.5% **f** 40.2%
5 **a** **i** 0.16 **ii** 0.55 **iii** 0.85
   b **i** 16% **ii** 55% **iii** 85%
6 **a** 0.23 **b** 0.98 **c** 0.9
   d 0.04 **e** 0.09 **f** 0.1
7 **a** $\frac{21}{100}$ **b** $\frac{1}{10}$ **c** $\frac{1}{2}$

   d $\frac{3}{4}$ **e** $\frac{4}{5}$ **f** $\frac{7}{10}$

8 **a** 25% **b** 0.25 **c** $\frac{1}{4}$

9 **a** 20% **b** 65% **c** 25%

What do you think?

1 $20\% = \frac{20}{100} = \frac{1}{5}$

2 $\frac{3}{5} = 60\%$, so Jackson did better as 60% > 55%

3 **a** 45% **b** $\frac{9}{20}$

Consolidate

1 **a** 21% **b** 86% **c** 6%
   d 60% **e** 10% **f** 1%
2 **a** 67% **b** 7% **c** 30%
   d 25% **e** 84% **f** 80%
3 **a** 0.85, 85% **b** 0.55, 55% **c** 0.325, 32.5%
4 **a** 0.41 **b** 0.99 **c** 0.8
   d 0.08 **e** 0.02 **f** 0.3
5 **a** $\frac{37}{100}$ **b** $\frac{3}{10}$ **c** $\frac{3}{5}$

   d $\frac{1}{4}$ **e** $\frac{7}{100}$ **f** $\frac{1}{20}$

6

Fraction	Decimal	Percentage
$\frac{4}{5}$	0.8	80%
$\frac{73}{100}$	0.73	73%
$\frac{9}{20}$	0.45	45%

7 **a** $\frac{1}{5} > 5\%$ **b** $0.9 > \frac{9}{100}$

   c $0.7 = 70\%$ **d** $40\% < \frac{3}{4}$

8 **a** 90% **b** 48% **c** 37.5%

9 $\frac{1}{4}$

10 45%

Stretch

1 **a** 0.333…, 33.33% or 33.$\dot{3}$%
   b 0.666…, 66.66% or 66.$\dot{6}$%

2 **a** $\frac{1}{8}$ **b** $\frac{111}{500}$ **c** $\frac{999}{1000}$

3

Fraction	Decimal	Percentage
$1\frac{13}{100}$	1.13	113%
$1\frac{1}{4}$	1.25	125%
$1\frac{1}{2}$	1.5	150%

4 Any percentage greater than 125% and less than 130%

Chapter 5.2

Are you ready? (A)

1 **a** $\frac{1}{5}$ **b** $\frac{3}{5}$ **c** $\frac{3}{10}$ **d** $\frac{5}{8}$

2 **a** 8 **b** 4 **c** 8 **d** 12
3 **a** 24 **b** 35 **c** 36 **d** 42
4 **a** 24 **b** 42 **c** 56

Practice (A)

1 **a** 3 **b** 7 **c** 7
   d 12 **e** 12 **f** 22
2 **a** 9 **b** 9 **c** 35
   d 40 **e** 66 **f** 91
3 **a** 188 **b** 236 **c** 708
   d 375 **e** 42.5 **f** 156.6
4 £26
5 3
6 **a** < **b** = **c** > **d** <
7 **a** More adults supported the red team
   b More children supported the blue team
8 £15

What do you think?

1 Sven is incorrect.
   Original cost = £48

   After $\frac{1}{4}$ increase = £60

   After $\frac{1}{4}$ decrease = £45

2 After 4 days Huda will be able to afford the coat as it will cost £131.22

Are you ready? (B)

1 **a** 5 **b** 8 **c** 4 **d** 2
2 **a** 3; 3 **b** 4; 4 **c** 7; 7
3 **a** 138 **b** 642 **c** 693.75
4 **a** 9 **b** 13 **c** 18 **d** 35

Practice (B)

1 30
2 **a** 12 **b** 24 **c** 36 **d** 78
3 **a** 3 **b** 6 **c** 8
4 15
5 **a** 12 **b** 20 **c** 36 **d** 28
6 **a** £84 **b** £36
7 **a** 5 **b** 15 **c** 25 **d** 28
8 18
9 **a** 9 **b** 25 **c** 49

Consolidate

1 **a** 8 **b** 7 **c** 12
   d 18 **e** 84 **f** 40
2 **a** 159 **b** 124 **c** 531
   d 1092 **e** 164.25 **f** 298.4
3 £51
4 4
5 **a** > **b** > **c** < **d** >
6 **a** 21 **b** 30 **c** 36 **d** 56
7 £33
8 **a** 10 **b** 9 **c** 22 **d** 54
9 £27

Stretch

1 £340
2 £350
3 £44.10

Chapter 5.3

Are you ready? (A)

1 **a** $\frac{1}{2}$ **b** $\frac{1}{4}$ **c** $\frac{1}{10}$

   d $\frac{1}{100}$ **e** $\frac{3}{4}$ **f** $\frac{1}{5}$

2 a 200 **b** 100 **c** 40
  d 4 **e** 300 **f** 80
3 a $\frac{1}{3}$ **b** $\frac{2}{3}$
4 a 10% **b** 80% **c** 85%

Practice (A)

1 a 12 **b** 150 **c** 6 **d** 90
  e 24 **f** 2.4 **g** 3 **h** 2.4
2 a 18 **b** 69 **c** 20
  d 80 **e** 40 **f** 160
3 a 6 **b** 18 **c** 54
  d 42 **e** 12 **f** 0.6
4 a 20 **b** 2 **c** 10
  d 30 **e** 90 **f** 94
5 a 372 **b** 868
6 £3960
7 a < **b** < **c** > **d** =
8 £1625

What do you think? (A)

1 a 475
  b Multiple methods, e.g. work out 75% + 20% or
    1% × 95

Are you ready? (B)

1 a 0.63 **b** 0.89 **c** 0.91
  d 0.04 **e** 0.4 **f** 1.09
2 a 552 **b** 480 **c** 72
  d 41.28 **e** 38.4 **f** 2.88

Practice (B)

1 a 0.54 **b** 0.76 **c** 0.08
  d 0.6 **e** 0.428 **f** 1.04
2 a 8800 **b** 226.8 **c** 137.16
  d 59.16 **e** 81.84 **f** 0.288
3 a 60% **b** 80% **c** 70% **d** 45%
4 £936
5 £5.85
6 a £65840 **b** £69680
7 2.43 m

Are you ready? (C)

1 a 0.76 **b** 0.7 **c** 0.06
2 a 1.04 **b** 1.25 **c** 1.8
3 29%
4 a 0.3 **b** 0.18 **c** 0.91

Practice (C)

1 a 80% **b** 0.8 **c** £384
2 a 0.6 **b** 0.75 **c** 0.25
  d 0.49 **e** 0.93 **f** 0.05
3 a Rhys has incorrectly written the answer – it would
    be £22.40
    Rhys has also used the incorrect multiplier, he
    should have used 0.72
  b £57.60
4 £202440
5 a 120% **b** 1.2 **c** 288
6 a 1.4 **b** 1.25 **c** 1.07
  d 1.51 **e** 1.075 **f** 1.502
7 80640
8 a Increase of 8% **b** Decrease of 13%
  c Increase of 40% **d** Decrease of 30%
  e Increase of 65% **f** Decrease of 96%

What do you think? (C)

1 No, Benji is incorrect.
  In April, the train ticket will cost £12.50 × 1.2 = £15
  In November, the train ticket will cost £15 × 0.8 = £12

Consolidate

1 a 28 **b** 33 **c** 205 **d** 40
2 a 27 **b** 280 **c** 320
  d 216 **e** 240 **f** 36

3 a 9600 **b** 14400
4 a 92% **b** 90% **c** 65% **d** 40%
5 a 9342 **b** 165.6 **c** 311.6
  d 76.26 **e** 51.84 **f** 124.08
6 £5460
7 a i 1.2 **ii** 1.45 **iii** 1.09
  b i 0.63 **ii** 0.92 **iii** 0.4
8 £199045
9 21624
10 £718.25

Stretch

1 No. If a represents the length and b represents the
  width then the new rectangle will have a length of
  $1.1a$ and a width of $1.1b$. The area will be $1.21ab$
  For example, if the length was 60 and the width 50,
  the new length and width would be 66 and 55 so the
  area would be 3630 and not 3300
2 After five bounces the ball will reach a height of
  0.949… m
3 £364.50

Chapter 5.4

Are you ready? (A)

1 a 18 **b** 28 **c** 6 **d** 216
2 a 1.3 **b** 0.6 **c** 1.08
  d 0.96 **e** 0.73 **f** 1.32
3 a 660 **b** 636 **c** 708 **d** 1020
4 a £96 **b** £75 **c** £32 **d** £31

Practice (A)

1 a £20 **b i** £420 **ii** £440 **iii** £460
2 £896
3 £1400
4 a £2688 **b** £384
5 £1251
6 £1962

What do you think? (A)

1 £3430
2 2%
3 6%

Are you ready? (B)

1 a 1.6 **b** 0.9 **c** 1.03
  d 0.98 **e** 0.81 **f** 1.21
2 a 1.44 **b** 1.16 **c** 5.77
3 a £288 **b** £463.05 **c** £577.35
4 £616

Practice (B)

1 a £8 **b** £408 **c** £8.16 **d** £416.16
2 £3009.83
3 a £3499.20 **b** £3779.14 **c** £4760.62
4 £11993.91
5 £12899.36
6 1.029 metres
7 £717

What do you think? (B)

1 5 minutes
2 4 years

Consolidate

1 a £64 **b i** £864 **ii** £992 **iii** £1184
2 £1680
3 £1152
4 £1608
5 £5314.68
6 a £561.80 **b** £631.24 **c** £796.92
7 £11631.99
8 11 years

Stretch

1 a Super Saver   **b** Gold Saver
2 Simple interest
3 £8399.57

Chapter 5.5

Are you ready?

1 a 0.7   **b** 0.06   **c** 0.82
  d 0.745   **e** 1.3   **f** 1.085
2 a 1.8   **b** 0.2   **c** 0.95
  d 1.03   **e** 1.138   **f** 0.915
3 a 30   30   **b** 45   45
  c 420   420   **d** 60   60

Practice

1 a 48   **b** 24   **c** 36
2 a 8   **b** 80   **c** 56   **d** 12
3 a 40   **b** 16   **c** 60   **d** 34
4 220 students
5 £300
6 £60
7 £575
8 £48
9 £20 000

What do you think?

1

Original amount	Percentage change	New amount
32	Increase by 10%	**35.2**
40	**Decrease by 6%**	37.6
110	Decrease by 40%	66
12	Decrease by 17%	**9.96**
16	**Increase by 60%**	25.6

Consolidate

1 a 170   **b** 17   **c** 102
2 a 92   **b** 46   **c** 9.2   **d** 32.2
3 a 80   **b** 24   **c** 136   **d** 36
4 70 counters
5 £20 000
6 £22.61
7 £243 243.24
8 £18 500
9 £12.00

Stretch

1 £5
2 29.27%
3 £168 000
4 £51 600

Percentages: exam practice

1

Decimal	0.25	0.5	0.75
Fraction	$\frac{1}{4}$	$\frac{1}{2}$	$\frac{3}{4}$
Percentage	25%	50%	75%

2 4
3 30
4 £600
5 0.125
6 a 75%   **b** $\frac{3}{20}$
7 £5061.89
8 £120

Block 6 Accuracy

Chapter 6.1

Are you ready?

1 a 500   **b** 5   **c** 0.5   **d** 0.05
2 a 3750   **b** 3800   **c** 4000
3 a 6.5   **b** 4.15
4 a A: 2.3   B: 2.82   C: 3.39   D: 3.75
  b A: 2   B: 3   C: 3   D: 4

Practice

1 a A: 5.08   B: 5.35   C: 5.72   D: 5.97
  b A: 5.1   B: 5.4   C: 5.7   D: 6.0
2 a i The answer has an extra 0 so is not rounded accurately to 1 d.p.
   ii The value after the first decimal place is 2, not 26
  iii The value has been rounded to the previous tenth, 1, rather than 2
  b i 7.9   **ii** 1.3   **iii** 4.2
3 a 2.3   **b** 10.5   **c** 7.2   **d** 0.1
  e 4.3   **f** 6.0   **g** 7.0   **h** 10.0
4 a A: 8.303   B: 8.329   C: 8.355   D: 8.381
  b A: 8.30   B: 8.33   C: 8.36   D: 8.38
5 a 2.34   **b** 2.35   **c** 8.03   **d** 0.15
  e 0.19   **f** 0.20   **g** 10.06   **h** 9.98

6

Rounded to nearest integer	Rounded to 1 d.p.	Rounded to 2 d.p.	Rounded to 3 d.p.
5.1783	5	5.2	5.18	5.178
11.2056	11	11.2	11.21	11.206
3.0095	3	3.0	3.01	3.010
8.9999	9	9.0	9.00	9.000

7 a 11.16   **b** 2.31   **c** 3.12   **d** 98.08
8 a £1.54 or 154p   **b** £2

Consolidate

1 a 7.8 and 7.9   **b** 7.8
  c i 7.2   **ii** 7.5   **iii** 8.0   **iv** 7.0
2 a 3.1   **b** 20.8   **c** 6.5   **d** 0.7
  e 5.7   **f** 4.0   **g** 12.0   **h** 1.0
3 a 4.83 and 4.84   **b** 4.84
  c i 4.86   **ii** 4.83   **iii** 4.80   **iv** 4.90
4 a 4.56   **b** 1.97   **c** 9.04   **d** 0.16
  e 0.30   **f** 0.29   **g** 100.01   **h** 20.00

5

Rounded to nearest integer	Rounded to 1 d.p.	Rounded to 2 d.p.	Rounded to 3 d.p.
9.90898	10	9.9	9.91	9.909
105.1055	105	105.1	105.11	105.106
$\frac{3}{8}$	0	0.4	0.38	0.375
π	3	3.1	3.14	3.142

6 a 8.15   **b** 5.94   **c** 2.98   **d** 1311.59
7 a 15 km   **b** 14 752 m   **c** 14 750 m   **d** 14 800 m

Stretch

1 Any three values from 8.45 to 8.4999…
2 10.35, 10.36, 10.37, 10.38, 10.39, 10.42, 10.43
3 a Any answer between 67.65 million and 67.74 million
  b Answers will vary depending on number chosen in part **a**, e.g. 67.65 million would be 67 650 000

Chapter 6.2

Are you ready?

1 a 260   b 30   c 0   d 0
2 a 256   b 26   c 3   d 0
3 a 256.1   b 25.6   c 2.6   d 0.3
4 a 0.256   b 25.613   c 2.561   d 0.257
5 a 300   b 325   c 325.18

Practice

1 a i 2   ii 4   iii 4   iv 6
   b i Thousands   ii Millions
     iii Thousandths   iv Ones
2 a i 4   ii 8   iii 3   iv 0
   b i Hundreds   ii Hundred thousands
     iii Ten-thousandths   iv Tenths
3 a 500   b 30   c 100
   d 0.09   e 0.005   f 0.1
4 a 450   b 31   c 400
   d 0.024   e 0.0031   f 0.080
5 a 100 000   b 11 000

6

	Rounded to 1 s.f.	Rounded to 2 s.f.	Rounded to 3 s.f.
23 763	20 000	24 000	23 800
1 056 198	1 000 000	1 100 000	1 060 000
0.2076	0.2	0.21	0.208
0.000 023 55	0.00002	0.000024	0.0000236

7 a 1 200 000   b 7.7   c 2.5   d 0.0000022

What do you think?

1 No, Sven has ignored the zero.
   3.04 rounded to 2 significant figures is 3.0
2 a 150   b 249
3 a 244 999   b 235 000

Consolidate

1 a i 3   ii 2   iii 2   iv 1
   b i 8   ii 0   iii 5   iv 5
2 a 300   b 40   c 200
   d 0.02   e 0.009   f 0.09
3 a 260   b 41   c 910
   d 0.051   e 0.0069   f 0.10
4 4.6 metres
5 0.009 kilograms
6 25°C
7 0.006 72 litres

8

	Rounded to 1 s.f.	Rounded to 2 s.f.	Rounded to 3 s.f.
105 672	100 000	110 000	106 000
23 091 124	20 000 000	23 000 000	23 100 000
0.3089	0.3	0.31	0.309
0.000 093 95	0.00009	0.000094	0.0000940

9 a 8.5   b 4.1
   c 1.8   d 0.000 000 18

Stretch

1 19.25 cm²
2 Disagree. Samira's statement is not accurate. Rounding 0.3519 to 1 significant figure gives 0.4, and rounding it to 1 decimal place gives 0.4 as well. However, these are not always the same. For example, rounding 0.0519 to 1 significant figure gives 0.05, while rounding it to 1 decimal place gives 0.1. It depends on the specific number being rounded.
3 Any number between 29 950 and 30 049

Chapter 6.3

Are you ready?

1 a 3000   b 400   c 200
   d 50   e 0.2   f 0.6
2 a 600   b 2950   c 0.8   d 3650
3 a 10 000   b 2   c 0.12   d 10

Practice

1 a 40 × 60   b 200 × 500   c 80 × 300   d 5 × 20
2 41 rounded to 1 significant figure = **40**
   75 rounded to 1 significant figure = **80**
   So 41 × 75 ≈ **40 × 80 = 3200**
3 0.875 rounded to 1 significant figure = **0.9**
   0.015 926 rounded to 1 significant figure = **0.02**
   So 0.875 + 0.015 926 ≈ **0.9 + 0.02 = 0.92**
4 a i 1000   ii 200
   b i 800   ii 1200   iii 200 000   iv 5
5 a i 3000   ii 1100   iii 240 000   iv 4
   b i 2638   ii 1175   iii 216 132   iv 4.13
6 a i 3   ii 1.7   iii 1.8   iv 40
   b i 2.88   ii 1.65   iii 2.14   iv 35.08
7 a Estimate £35 to £42   b £36.68
8 a 30   b 54   c 14   d 140

What do you think?

1 Junaid's method is more appropriate.
   Using Ali's method, 18 would be rounded to 0 so the estimate for 18 × 287 ≈ 0 × 300 = 0

Consolidate

1 a 37 rounded to 1 significant figure = **40**
   88 rounded to 1 significant figure = **90**
   So 37 × 88 ≈ **40 × 90 = 3600**
   b 3256
2 a i 4100   ii 1600   iii 350 000   iv 10
   b i 4385   ii 1573   iii 350 892   iv 11.21
3 a i 5   ii 1.5   iii 3.6   iv 30
   b i 5.48   ii 1.46   iii 4.00   iv 29.97
4 7
5 a 84   b 40   c 25
6 £230
7 a 3600   b 8 packs = 4000 bricks, 8 packs ≈ £800
8 20 litres
9 £32 000

Stretch

1 a 10 m²   b 7.5 cm²   c 500 cm²   d 2400 mm²
2 5.6 kg
3 29 > 25, so √29 > 5
   29 < 36, so √29 < 6
   So √29 is between **5** and **6**
4 a i 8 and 9   ii 7 and 8   iii 9 and 10   iv 11 and 12
   b i 8.49   ii 7.14   iii 9.70   iv 11.79

Chapter 6.4

Are you ready?

1 a 400   b 300   c 3300
2 a 4.17   b 0.04   c 10.21
3 a 4000   b 0.5   c 40
4 1512   2018   2399
5 0.39   0.415   0.351

Practice

1 a 5.2   4.51   5.09   4.897
   b $4.5 \leqslant x < 5.5$
2 $12.5 \leqslant a < 13.5$
3 a 3.15   3.167   3.249
   b $3.15 \leqslant y < 3.25$
4 $22.55 \leqslant b < 22.65$
5 a 75   b 84

6 a 6.50   **b** 7.49
7 a $59.5\,\text{m} \leqslant f < 60.5\,\text{m}$ **b** $55\,\text{m} \leqslant f < 65\,\text{m}$
8 a $450 \leqslant n < 550$ **b** $495 \leqslant n < 505$
   c $499.5 \leqslant n < 500.5$

Consolidate

1 a 6013.5   6499   5500.5
   b $5500 \leqslant x < 6500$
2 a 0.251   0.245   0.2549
   b $0.245 \leqslant y < 0.255$
3 $35 \leqslant c < 45$
4 a 67 500 000   **b** 68 499 999
5 a 0.750   **b** 0.849
6 a $9.5\,\text{kg} \leqslant b < 10.5\,\text{kg}$ **b** $9.5\,\text{kg} \leqslant b < 15\,\text{kg}$

Stretch

1 a 40   **b** 31
2 a 39   **b** 140
3 £2772

Accuracy: exam practice

1 a 500   **b** 10 600   **c** 3000
2 a 31.7   **b** 32
3 4
4 a 3.605 762 257   **b** 3.61
5 300
6 800
7 a 24.5 kg   **b** 25.4$\dot{9}$ kg

Number: exam practice

1 a 3630   **b** 936   **c** 1736
2 0.55
3 a 2.770 517 777   **b** 2.8
4 194
5 £3.60
6 $\frac{67}{100}$
7 3000 m
8 −4°C
9 £5.75
10 66
11 a −6   **b** 4
12 500 ml glass
13 2×5^3
14 $4\frac{7}{15}$
15 2.7×10^7
16 a 189 500   **b** 190 499
17 £7369.50

Block 7 Understanding algebra

Chapter 7.1

Are you ready? (A)

1 a $4y$ **b** $4y$ **c** $\frac{4}{y}$
   d $\frac{y}{4}$ **e** y^2 **f** ay
2 a $\frac{1}{2}$ **b** $\frac{1}{3}$ **c** $\frac{3}{4}$

Practice (A)

1 a $p + 6$ **b** $p - 3$ **c** $2p$
   d $\frac{p}{3}$ **e** $\frac{p}{2} - 2$ **f** $2p + 10$
2 a i $2m$ **ii** $5r$ **iii** $2m + 5r$
   b $5r - 3m$
3 a i $x + 3$ **ii** $x + k$ **iii** $x + w$
   b $x - 6$
4 a £$(b - 10)$ **b** £$(12b - 120)$ **c** £$0.5b$
5 a $h + g$ **b** $h + g + 3$ **c** hg **d** $\frac{h}{g}$

Are you ready? (B)

1 a $2y$ **b** $3y$ **c** y **d** $3y$
2 a −1   **b** −7   **c** 1   **d** 7   **e** −1
3 $2w$, $-5w$ and w

Practice (B)

1 a $6g$ **b** $8h + 6$ **c** $6k + 5p$
   d $7u + 5$ **e** $10j + 3r$ **f** $2e + 2a$
2 a $5a + 6b$ **b** $a + 6b$ **c** $5a + 4b$
   d $9c + d$ **e** $-c - d$ **f** $c - d$
3 a $10e^2 + 13f$ **b** $2e^2 + 8f$ **c** $10ef + 5f$
   d $9x^2 - 2x$ **e** $3x^2 + 2x$ **f** $2xy + 3x + 3y$
4 $(6h + 4g)$ cm
5 a $gh^2 + 5gh$ **b** $3gh^2 + 5g^2h$ **c** $7gh^2 - 4gh$

Consolidate

1 a $g - 5$ **b** $g + 8$ **c** $2g + 1$
   d $2g - 8$ **e** $\frac{g}{4}$ **f** $\frac{g}{3} - 3$
2 $7n$
3 a $8y$ **b** $25y$ **c** xy
4 a $3m + 5k$ **b** $m + 3k$ **c** $3m + k$
   d $2m - k$ **e** $m + k$ **f** $-4m - 8k$
   g $4m^2 - 3m$ **h** $m^2 - 5m$ **i** $-3m$

Stretch

1 Many possible answers, e.g. $7x + 5 - x$, $3x + 2 + 3x + 3$, $1 + 8x + 4 - 2x$
2 a $6hg$ cm²   **b** $9y^2$ cm²
3 a $(8a - 2b)$ cm   **b** $20k$ cm

Chapter 7.2

Are you ready? (A)

1 a 5^3 **b** h^2 **c** 6^5 **d** k^4
2 a $5u$ **b** mp **c** $6wg$ **d** $4h^2$

Practice (A)

1 a g^5 **b** h^{10} **c** k^4
   d m^9 **e** n^8 **f** p^{10}
2 a b^2 **b** c^6 **c** d^8
   d e^6 **e** f^5 **f** g^4
3 a h^6 **b** k^{10} **c** m^{16} **d** p^{35}

What do you think?

1 Emily is incorrect. You need to multiply the coefficients and add the powers so the answer should be $10y^8$.

Are you ready? (B)

1 a g^5 **b** $6a^2$ **c** $4b^2$ **d** $3c$
2 a $\frac{1}{16}$ **b** 3   **c** 11   **d** −3

Practice (B)

1 a $2x^8$ **b** $4y^5$ **c** $6w^7$
   d $35u^8$ **e** $15h^7$ **f** $6g^6$
2 a $3b^2$ **b** $2c^5$ **c** $5d^4$
   d $4e^4$ **e** $6f^5$ **f** $1.5g^8$ or $\frac{3}{2}g^8$
3 a $9h^8$ **b** $64n^6$ **c** $32w^{25}$ **d** $125p^{24}$
4 a 1   **b** 1   **c** $\frac{1}{25}$ **d** $\frac{1}{8}$
5 a g^2 **b** h^{-1} **c** k^{-2}
   d a^2 **e** b^{-3} **f** x^{-2}
6 a y^7 **b** w^7 **c** c^{-5}
   d d^{-6} **e** b^{-3} **f** g^5

Consolidate

1 a g^7 **b** k^6 **c** m^8
   d b^4 **e** c^8 **f** g^5
2 a h^6 **b** k^{28} **c** m^{40}
   d $16g^{10}$ **e** $27n^{12}$ **f** $16w^{32}$
3 a $10m^7$ **b** $5y^{10}$ **c** $15w^{10}$
   d $2b^5$ **e** $3c^6$ **f** $9g^3$

4 a 1 **b** 1 **c** $\frac{1}{64}$

5 a g^2 **b** k^{-4} **c** a^{-1}
d y^7 **e** c^{-4} **f** b^{-2}

Stretch

1 a y **b** x^2 **c** u^2
d w **e** $4w^4$ **f** g^{-2}

2 a $24g^3$ **b** $64y^{-8}$ **c** $3x^{12}$ **d** $3h$

3 a x^6 **b** y^{-20} **c** $16w^{-6}$ **d** $\frac{1}{8u^{12}}$ or $\frac{1}{8}u^{-12}$

Chapter 7.3

Are you ready?

1 a 5 multiplied by x **b** 5 add x
c x divided by 5 **d** x subtract 5
e x subtracted from 5
2 $2x$ means 2 multiplied by x and x^2 means x multiplied by x.
3 a 9 **b** 9 **c** −12 **d** 12 **e** −48

Practice

1 a 35 **b** 6 **c** 150 **d** 25
2 a 40 **b** 32 **c** 8 **d** 53 **e** 14
3 a 8.46 **b** 54 **c** 38 **d** 7.54
e 0.3 **f** 1.16 **g** 41.16 **h** −0.3
4 a 105 **b** 55 **c** 2000 **d** 3.2
e 0.3125 **f** 20 **g** 5 **h** 285
5 a i 35 **ii** 5 **iii** −5 **iv** −35
b i −5 **ii** −35 **iii** 35 **iv** 5
6 a −2 **b** −14 **c** −48 **d** 30
e −4 **f** −16 **g** 10 **h** 7
7 a 36 **b** 48 **c** 36 **d** 192
e 100 **f** 75 **g** −45 **h** −375

What do you think?

1 a Various possible answers, e.g. $-a - b$
b Various possible answers, e.g. $-ab$
2 a i 500 and 2500 **ii** 12 500 and 62 500
iii 20 and 100
b $e = 0$

Consolidate

1 a 39 **b** 12 **c** 108 **d** 33
2 a 50 **b** 40 **c** 10 **d** 50 **e** 5
3 a 42 **b** 18 **c** 360 **d** 2.5
e 0.4 **f** 6 **g** −6 **h** −24
4 a −2 **b** −18 **c** −80 **d** 54
e −5 **f** −20 **g** 13 **h** 9
5 a 75 **b** 180 **c** 150 **d** 432
e 48 **f** 80 **g** −100 **h** −128

Stretch

1 a 81 **b** 64 **c** 9 **d** 8
2 a 2304 **b** 20 992 **c** 288
d 16 **e** 2880 **f** 20.5

Chapter 7.4

Are you ready?

1 a $8x + 6$ **b** $18w + 4$
2 a False **b** True **c** True **d** False
3 a $20\,\text{cm}^2$ **b** $10\,\text{cm}^2$

Practice

1 $2w + 5a^2 - 3$ is an expression; $V = lwh$ is a formula; $6 + 5h = 21$ is an equation; $15 + 10y \equiv 10y + 15$ is an identity
2 $3p + 5r$
3 a $3y$ **b** $3x$ **c** $8g$ **d** $4h$
4 a $8a\,\text{cm}$ **b** $3a^2\ \text{cm}^2$

Consolidate

1 $5g \times 9h \equiv 45gh$ is an identity; $a^2 + b^2 = c^2$ is a formula; $5x + 20 = 3x + 22$ is an equation; $5g + 2g^2$ is an expression
2 a 24 **b** 36 **c** −36 **d** −216
3 $6abc$

Stretch

1 a πk^2 **b** $9\pi g^2$ **c** $16\pi x^2$
2 $a = 23$, $b = 3$

Understanding algebra: exam practice

1 a $4y$ **b** $7a + 2b$ **c** $3t^2 + 7t$
2 $x - 6$
3 $4t + 12c$
4 95
5 a d^3 **b** $20fg$
6 a $4x$ **b** x^2
7 £850
8 a m^6 **b** $15x^6y^4$

Block 8 Working with brackets

Chapter 8.1

Are you ready? (A)

1 a $6a$ **b** $6b$ **c** $6c$
d $-10d$ **e** $6e$ **f** $15f^2$
2 a $5a + 9$ **b** $6c + 7$ **c** $2f + 3$ **d** $5 - g$

Practice (A)

1 a $3a + 12$ **b** $5c - 10$ **c** $12 + 4e$
d $6g + 8$ **e** $12h - 6$ **f** $20 - 15j$
g $21x + 35$ **h** $16 - 32k$ **i** $90 + 81y$
2 a $2b + 6$ **b** $-2b - 6$ **c** $-2b + 6$
d $12g - 6$ **e** $-12g - 6$ **f** $-12g + 6$
3 a $6g + 20$ **b** $2h + 5$ **c** $8r + 8$
d $3k - 10$ **e** $11y + 19$ **f** $2p + 20$
4 a $8c + 28$ **b** $10x + 1$ **c** $17e + 4$
d $9t + 24$ **e** $8d + 15$ **f** 5
5 a $x^2 + 8x$ **b** $3g^2 - 2g$ **c** $10k + 2k^2$
d $t^3 + gt$ **e** $3w^3 + 2w^2y$ **f** $2e^2f - 3ef^3$

Are you ready? (B)

1 a 4 **b** 6 **c** 7 **d** 8
2 a 3 **b** $3y$ **c** $3x$ **d** $3x$

Practice (B)

1 a $2(x + 4)$ **b** $4(h - 4)$ **c** $8(3y + 1)$
d $8(p - 1)$ **e** $3(2 + k)$ **f** $6(2 - t)$
g $2(2m + 7)$ **h** $3(5 + 4w)$ **i** $2(3u - 14)$
2 a $4(3x + 4y + 5)$ **b** $5(3h - 4 + 7g)$ **c** $8(2 - 5w - 3g)$
3 a $x(x + 1)$ **b** $y(1 - y)$ **c** $f(3 + f)$
d $g(5g + 2)$ **e** $u(8 - 3u)$ **f** $n(4n^2 + 3)$
4 a $5(1 + 3p^2)$ **b** $2n(1 + 2n)$ **c** $2x(x - 4)$
d $7y(2 + y)$ **e** $5g(g + 2)$ **f** $2w^2(3w - 1)$
5 a $8h(h + 2f)$ **b** $6ab(b + 4a)$
c $4x(3xy + 5y + 2x^2)$ **d** $5cd(3d + 2cd + 5)$

What do you think?

1 Tiff is correct. The highest common factor of $14xy^2$ and $-28x^2y$ is $14xy$. Kath has not fully factorised the expression.
2 There are no common factors of $10x$ and $-7y$ apart from 1
3 a Various possible answers, e.g. $2a(8 + 5b)$
b Various possible answers, e.g. $3a(4 + 6b) + 4a(1 - 2b)$

Consolidate

1 a $4b + 12$ **b** $8b + 12$ **c** $8b - 12$
d $8b + 12c$ **e** $8b - 12c + 20$ **f** $-4b - 12$
g $b^2 + 3b$ **h** $2b^2 - 5b$ **i** $20b - 4b^3$

Answers

2 a $9g + 38$ **b** $8h + 16$ **c** $18w + 6$
d $y + 6$ **e** $18p + 5$ **f** $16k - 4$
3 a $3(c + 3)$ **b** $5(3 - d)$ **c** $7(e - 1)$
d $2(3f + 8)$ **e** $3(4g + 5h + 9)$ **f** $6(3 - 7j - 4k)$
4 a $m(m + 1)$ **b** $n(8 + 3n)$ **c** $4(1 + 3p^2)$
d $3r(1 + 3r)$ **e** $2u(5 + 3u)$ **f** $2w(7y - 3w)$

Stretch

1 a 3500 **b** 470 **c** 2700
2 $(24x + 14)$ cm or $2(12x + 7)$ cm

Chapter 8.2

Are you ready?

1 a $2y + 6$ **b** $y^2 + 3y$ **c** $2y - 6$ **d** $y^2 - 3y$
2 a $11y$ **b** $-y$ **c** y **d** $-11y$
3 a $-2y$ **b** $2y$ **c** $6y$ **d** $-6y$
4 a $x^2 + 7x + 10$ **b** $x^2 - 3x - 10$ **c** $x^2 + 3x - 10$

Practice

1 a $x^2 + 7x + 12$ **b** $x^2 + 7x + 10$ **c** $y^2 + 8y + 15$
d $y^2 + 5y + 4$ **e** $w^2 + 12w + 35$ **f** $w^2 + 10w + 9$
2 a $x^2 + 2x - 15$ **b** $x^2 + 3x - 18$ **c** $p^2 + 2p - 35$
d $p^2 - 2p - 15$ **e** $g^2 - g - 42$ **f** $g^2 - 7g - 8$
3 a Both expressions contain terms y^2 and 21 but one
expression contains $+10y$ and the other contains $-10y$.
b i $x^2 - 5x + 6$ **ii** $h^2 - 11h + 28$ **iii** $g^2 - 13g + 40$
4 a $x^2 - 4$ **b** $x^2 - 49$ **c** $x^2 - 1$
There is no term in x.
5 a Benji is incorrect as $(y + 5)^2 \equiv (y + 5)(y + 5) \equiv$
$y^2 + 10y + 25$
b i $y^2 + 6y + 9$ **ii** $x^2 + 2x + 1$
iii $w^2 - 8w + 16$ **iv** $r^2 - 16r + 64$
6 a Marta is wrong as $2x \times 3 = 6x$ and $4 \times x = 4x$ so the
coefficient of x is 10
$4 \times 3 = 12$ so the correct expansion of
$(2x + 4)(x + 3)$ is $2x^2 + 10x + 12$
b i $2y^2 + 7y + 3$ **ii** $3x^2 + 10x + 8$ **iii** $6w^2 + 17w + 5$
7 a $3w^2 - 10w - 8$ **b** $5p^2 - 23p - 10$ **c** $12y^2 + 5y - 3$
d $6g^2 - 13g + 6$ **e** $4x^2 + 12x + 9$ **f** $25h^2 - 70h + 49$

Consolidate

1 a $g^2 + 4g + 3$ **b** $x^2 + 10x + 24$ **c** $y^2 + 4y - 12$
d $p^2 + p - 2$ **e** $h^2 - 3h - 4$ **f** $k^2 - 4k - 21$
2 a $x^2 - 6x + 8$ **b** $u^2 - 9u + 20$ **c** $y^2 - 4y + 3$
d $t^2 - 9$ **e** $w^2 - 81$ **f** $n^2 + 2n - 99$
3 a $x^2 + 2x + 1$ **b** $x^2 - 20x + 100$
4 a $2x^2 + 11x + 12$ **b** $3u^2 + 2u - 5$ **c** $4y^2 - 13y - 12$
d $8h^2 - 20h + 8$ **e** $9w^2 + 12w + 4$ **f** $36g^2 - 12g + 1$

Stretch

1 a $(y^2 + 13y + 40)$ cm² **b** $(3a^2 - 5a - 28)$ cm²
2 $gk + gm + hk + hm$
3 $a = 2, b = -1, c = 1, d = 3$
4 a $e^2 - 4$ **b** $r^2 - 25$ **c** $b^2 - 81$
$(x - a)(x + a) \equiv x^2 - a^2$
5 a 2400 **b** 6800 **c** 896

Chapter 8.3

Are you ready?

1 a $4(g + 3)$ **b** $g(g - 4)$ **c** $4g(g - 2)$
2 a 1, 2, 5, 10 **b** 1, 2, 3, 4, 6, 12
c 1, 2, 4, 5, 10, 20 **d** 1, 2, 3, 4, 6, 8, 12, 24
3 a 2 and 3 **b** 5 and −2
c 3 and −6 **d** −4 and −5

Practice

1 a $(x + 5)(x + 1)$ **b** $(x + 5)(x + 2)$ **c** $(x + 3)(x + 5)$
d $(y + 3)(y + 4)$ **e** $(w + 10)(w + 2)$ **f** $(p + 4)(p + 5)$
2 a $(x + 7)(x - 1)$ **b** $(x + 8)(x - 4)$ **c** $(x + 5)(x - 2)$
d $(g + 5)(g - 3)$ **e** $(k + 6)(k - 3)$ **f** $(h + 9)(h - 2)$
3 a $(x - 3)(x + 1)$ **b** $(x - 6)(x + 1)$ **c** $(x - 5)(x + 3)$
d $(y - 6)(y + 2)$ **e** $(a - 8)(a + 2)$ **f** $(c - 5)(c + 4)$

4 a The coefficient of x is −9 so
$x^2 - 9x + 20 \equiv (x - 4)(x - 5)$
b i $(x - 6)(x - 3)$ **ii** $(y - 6)(y - 4)$
iii $(w - 5)(w - 6)$ **iv** $(u - 12)(u - 2)$
5 a $(x + 4)^2$ **b** $(y + 5)^2$ **c** $(w - 2)^2$ **d** $(p - 6)^2$
Each expression factorises to an identical pair of
brackets.
6 a $(x + 3)(x - 3)$ **b** $(y + 7)(y - 7)$ **c** $(w + 10)(w - 10)$
Only the sign is different in each pair of brackets.
7 a $(p + 3)(p + 8)$ **b** $(q + 9)(q + 4)$ **c** $(r - 3)(r - 4)$
d $(s - 3)(s + 2)$ **e** $(u - 10)(u + 2)$ **f** $(v - 8)^2$
g $(w - 10)(w + 3)$ **h** $(x - 5)(x - 6)$ **i** $(y + 6)(y - 5)$

Consolidate

1 a $(x + 1)(x + 7)$ **b** $(y + 2)(y + 7)$ **c** $(x + 3)(x + 7)$
d $(y + 4)(y + 7)$ **e** $(w + 20)(w + 1)$ **f** $(p + 6)^2$
2 a $(x + 8)(x - 1)$ **b** $(x + 8)(x - 2)$ **c** $(x + 7)(x - 3)$
d $(y + 10)(y - 3)$ **e** $(w - 6)(w + 5)$ **f** $(a - 12)(a + 2)$
g $(g - 9)(g + 4)$ **h** $(p - 9)(p + 5)$ **i** $(u - 9)^2$
3 a $(x + 1)(x - 1)$ **b** $(g + 8)(g - 8)$ **c** $(k + 12)(k - 12)$

Stretch

1 $(x + 5)$ cm
2 $a = 6, b = 2, c = -56, d = 8$
3 $(8x + 3y)(8x - 3y)$
4 $x^2 - 25 = 24$; subtracting 24 from each side of the
equation gives $x^2 - 49 = 0$, which is equivalent to
$(x + 7)(x - 7) = 0$

Chapter 8.4

Are you ready?

1 a $3y$ **b** $3y^2 + y$ **c** y **d** $2y + 4w$
2 a $8y + 11$ **b** $7x - 17$
3 a $6(y + 3)$ **b** $8(3x - 2)$ **c** $5w(3w - 7)$

Practice

1 a False, as $5x + x \equiv 6x$
b False, as x and y are not like terms
c True **d** True
e True **f** False, as $5y - y \equiv 4y$
g False, as xy and y are not like terms
h True
i False as $5x + 4y + x \equiv 6x + 4y$
2 a $5g + 2g \equiv 7g$ **b** $10h - 3h \equiv 7h$
c $9k + 2k - k \equiv 10k$ **d** $6p^2 + 2p + 3p^2 \equiv 9p^2 + 2p$
e $5w - m + 5w + 2m \equiv m + 10w$
f Various possible answers, e.g. $6f + 4n - 2f \equiv 4f + 4n$
3 a $8(2u + 3) \equiv 16u + 24 \equiv 2(8u + 12) \equiv 2(12 + 8u)$
b $3(2y + 5) - 3 \equiv 6y + 15 - 3 \equiv 6y + 12 \equiv 6(y + 2)$
c $3(4 + 2h) + 8(h - 1) \equiv 12 + 6h + 8h - 8 \equiv$
$14h + 4 \equiv 2(7h + 2)$
4 a $a = 7, b = 11$ **b** $a = 7, b = 3$
c $a = 1, b = 8$ **d** $a = 2, b = 6$

Consolidate

1 $4p - p$ $p + p + p$ $3 \times p$ $6p \div 2$ $p \times 3$
2 a False, as $3m + 3m \equiv 6m$
b False, as $2m + 3k + 4m + 5k \equiv 6m + 8k$
c True **d** True
e False, as $4m$ and 4 are not like terms
f True
3 a $5t - t \equiv 4t$
b $7g^2 + 2g + 11g^2 \equiv 18g^2 + 2g$
c $3f - e + 2f - 5f \equiv -e$
d Various possible answers, e.g. $a + 2a - 2b \equiv 3a - 2b$
4 $6(3x + 10) \equiv 18x + 60 \equiv 18x + 54 + 6 \equiv 9(2x + 6) + 6 \equiv$
$6 + 9(2x + 6)$
5 a $a = 9, b = 2$ **b** $a = 1, b = 7$

Stretch

1 a $4(x + 3) \equiv 4x + \mathbf{12}$
b $3(y + 7) \equiv 3y + 21$
c $3(\mathbf{2}w - 4) + 6(2w - \mathbf{2}) \equiv 18w - 24$
2 $5(8 + y) \equiv 40 + 5y$
3 $y = 0$ and $y = 2$

Working with brackets: exam practice
1 a $2a + 12$ **b** $y^2 - 4y$ **c** $24 + 20x$
2 a $3(w - 2)$ **b** $b(b + 7)$
3 $4(4m + 3n)$
4 $3(2x + 5)$ or $6x + 15$
5 a $7t + 41$ **b** $y - 10$
6 a $x^2 - 3x - 54$ **b** $3x^2 - 5x + 2$
7 $a = 6, b = 2$

Block 9 Functions and linear equations

Chapter 9.1

Are you ready? (A)
1 a $3y$ **b** p^2 **c** $\frac{h}{4}$ **d** $\frac{7}{w}$
2 a $2y$ **b** $\frac{y}{5}$ **c** $y + 5$ **d** $y - 5$

Practice (A)
1 a 70 **b** 600 **c** 50 **d** 6
e 60.6 **f** 36 **g** 59.4 **h** 100
2 a i 118.3 **ii** 32.1 **iii** −3.2
iv 3518.2 **v** 381.8
b i 1826.5 **ii** 294.9 **iii** 259.6
iv 3781 **v** 119
3 a $8y$ **b** $y + 8$ **c** $y - 8$
d $\frac{y}{8}$ **e** $8 - y$
4 a i 1 **ii** 42 **iii** $\frac{7}{6}$
iv $7g$ **v** $7h$ **vi** $\frac{7}{h}$
b i $h - 6$ **ii** $6h$ **iii** $\frac{h}{6}$
iv hg **v** h^2 **vi** 1
c i $5h - 6$ **ii** $30h$ **iii** $\frac{5h}{6}$
iv $5hg$ **v** $5h^2$ **vi** 5

Are you ready? (B)
1 a $6e$ **b** $5c$ **c** $\frac{7}{f}$
d $\frac{g}{8}$ **e** h^2 **f** jk
2 a $\div 3$ **b** $- 4$ **c** $\times 5$ **d** $+ 6$
3 Square rooting

Practice (B)
1 a 54 **b** 66 **c** 10 **d** 360
2 a 142 **b** 30.1 **c** 10.773
d 15.6 **e** 83.23 **f** 0
3 a $w - 9$ **b** $\frac{w}{9}$ **c** $9w$ **d** $w + 9$
4 a $2g$ **b** $8g + 4$ **c** $8g - 4$ **d** $32g$

What do you think?
1 a $m + 2$ **b** $m + 12$ **c** $5(m + 7)$ **d** $\frac{m + 7}{5}$
2 a 6 **b** 1.5 **c** y **d** $2y$

Are you ready? (C)
1 a 24 **b** 3
2 a $3a - 5$ **b** $5(b + 3)$ **c** $\frac{c}{3} + 5$ **d** $\frac{d + 5}{3}$

Practice (C)
1 a 6 **b** 8 **c** 24 **d** 4
2 a Input → ×8 → −5 → Output
b i 51 **ii** 124.6 **iii** $8k - 5$
3 e.g. Inputting 10 gives 52 and 25, so Marta is wrong as these are not the same outputs (true only for input 1).
4 a $3p + 4$ **b** $4(p + 3)$ **c** $\frac{p - 3}{4}$ **d** $\frac{p}{4} + 3$
5 a 12 **b** 8 **c** 4 **d** 24

6 a Input → −7 → ×4 → Output
b 17 **c** 8 **d** 12.25
7 a 12 **b** 79 **c** 0
d y **e** $15y$ **f** $4y + 1$

Consolidate
1 a 55 **b** 250 **c** 45 **d** 10
e 50.5 **f** 25 **g** 49.5 **h** 100
2 a i $t - 7$ **ii** $t + 7$ **iii** $7t$
iv $\frac{t}{7}$ **v** $7 - t$ **vi** t^2
b i $7t - 7$ **ii** $7t + 7$ **iii** $49t$
iv t **v** $7 - 7t$ **vi** $7t^2$
3 a 45 **b** 55 **c** 10 **d** 250
4 a i $t - 7$ **ii** $7t$ **iii** $t + 7$ **iv** $\frac{t}{7}$
b i $7t - 7$ **ii** $49t$ **iii** $7t + 7$ **iv** t
5 a i 20 **ii** 25 **iii** 280 **iv** 5
b i 50 **ii** 40 **iii** 5 **iv** 280
6 a Input → ÷10 → +17 → Output
b i 25 **ii** 31.3 **iii** $\frac{f}{10} + 17$
c i 630 **ii** 1260 **iii** $10(f - 17)$
7 a i $5w + 2$ **ii** $2(w + 5)$
iii $\frac{w - 5}{2}$ **iv** $\frac{w}{5} + 2$
b i $2y + 4$ **ii** $5y + 6$
iii $20y + 49$ **iv** $5(10y + 20)$

Stretch
1 a i 20 **ii** Various possible answers, e.g. 1
b i 4 **ii** Various possible answers, e.g. 1
2 a $e, 4$ **b** $2f, 5$ **c** $g, 8$
3 a Missing input: $w + 2$
Missing operation: × 4
Missing output: $12y - 7$
b Missing inputs: (top) $2x$ and (bottom) $w + 2$
Missing operation: + 8
Missing output: $5(4y + 8)$

Chapter 9.2

Are you ready? (A)
1 a $+ 5$ **b** $- 5$ **c** $\times 5$ **d** $\div 5$
2 a 70 **b** 30 **c** 1000
d $\frac{5}{2} = 2.5$ **e** $\frac{2}{5} = 0.4$
3 7

Practice (A)
1 a $g = 16$ **b** $b = 5$ **c** $h = 24$ **d** $x = 80$
2 a $y = 46$ **b** $w = 273$ **c** $c = 204$
d $r = 1.9$ **e** $u = 2981$ **f** $g = 0.166$
3 a $k = 50$ **b** $d = 101$ **c** $c = 25.1$
4 a $a = 3$ **b** $a = 147$ **c** $t = 1.8$ **d** $t = 180$
e $q = 6.2$ **f** $q = 99.2$ **g** $u = 45$ **h** $u = 7.2$
5 a $w = 32$ **b** $w = 68$ **c** $w = 32$
d $y = 24.6$ **e** $y = 13.4$ **f** $y = 84.5$
6 a $a = 72$ **b** $a = 2$ **c** $c = 9$ **d** $c = 2025$
e $w = 6$ **f** $y = 2.5$ **g** $x = 16$ **h** $f = 1$

What do you think? (A)
1 No, because you can subtract 75 from both sides so $y = 0$, and you can divide both sides by 75 so $w = 1$
2 a $p = 32$ (add 24 to both sides of the equation)
$p = 16$ (add p to both sides of the equation and then subtract 8 from both sides)
b $p = 192$ (multiply both sides by 24)
$p = 3$ (multiply both sides by p and then divide both sides by 8)

Are you ready? (B)

1 a $k = 7$ **b** $k = 63$ **c** $k = 24$ **d** $18 = k$
2 a $a = 17$ **b** $a = 13$ **c** $a = 30$ **d** $a = 7.5$
3 a $3x + 12$ **b** $10x - 35$

Practice (B)

1 a $y = 5$ **b** $p = 10$ **c** $k = 4$
 d $u = 1$ **e** $e = 29$ **f** $w = 6.5$
2 a $q = -1$ **b** $x = -3$ **c** $r = -3$
 d $w = 4$ **e** $r = -3$ **f** $g = -9.6$
3 a $x = 3$ **b** $r = 8$ **c** $w = 13$
 d $t = -1$ **e** $y = -3$ **f** $p = -1$
4 a $w = 24$ **b** $p = 60$ **c** $u = 12$ **d** $r = 135$
5 a $w = 27$ **b** $p = 51$ **c** $u = 26$ **d** $r = 103$
6 a Preferred method stated
 b i $x = 3$ **ii** $x = 5.5$ **iii** $x = -1$

What do you think? (B)

1 a Huda's and Lida's numbers are different as the
 operations have different inverses.
 b $\frac{x - 5}{3} = 9$ $x = 32$
 c $\frac{x}{5} - 3 = 9$ $x = 60$
 d $3(x + 5) = 9$ $x = -2$

Consolidate

1 a $p = 10$ **b** $b = 3$ **c** $h = 20$ **d** $x = 75$
2 a $y = 32$ **b** $k = 118$ **c** $w = 184$
 d $f = 186$ **e** $r = 0.7$ **f** $h = 95.4$
3 a $x = 3$ **b** $x = 192$ **c** $u = 77.5$
 d $u = 310$ **e** $s = 11$ **f** $y = 22$
 g $b = 162$ **h** $b = 2$ **i** $k = 2.5$
4 a $y = 8$ **b** $p = 7$ **c** $r = -2$
 d $p = 1.5$ **d** $w = 3.5$ **e** $q = 5.2$
 f $m = 2$ **g** $b = 7$ **h** $x = -4$
5 a $w = 4$ **b** $p = 7$ **c** $v = 15$
 d $c = -2$ **e** $m = 3.5$ **f** $a = 2$
6 a $y = 24$ **b** $g = 87$ **c** $k = 205$
 d $m = 177$ **e** $y = 32$ **f** $x = -48$

Stretch

1 a $p = 78$ **b** $p = 39$ **c** $p = 27$ **d** $p = 75$
2 a $p = 2$ **b** $p = -1$ **c** $p = 1$ **d** $p = 17$
3 a $3y - 8 + 2(34) = 180$ **b** $4p + 256 = 360$
 $3y - 60 = 180$ $p = 26$
 $y = 80$

Chapter 9.3

Are you ready?

1 a $y = 15$ **b** $x = 324$ **c** $m = -12$ **d** $r = 8$
2 a $5a - 10$ **b** $28 + 7y$ **c** $32 - 4c$ **d** $10g - 15$
3 a $g = 4$ **b** $w = 51$ **c** $u = 1$ **d** $x = \frac{7}{4}$

Practice

1 $x = 5$
2 a $y = 5$ **b** $w = 2$ **c** $p = 2$
 d $m = 5$ **e** $t = 7$ **f** $k = 11$
3 a $y = 6$ **b** $u = 2.5$ **c** $p = 20$
4 Filipo has written 8 instead of -8.
 The solution is $x = -3$
5 a $b = 2$ **b** $k = 2$ **c** $m = -5$
 d $t = -3$ **e** $a = 7$ **f** $w = 0.4$
6 a $p = 2$ **b** $c = 3$ **c** $x = 3$
 d $y = 1$ **e** $n = -2$ **f** $g = 3.5$

What do you think?

1 a Flo's strategy is more sensible as she will end up
 with a positive coefficient of y.
 b $y = 3$
 c Yes, Benji's method works.
2 a $u = 5$ **b** $e = 4$ **c** $m = -1$

Consolidate

1 $y = 4$
2 a i Subtract $3a$ from both sides of the equation.
 ii Subtract w from both sides of the equation.
 iii Subtract m from both sides of the equation.
 iv Subtract $3x$ from both sides of the equation.
 b i $a = 5$ **ii** $w = 3$ **iii** $m = 6$ **iv** $x = 14$
3 a $b = 9$ **b** $e = 1$ **c** $p = 6$
 d $r = 2$ **e** $g = 1.5$ **f** $a = -4$
4 a $q = 4$ **b** $y = 4.7$ **c** $e = 2$
 d $h = 0.5$ **e** $m = 4$ **f** $u = -5$

Stretch

1 a $y = -13$ **b** $y = -14$ **c** $y = -\frac{11}{5}$ **d** $y = 10$
2 a $x = \frac{3}{2}$ **b** 6.5 cm

Functions and linear equations: exam practice

1 a 45 **b** 1
2 a $y = 2$ **b** $m = 9$ **c** $x = 45$
3 a $p = 7$ **b** $q = 80$
4 19
5 a $a = 8$ **b** $b = \frac{11}{3}$
6 46 cm

Block 10 Formulae

Chapter 10.1

Are you ready?

1 a 40 **b** 8 **c** 384 **d** 1.5
 e $\frac{2}{3}$ **f** 40 **g** -8 **h** -8
2 a 48 **b** 100 **c** 9 **d** 180

Practice

1 a $B = 15$ **b** $B = 9$ **c** $B = 36$ **d** $B = 4$
2 a $R = 37$ **b** $R = 57$ **c** $R = 7.5$ **d** $R = 225$
3 a $A = 29$ **b** $A = 24$ **c** $A = 9$ **d** $A = -10$
4 a $K = 23$ **b** $K = 40$ **c** $K = 42$
 d $K = 35$ **e** $K = 162$ **f** $K = 225$
5 a 86°F **b** 78.8°F
6 a $T = 1$ **b** $T = 7$ **c** $T = 6$
 d $T = 9$ **e** $T = 18$ **f** $T = 2$
7 a $A = 80 \text{ mm}^2$ **b** $w = 3.5 \text{ mm}$

Consolidate

1 a $T = 2$ **b** $T = 42$ **c** $T = 0$ **d** $T = -6$
2 a $l = 3$ **b** $l = 7.5$ **c** $l = 0$ **d** $l = -2$
3 a $S = 29$ **b** $S = 36$ **c** $S = 44$
 d $S = 28$ **e** $S = 45$ **f** $S = 384$
4 a $A = 4$ **b** $A = 13$ **c** $A = 8$
 d $A = 12$ **e** $A = 32$ **f** $A = 2$
5 a 785 cm³ **b** 16 m

Stretch

1 a $A = 20 \text{ cm}^2$ **b** $h = 8 \text{ cm}$ **c** $b = 7 \text{ cm}$
2 a i $P = 36 \text{ mm}$ **ii** $w = 7 \text{ mm}$
 b Various possible answers where l and w sum to 25,
 e.g. $l = 20 \text{ cm}$ and $w = 5 \text{ cm}$.
3 $0 \times 4 = 0$ not 4 $2 \times 4^2 = 2 \times 16$ not 8^2 $8^2 = 64$ not 16
 The correct solution is $s = 16$
4 a £1552.50 **b** £1663.08 **c** £2513.02

Chapter 10.2

Are you ready?

1 a $A = 495$ **b** $A = 19.8$ **c** $A = 6.3$
2 a $n = 12$ **b** $x = 18$
3 $(6x + 6)$ cm

Practice

1 $A = 3x + 1$
2 a $C = 16n + 78$ **b** 25 miles
3 $T = 15c + 24a$

4 a $P = (10y + 20)$ cm   **b** $P = 60$ cm
5 a $C = 0.85m + 1.3$
    b £14.90   **c** 22 miles

Consolidate

1 a $P = 3x + 2y$ **b** 18 cm   **c** 11 cm
2 a $C = 30h + 75$ **b** £165   **c** 5.5 hours
3 a $C = 0.8m + 5$ **b i** £9 **ii** £25.80   **c** 35 miles

Stretch

1 $A = 18g$

2 a $s = \frac{d}{t}$ **b** $s = 6$ mph
    c $s = 3.75$ m/s   **d** $d = 60$ km

Chapter 10.3

Are you ready?

1 a $+8$ **b** $\times 7$ **c** $\div 9$ **d** -6
2 a $A = 20$ cm² **b** $b = 12$ cm
3 a $x = 6$ **b** $x = 144$ **c** $x = 141$ **d** $x = 126$
4 a 3 (or -3)   **b** 81

Practice

1 a P **b** l **c** F **d** m
  e v **f** t **g** s **h** u
2 a $x = a - 5$ **b** $x = a + 5$ **c** $x = \frac{a}{5}$ **d** $x = 5a$
3 a $r = t - k$ **b** $r = t + k$ **c** $r = \frac{t}{k}$ **d** $r = tk$
4 a $w = \frac{g - 8}{3}$ **b** $w = \frac{g - h}{3}$ **c** $w = \frac{g + 8}{3}$
  d $w = \frac{g + h}{3}$ **e** $w = 3(g + h)$ **f** $w = 3g + h$
  g $w = 3(g - h)$ **h** $w = 3g - h$
5 $d = \frac{c}{3e}$
6 a $f = \frac{H}{5g}$ **b** $f = \frac{H}{rg}$
  c $f = \frac{H - 2r}{3}$ **d** $f = \frac{H - gk}{r}$
7 a $d = \frac{C}{\pi}$ **b** $b = \frac{2A}{h}$
8 a $a = \sqrt{P}$ **b** $a = \sqrt{P - w}$ **c** $a = \sqrt{\frac{P}{w}}$
  d $a = \sqrt{5P - w}$ **e** $a = P^2$ **f** $a = (P + w)^2$
  g $a = (Pw)^2$ **h** $a = \left(\frac{P}{w}\right)^2$

What do you think?

1 a Filipo has missed the negative coefficient of p and has written $3p$ instead of $-3p$.
  b

$$g = h - 3p$$
$+3p \quad\quad +3p$
$$g + 3p = h$$
$-g \quad\quad -g$
$$3p = h - g$$
$\div 3 \quad\quad \div 3$
$$p = \frac{h - g}{3}$$

2 a $v = 2 - k$ **b** $v = m - k$ **c** $v = 2m - k$
  d $v = \frac{m - k}{2}$

Consolidate

1 a $u = \frac{d}{3}$ **b** $u = 3d$ **c** $u = d - 3$ **d** $u = d + 3$
  e $u = d - e$ **f** $u = d + e$ **g** $u = \frac{d}{e}$ **h** $u = de$

2 a $t = \frac{p - k}{5}$ **b** $t = \frac{p + k}{5}$ **c** $t = 5p + k$
  d $t = 5(p + k)$ **e** $t = 5(p - k)$ **f** $t = 5p - k$
3 a $y = \frac{a}{3x}$ **b** $y = \frac{a}{rx}$
  c $y = \frac{a - 2x}{3}$ **d** $y = \frac{a + xw}{r}$
4 a $y = 4 - 6x$ **b** $y = x + 4$
  c $y = 2 - 3x$ **d** $y = \frac{1}{2}x - 2$
5 $r = \sqrt{\frac{A}{\pi}}$

Stretch

1 a $h = \frac{2A}{a + b}$ **b** $a = \frac{2A}{h} - b$
2 $C = \frac{5(F - 32)}{9}$

Formulae: exam practice

1

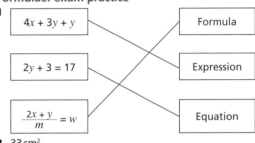

2 33 cm²
3 $t = c + m$
4 $T = 12c + 30a$
5 2144.66 m³
6 a 3 miles   **b** 54 feet

Block 11 Inequalities

Chapter 11.1

Are you ready?

1 a $-2, 3, 8$ **b** $-6, -3, 5$ **c** $-7, -2, 2, 7$
2 a Any number less than 4
  b Any number greater than 4
  c Any number less than -4
3 5   4   6   41   65

Practice

1 a i x is greater than 6   **ii** y is less than 5
    iii a is less than or equal to -3
  b i $x < 2$ **ii** $a \geqslant 5$ **iii** $8 > y$
2 $4 < b$
3 a i Any integer greater than 5
    ii Any integer less than -3
    iii 4 or any integer less than 4
  b i For example 7.9
    ii For example -6.1
    iii For example -3.5
4 a $-2, -1, 0, 1$ **b** $-4, -3, -2, -1, 0, 1, 2$
  c $-4, -3, -2, -1, 0, 1, 2$ **d** $-2, -1, 0, 1, 2, 3$
  e $-1, 0, 1, 2$ **f** $0, 1, 2$

Consolidate

1 a Any number greater than 10
  b Any number less than 10
  c Any number less than -10
  d Any number greater than -10
2 $y > 3$ $3 < y$ $4 \leqslant y$ $y \leqslant 4$ $5 > y$
3 a $-4, -3, -2, -1, 0, 1$ **b** $0, 1, 2, 3, 4$
  c $-3, -2, -1, 0, 1, 2, 3$ **d** $0, 1, 2, 3$
  e $-3, -2, -1, 0, 1, 2$ **f** $-3, -2, -1, 0, 1, 2, 3$
4 $-2 \leqslant p \leqslant 3$ or $-3 < p < 4$

Stretch

1 a i 4 **ii** 7 **iii** 2
b i 3 **ii** 6 **iii** 1
2 $w > 3$ includes any number greater than 3, such as 3.1, whereas $w \geqslant 4$ includes 4 and any number greater than 4
3 a 24 and 32 **b** 25 and 36 **c** 29 and 31 **d** 30
4 $x = 2$ and $y = 1$ therefore $x^2 = 4$ and $y^2 = 1$ $x^2 > y^2$
$x = 2$ and $y = -3$ therefore $x^2 = 4$ and $y^2 = 9$ $x^2 < y^2$
$x = 2$ and $y = -2$ therefore $x^2 = 4$ and $y^2 = 4$ $x^2 = y^2$

Chapter 11.2

Are you ready? (A)

1 a $y = 9$ **b** $y = 2$ **c** $y = 1$
d $y = 15$ **e** $y = 12$ **f** $y = 14$
2 Various possible answers, for example:
a 2, 1, 0 **b** 7, 8, 9 **c** −1, 0, 1
3 Various possible answers, for example:
a 2 **b** 4 **c** −4 **d** −3

Practice (A)

1 a $g > 8$ **b** $a \leqslant 5$ **c** $b < 12$ **d** $h \geqslant 20$
2 a $x > 6$ **b** $a \leqslant 5$ **c** $g > 6$
d $f \leqslant 2$ **e** $19 < b$ **f** $6 \geqslant k$
3 a i 3 **ii** 6 **iii** 5
b i 2 **ii** −4 **iii** 2
4 a $x \leqslant 12$ **b** $x \leqslant 21$ **c** $m > 84$ **d** $m > 75$
5 a Possible answers:
  i Expand the brackets or divide by 4
  ii Expand the brackets or divide by 2
  iii Expand the brackets or divide by 5
  iv Expand the brackets or divide by 4
  v Expand the brackets or divide by 7
  vi Expand the brackets or divide by 5
b i $x > 3$ **ii** $r \leqslant 10.5$ **iii** $12 \geqslant w$
  iv $t \leqslant -\frac{3}{4}$ **v** $-3 < y$ **vi** $p \geqslant 0$
c i Smallest integer is 4 **ii** Greatest integer is 10
  iii Greatest integer is 12 **iv** Greatest integer is −1
  v Smallest integer is −2 **vi** Smallest integer is 0
6 a $y > 3$ **b** 4

What do you think?

1 a Possible answer: substituting in $x = 4$ gives $3 - 4 = -1$, and −1 is not greater than 1 so Abdullah is incorrect.
b 2 placed in the left box and x placed in the right box.
2 a $5 > y$ **b** $3 < x$ **c** $x \leqslant 2$

Are you ready? (B)

1 a $y = 11$ **b** $x = 1$ **c** $p = 4.6$
2 a $a = 4$ **b** $b = 4.7$ **c** $f = 2.5$

Practice (B)

1 a 6 placed in the box **b** $4g$ placed in the box
c $5w$ placed in the left box and 19 placed in the right box
d 21 placed in the left box and $7m$ placed in the right box
2 a Possible answers:
  i Subtract $2y$ from both sides of the inequality.
  ii Subtract y from both sides of the inequality.
  iii Subtract $3y$ from both sides of the inequality.
  iv Subtract $4y$ from both sides of the inequality.
b i $y < 4$ **ii** $y < 3$ **iii** $y < 6$ **iv** $y < 12$
3 a $y > 8$ **b** $w \leqslant 1.5$ **c** $2 \geqslant p$
d $10 < r$ **e** $y > 6$ **f** $45 \leqslant p$
4 a Possible answer: all have terms $2x$, $6x$, 2 and 18 and the same inequality symbol but the order of terms and signs are different in each inequality.
b i $4 < x$ **ii** $5 < x$ **iii** $4 < x$

---

5 a i $y < 2$ **ii** $u \geqslant 3$ **iii** $18 \leqslant p$
  iv $b \leqslant 1$ **v** $k \geqslant 3$ **vi** $t \leqslant -4.5$
b i Greatest integer is 1 **ii** Smallest integer is 3
  iii Smallest integer is 18 **iv** Greatest integer is 1
  v Smallest integer is 3 **vi** Greatest integer is −5
6 a $2 < x$ **b** $g > 3$ **c** $3 \leqslant y$
d $2 \geqslant a$ **e** $4.2 < n$ **f** $2 \geqslant g$

Consolidate

1 a $b > 12$ **b** $7 < c$ **c** $d \leqslant 6$ **d** $e \geqslant 49$
e $f > 5$ **f** $2 \geqslant g$ **g** $h > 8$ **h** $50 \leqslant k$
2 a $w < 9$ **b** $x \leqslant 3$ **c** $y \geqslant 4.5$
d $a \leqslant -1$ **e** $g > -5$ **f** $m < -7$
3 a 1 **b** 2 **c** 5 **d** −9
4 a $b > 4$ **b** $2 \geqslant g$ **c** $x > 0.5$
d $y \geqslant -2.5$ **e** $16 > p$ **f** $e \leqslant -1$
5 a $4y - 6 < 75$ **b** 20

Stretch

1 Let the number be x.
$2x - 5 > 25$, so $x > 15$
2 Let the width be x.
a $4x + 10 < 50$ **b** $4x + 10 \leqslant 50$
3 $w < \frac{110}{3}$
4 a $9 < x$ **b** $y \geqslant 19.5$ **c** $-\frac{19}{7} \leqslant w$
5 a $2 \leqslant n < 5$ **b** $-1 < n \leqslant 4$

Chapter 11.3

Are you ready?

1 a 2, 3, 4, 5 **b** −3, −2, −1, 0, 1, 2 **c** −3, −2, −1, 0, 1, 2
2 a $k > 8$ **b** $n \leqslant 7$ **c** $h > 5$
3 a $p > 2$ **b** $r \leqslant 11.5$ **c** $10 \geqslant e$

Practice

1 a
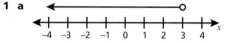

b
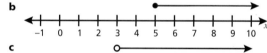

c
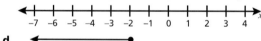

d
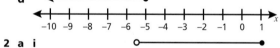

2 a i

ii

iii

b i −2, −1, 0, 1, 2, 3, 4
  ii −4, −3, −2, −1, 0, 1, 2
  iii −4, −3, −2
3 a $x \geqslant -2$ **b** $x < 3$ **c** $-2 < x < 5$ **d** $-5 \leqslant x < -1$
4 a
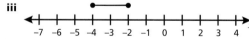

b
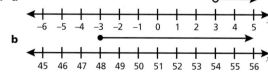

c

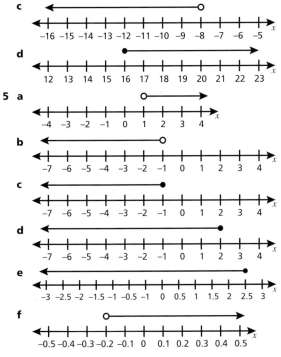

d

5 a

b

c

d

e

f

c

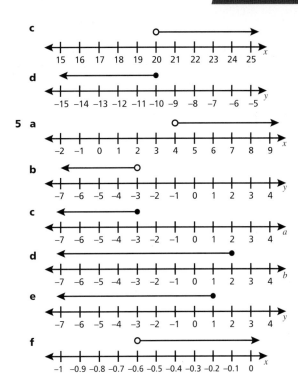

d

5 a

b

c

d

e

f

What do you think?

1 No, because if x is less than -1 then it is not possible for x to be greater than or equal to 2

Consolidate

1 a

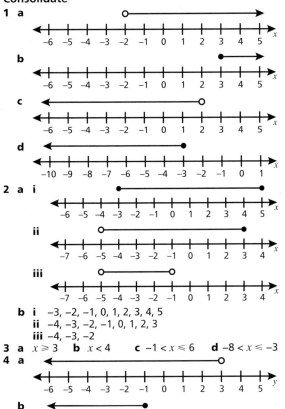

b

c

d

2 a i

ii

iii

b i $-3, -2, -1, 0, 1, 2, 3, 4, 5$
ii $-4, -3, -2, -1, 0, 1, 2, 3$
iii $-4, -3, -2$

3 a $x \geqslant 3$ **b** $x < 4$ **c** $-1 < x \leqslant 6$ **d** $-8 < x \leqslant -3$

4 a

b

Stretch

1 a Jackson is incorrect as the number line shows x is greater than -4 but less than or equal to 3
 b $-4 < x \leqslant 3$

2 $2x + 10 \geqslant 6$ $\qquad 6 \leqslant 3x + 12$ $\qquad 5x + 3 \geqslant -7$

3

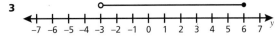

Inequalities: exam practice

1 D
2 y is less than or equal to 2
3 3, 4, 5, 6
4 a $x > 2$ **b** $y < -2$
5 a

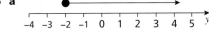

 b $-3 < x \leqslant 4$ **c** $m < 5$
6 3
7 a $-2, -1, 0, 1, 2, 3$
 b

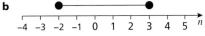

Block 12 Linear graphs

Chapter 12.1

Are you ready?

1 a The first number in each pair of coordinates is 4
 b In each pair of coordinates, the x value is equal to **4**
2 Example answers: (0, –3) (1, –3) (2, –3)
3 a –1 **b** –7 **c** 7 **d** 7

4

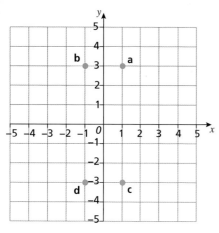

Practice

1 a Coordinate grid drawn.
  b i Example answers: (0, 4) (−2, 4) (8, 4)
    ii

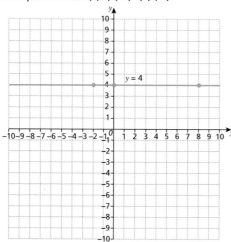

  iii Example answer: (6, 4)
    This confirms the line has the equation $y = 4$ as
    the y value is 4

  c

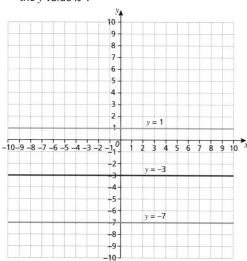

2 a Coordinate grid drawn.
  b i Example answers: (4, 0) (4, −2) (4, 8)
    ii

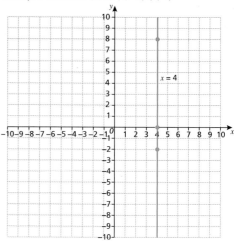

  c

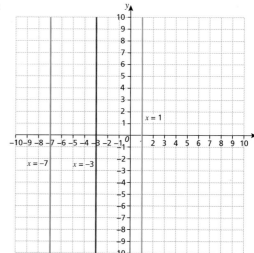

  d Same: they are all vertical straight lines
    Different: they all intersect the x-axis at
    different points
  e i Example answers: (2, 2) (5, 5) (−7, −7)
    ii

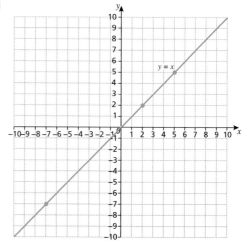

3 a

x	−2	−1	0	1	2
y	−3	−1	1	3	5

b

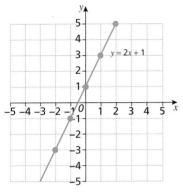

4 a i $y = x - 4$

x	−2	−1	0	1	2
y	−6	−5	−4	−3	−2

ii $y = 4x - 1$

x	−2	−1	0	1	2
y	−9	−5	−1	3	7

iii $y = 4(x - 1)$

x	−2	−1	0	1	2
y	−12	−8	−4	0	4

b

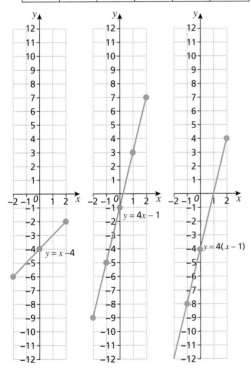

5 a Example answer: both have x, y, 3 and 2 but one has a positive coefficient of x and one has a negative coefficient of x.

b i $y = 3x - 2$

x	−3	−2	−1	0	1	2	3
y	−11	−8	−5	−2	1	4	7

$y = 2 - 3x$

x	−3	−2	−1	0	1	2	3
y	11	8	5	2	−1	−4	−7

ii

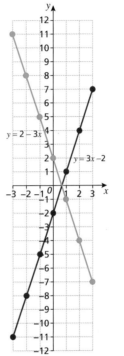

6 a i $y = -5 + 2x$

x	−2	−1	0	1	2
y	−9	−7	−5	−3	−1

ii $y = 2(x + 5)$

x	−2	−1	0	1	2
y	6	8	10	12	14

iii $y = 5 - 2x$

x	−2	−1	0	1	2
y	9	7	5	3	1

iv $y = \frac{1}{2}x + 5$

x	−2	−1	0	1	2
y	4	4.5	5	5.5	6

b

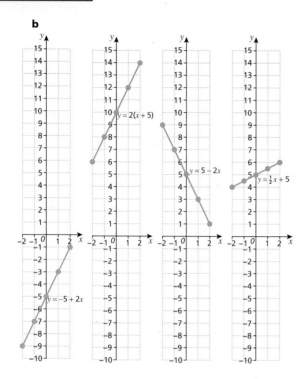

b

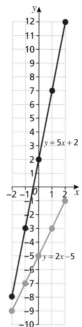

What do you think?

1 **a** Yes, Junaid is correct in this case as the x values increase by 1, the y values increase by 3

 b Junaid's method won't work for the table of values given because the x values are not increasing by 1

Consolidate

1 **a** $y = 3$ **b** $x = -2$ **c** $y = x$ **d** $y = -x$

2 **a** **i** $y = 5x + 2$

x	−2	−1	0	1	2
y	−8	−3	2	7	12

 ii $y = 2x - 5$

x	−2	−1	0	1	2
y	−9	−7	−5	−3	−1

3 **a** **i** $y = -4 + 3x$

x	−3	−2	−1	0	1	2	3
y	−13	−10	−7	−4	−1	2	5

 ii $y = 3(x + 4)$

x	−3	−2	−1	0	1	2	3
y	3	6	9	12	15	18	21

 iii $y = 4 - 3x$

x	−3	−2	−1	0	1	2	3
y	13	10	7	4	1	−2	−5

 iv $y = \frac{6x + 1}{2}$

x	−3	−2	−1	0	1	2	3
y	−8.5	−5.5	−2.5	0.5	3.5	6.5	9.5

b

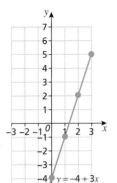 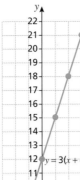

$y = -4 + 3x$

$y = 3(x + 4)$

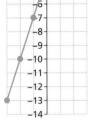 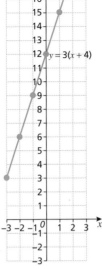

$y = 4 - 3x$

$y = \dfrac{6x + 1}{2}$

Stretch

1 (1, 1) and (3, −13)

2 a i $x + y = 6$

x	−2	−1	0	1	2
y	8	7	6	5	4

ii $x - y = 6$

x	−2	−1	0	1	2
y	−8	−7	−6	−5	−4

iii $2x + y = 6$

x	−2	−1	0	1	2
y	10	8	6	4	2

iv $3x - 2y = 6$

x	−2	−1	0	1	2
y	−6	−4.5	−3	−1.5	0

b

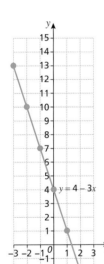

$x + y = 6$

$x - y = 6$

$2x + y = 6$

$3x - 2y = 6$

3 Graphs A and B

Chapter 12.2

Are you ready?

1

x	−2	−1	0	1	2
y	−8	−5	−2	1	4

2

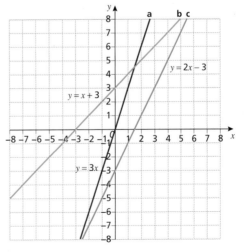

$y = 2x - 3$

$y = x + 3$

$y = 3x$

Practice

1 a Positive   **b** Negative   **c** Positive   **d** Negative

2 a

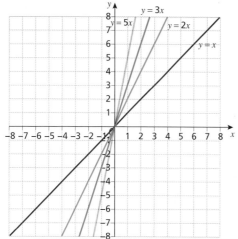

b As the gradient increases, the line gets steeper.

3 a

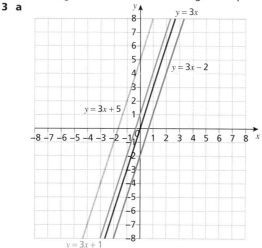

b i The lines are going in exactly the same direction and never meet.
   ii The gradient in each equation is 3
c Various possible answers, e.g. $y = 3x + 10$

4 a

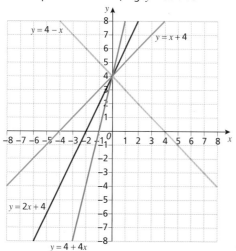

b i They all intercept the y-axis at the same point.
  ii The y-intercept of each line is 4
c Various possible answers, e.g. $y = 3x + 4$

5 a i 4   **ii** (0, 1)   **b i** 5   **ii** (0, –3)
 c i –3   **ii** (0, 2)   **d i** –2   **ii** (0, –5)
 e i 8   **ii** (0, 0)   **f i** 0   **ii** (0, 6)
 g i $\frac{1}{3}$ **ii** (0, 9)   **h i** $\frac{1}{4}$ **ii** (0, –7)

6 a Various possible answers, e.g. $y = 7x$, $y = 7x - 1$,
   $y = 7x + \frac{1}{2}$
 b Various possible answers, e.g. $y = x - 12$, $y = -2x - 12$,
   $y = \frac{1}{2}x - 12$
 c Various possible answers, e.g. $y = 1 - 3x$, $y = -1 - 3x$,
   $y = \frac{1}{2} - 3x$
 d Various possible answers, e.g. $y = 10 - x$, $y = x + 10$,
   $y = 10 + \frac{1}{2}x$

7 a $y = 3x + 4$ **b** $y = -2x$
 c $y = x - 7$ **d** $y = 0.5x - 1$

8 a i 3   **ii** (0, 6)   **b i** –2   **ii** (0, 8)
 c i –4   **ii** (0, 0.5)   **d i** $-\frac{1}{5}$ **ii** $\left(0, \frac{3}{4}\right)$

What do you think?

1 a Line A is not parallel to the others as it is going in a different direction.
 b Line A must have the same y-intercept as line D and a positive gradient greater than $\frac{1}{2}$. Line A could be $y = 4x - 3$

Line B must have a gradient of $\frac{1}{2}$ and a positive y-intercept. Line B could be $y = \frac{1}{2}x + 5$

The equation of line C is $y = \frac{1}{2}x$

Consolidate

1 a 9   **b** 1   **c** –5   **d** $\frac{1}{5}$

2 a (0, 14)   **b** (0, 7)   **c** (0, –18)   **d** $\left(0, \frac{1}{2}\right)$

3 a Various possible answers, e.g. $y = x + 5$, $y = 2x + 5$,
   $y = 3x + 5$
 b Various possible answers, e.g. $y = 8x$, $y = 8x + 1$,
   $y = 8x - 10$
 c Various possible answers, e.g. $y = 4x$, $y = 15x$,
   $y = 3.5x$

4 a $y = 5x + 1$ **b** $y = 12x - 5$
 c $y = -8x$ **d** $y = \frac{1}{3} - x$

Stretch

1 $y = 5x - 3$
2 $y = 9x + 12$

3 a i $\frac{1}{2}$ **ii** (0, 4)   **b i** 2   **ii** $\left(0, \frac{5}{3}\right)$
 c i 2   **ii** $\left(0, -\frac{3}{5}\right)$ **d i** $-\frac{1}{2}$ **ii** $\left(0, \frac{7}{4}\right)$

Chapter 12.3

Are you ready? (A)

1 a i 3   **ii** (0, 5)   **b i** 4   **ii** (0, –1)
 c i –2   **ii** (0, 3)   **d i** –5   **ii** (0, 0)
2 a i (0, –1)   **ii** $y = 3x - 1$
 b i (0, 3)   **ii** $y = -2x + 3$

Practice (A)

1 a 2   **b** 3   **c** –2   **d** –1
2 a i (0, 1)   **ii** 2   **iii** $y = 2x + 1$
 b i (0, –2)   **ii** 3   **iii** $y = 3x - 2$
 c i (0, 1)   **ii** –1   **iii** $y = -x + 1$
 d i (0, –2)   **ii** –2   **iii** $y = -2x - 2$

3 a $y = 4x + 3$ **b** $y = -3x - 2$
  c $y = 0.5x - 7$ **d** $y = -2x - 10$
4 a i £9   **ii** £4.50
  b £1.50
  c The gradient is 1.5, meaning the charge is £1.50 per mile. The y-intercept is 3, meaning there is an initial charge of £3 before any distance is covered in the taxi.
5 a $x = 1$ **b** $x = -2$
6 a $y = 2x + 1$ **b** $y = 5x + 5$ **c** $y = 10x - 5$

Are you ready? (B)

1 180 km
2 2.5 hours

Practice (B)

1 a i 20 miles **ii** 30 miles   **b i** 1 hour **ii** 5 hours
  c 1.5 hours **d** 10 mph   **e** 3.5 hours **f** 10 mph
2 a 30 miles   **b** 2 hours   **c** 4:30 pm
  d i 30 mph   **ii** 60 mph   **iii** 40 mph
  e 40 mph
3 a

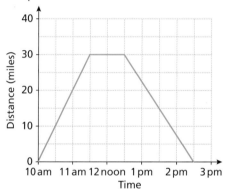

  b i 60 miles   **ii** 15 mph

Consolidate

1 a $y = 3x - 2$ **b** $y = -2x - 3$
  c $y = 4x - 3$ **d** $y = -0.5x + 1$
2 a £1.50
  b i £2   **ii** £7.50
  c £0.60
  d The gradient is the cost per mile. The y-intercept of the line is the cost for travelling 0 miles.
3 a 2.5 km   **b** 30 minutes
  c 1.5 km/h   **d** 1 hour and 20 minutes

Stretch

1 a $y = \frac{x}{3} - 1$ **b** $y = -\frac{x}{4} + 2$
  c $y = 4x + 4$ **d** $y = \frac{x}{2} - 6$
2 $y = 2x - 24$
3 a $y = 2x + 4$ **b** $y = 6 - 2x$
4 a There is a linear relationship between the number of hours and the cost.
  b i $y = 10x + 45$
    ii The gradient is the cost per hour. The y-intercept of the line is the fixed cost of £45.

Linear graphs: exam practice

1 a

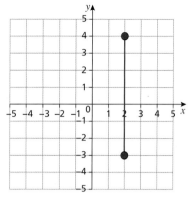

  b $x = 2$
  c

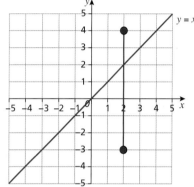

  d Example answers: $(-1, -4)$ or $(3, -4)$
2 a

x	-2	-1	0	1	2
y	-2	1	4	7	10

  b

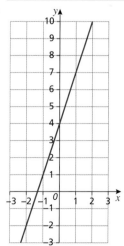

3 a 7   **b** 3 or (0, 3)
  c Example answer: $y = 7x + 5$
4 Example answers: (0, 3), (1, 2), (−1, 4)

Block 13 Non-linear functions

Chapter 13.1

Are you ready?

1 a 9   **b** 6   **c** 5
  d 3   **e** 15   **f** 3

2 a 9   **b** −6   **c** −13
  d 3   **e** 3   **f** 15

3

x	−3	−2	−1	0	1	2	3
y	−9	−7	−5	−3	−1	1	3

Practice

1 a

x	−3	−2	−1	0	1	2	3
y	9	4	1	0	1	4	9

b

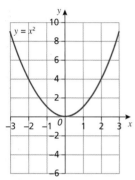

2 a

x	−3	−2	−1	0	1	2	3
y	6	1	−2	−3	−2	1	6

b

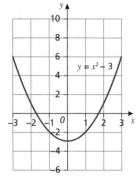

c Example answer: Both are parabolas but $y = x^2 - 3$ crosses the y-axis at (0, −3) whereas $y = x^2$ crosses the y-axis at (0, 0).

d Discussion of graphs as a class

e

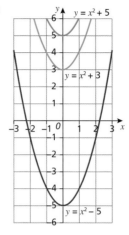

3 a $x = 2.5$ $y = -6.25$

b

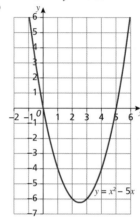

4 a

x	−5	−4	−3	−2	−1	0	1	2
y	10	4	0	−2	−2	0	4	10

b

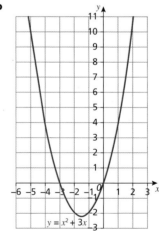

5 a $y = 0$ **b** $y = -3$ **c** $y = -3$
  d $x = 1$ **e** $y \approx -1.6$ **f** $x \approx -0.7$ and $x \approx 2.7$

6 a

x	−3	−2	−1	0	1	2	3
y	16	8	2	−2	−4	−4	−2

b

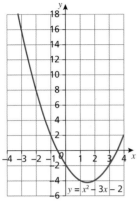

$y = x^2 - 3x - 2$

c   i $y \approx -3.3$ **ii** $x \approx -1.2$ and $x \approx 4.2$

What do you think?

1

x	−3	−2	−1	0	1	2	3
y	−9	−4	−1	0	−1	−4	−9

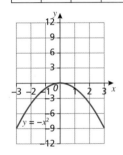

$y = -x^2$

2

Linear graphs	Quadratic graphs
$y = 4x + 1$	$y = 4x^2$
$y = 4 + x$	$y = x^2 + 4x + 1$
$y = -4 - x$	$y = 4 - x^2$
$y = x - 4$	$y = (x + 4)^2$

Consolidate

1 a

x	−3	−2	−1	0	1	2	3
y	5	0	−3	−4	−3	0	5

b

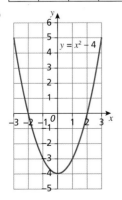

$y = x^2 - 4$

2 a

x	−6	−5	−4	−3	−2	−1	0	1
y	6	0	−4	−6	−6	−4	0	6

b

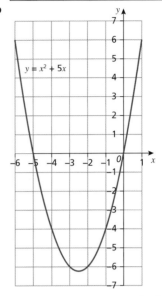

$y = x^2 + 5x$

3 a

x	−1	0	1	2	3	4	5	6
y	9	3	−1	−3	−3	−1	3	9

b

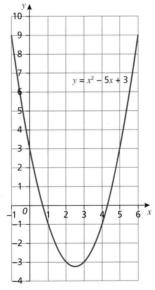

$y = x^2 - 5x + 3$

c   i $y \approx -3.2$ **ii** $x \approx 0.4$ or $x \approx 4.6$

Stretch

1 a

x	−3	−2	−1	0	1	2	3
y	21	11	5	3	5	11	21

b

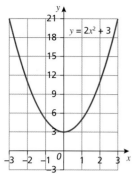

2 a

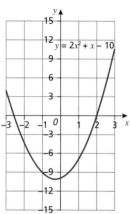

b

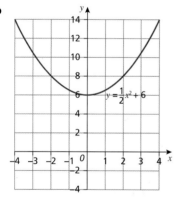

3 $y = x^2 - 7$ is graph 3 $y = x^2 - 7x$ is graph 1

$y = -x^2 + 7$ is graph 4 $y = x^2 + 7x - 7$ is graph 2

Chapter 13.2

Are you ready?

1 a 8   **b** 4   **c** 14   **d** 9

2 a −8   **b** −12   **c** −14   **d** −19

3 a 2   **b** 3   **c** −1   **d** 12

Practice

1 a

x	−2	−1	0	1	2
y	−12	−5	−4	−3	4

b

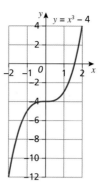

2 a

x	−2	−1	0	1	2
y	−12	−3	0	3	12

b

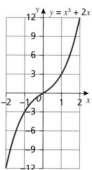

3 a

x	−4	−2	−1	−0.5	0.5	1	2	4
y	−1	−2	−4	−8	8	4	2	1

b

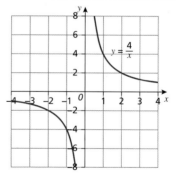

4 a

x	−2	−1	0	1	2
y	−16	−2	0	2	16

b

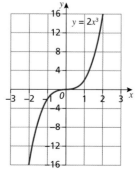

5 a

x	−4	−2	−1	−0.5	0.5	1	2	4
y	−0.5	−1	−2	−4	4	2	1	0.5

b

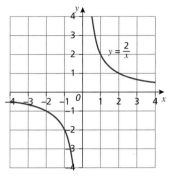

$y = \dfrac{2}{x}$

c $x \approx 1.3$

Consolidate

1 a

x	−2	−1	0	1	2
y	−9	−1	1	3	11

b

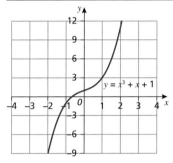

$y = x^3 + x + 1$

2 a

x	−5	−2	−1	−0.5	0.5	1	2	5
y	−1	−2.5	−5	−10	10	5	2.5	1

b

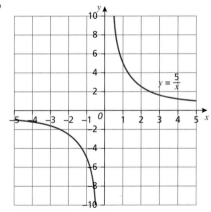

$y = \dfrac{5}{x}$

3 Graph 1 is $y = x + 2$ Graph 2 is $y = \dfrac{2}{x}$

Graph 3 is $y = x^2$ Graph 4 is $y = 2x^3$

Stretch

1 a

x	−2	−1	0	1	2
y	8	1	0	−1	−8

b

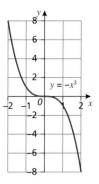

$y = -x^3$

2 a

x	−3	−2	−1	−0.5	0.5	1	2	3
y	−1	−1.5	−3	−6	6	3	1.5	1

b

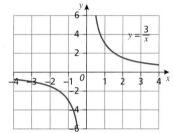

$y = \dfrac{3}{x}$

c 1.5

d

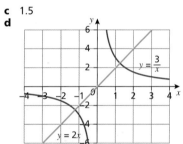

$y = \dfrac{3}{x}$

$y = 2x$

e (1.2, 2.5) and (−1.2, −2.5)

3 Graph 1 is $y = 2x + 4$ Graph 2 is $y = -x^2 + 6$

Graph 3 is $y = -x^3$ Graph 4 is $y = \dfrac{1}{x}$

Non-linear functions: exam practice

1 A and C

2 a

x	−3	−2	−1	0	1	2	3
y	9	4	1	0	1	4	9

b

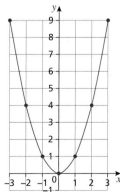

c Accept answers in the range of 6 to 6.3

d 49

3

Equation	Letter
$y = -x^3 + 3$	B
$y = -3x - 1$	D
$y = -\dfrac{3}{x}$	A
$y = (x - 3)^2$	C

Block 14 More complex equations

Chapter 14.1

Are you ready?

1 a $4(y + 3)$ **b** $g(g - 7)$ **c** $5t(t - 2)$
2 a $x = -4$ **b** $x = 5$
3 a 4 and 1 **b** 5 and 8 **c** −3 and 5

Practice

1 a $x = 3, x = -1$ **b** $x = 4, x = 9$
  c $y = -5, y = 8$ **d** $y = -7, y = -4$
  e $w = 0, w = 2$ **f** $w = 0, w = -7$
2 a $x = -2, x = -3$ **b** $y = -5, y = -3$
  c $g = -4, g = -7$ **d** $a = -6, a = -2$
3 a $m = 2, m = -5$ **b** $b = 5, b = -7$
  c $h = -3, h = 1$ **d** $w = -8, w = 3$
4 a $k = 6, k = -2$ **b** $u = 4, u = -1$
  c $q = 8, q = -5$ **d** $f = 6, f = -5$
5 Tiff is incorrect as $x^2 - 10x + 24 = 0$ can be written as $(x - 4)(x - 6) = 0$ so the solution is $x = 4$ or $x = 6$
6 a $x = 9, x = 2$ **b** $y = 12, y = 2$ **c** $u = 8, u = 3$
7 a $x = -5, x = 3$ **b** $d = -3, d = 9$
  c $h = 8, h = 2$ **d** $y = -5, y = -4$
8 a $x = 5, x = -5$ **b** $y = 8, y = -8$
  c $w = 11, w = -11$
    Roots have the same value but opposite signs.

Consolidate

1 a $x = -7, x = -1$ **b** $q = -9, q = -4$
  c $y = -8, y = 2$ **d** $w = -10, w = 3$
  e $p = 12, p = -2$ **f** $h = 9, h = 5$
2 a $x = 1, x = -1$ **b** $r = 9, r = -9$
  c $g = 13, g = -13$
3 a $e = 9, e = -4$ **b** $k = -3, k = -12$
  c $m = 5, m = 8$

Stretch

1 $x^2 + 8x + 16 = 0$ can be rewritten as $(x + 4)(x + 4) = 0$ and the only solution is $x = -4$
2 a Jackson has not made the equation equal to zero before attempting to solve.
  b $y = -8, y = 2$
3 Length = 9 cm, width = 4 cm

Chapter 14.2

Are you ready?

1 a $x = 4$ **b** $x = -8$ **c** $x = -6$ **d** $x = 3$
2 a $2a$ **b** a **c** $5a$
3 $6x + 2y$

Practice

1 a $x = 5, y = 6$ **b** $x = 4, y = 1$
  c $x = 3, y = 4$ **d** $x = 8, y = 6$
2 a $a = 6, b = 5$ **b** $c = 4, d = 7$
  c $e = 3, f = 5$ **d** $g = 5, h = 1$
3 a $x = -2, y = -2$ **b** $x = 4, y = 4$
  c $x = 1.5, y = 5.5$
4 a It should be $-4x = 4$, so $x = -1$
  b $x = -1, w = 2$
  c i $k = 8, n = 3$ **ii** $p = 4, t = 6$
    iii $m = -2, q = 3$ **iv** $r = 8, w = -1$
5 a $x = 4, y = 3$ **b** $m = 8, p = 2$
  c $w = 11, k = 5$ **d** $h = 6, t = 4$

6 a He has subtracted $3b$ instead of subtracting $-3b$
  b $a = 10, b = 3$
7 a $x = 5, y = 4$ **b** $f = 8, g = 3$
  c $g = 5, h = 6$ **d** $m = 4, k = -1$
  e $w = 8, p = -2$ **f** $v = 5, n = -3$
8 A, B, C and E

Consolidate

1 a $x = 4, y = 2$ **b** $x = 3, y = 7$
  c $x = 1, y = 2$ **d** $x = 3, y = 0$
2 a $a = 3, b = 2$ **b** $e = -6, f = 4$
  c $k = 1, n = 2$ **d** $r = -3, w = 3$
3 a $m = 7, p = 4$ **b** $h = 1, t = 3$
  c $x = 9, y = 2$ **d** $f = 1, g = -2$
  e $g = 6, h = -4$ **f** $n = 10, k = 3$
4 When $a + 2y = 13$ and $a - y = 1$
    $a = 5, y = 4$
  When $a + 2y = 13$ and $2a + 3y = 21$
    $a = 3, y = 5$
  When $a - y = 1$ and $2a + 3y = 21$
    $a = 4.8, y = 3.8$

Stretch

1 a $x = 5, y = -2$ **b** $x = 0.5, y = 4$ **c** $x = 5, y = 2$
2 a A cup of tea costs £1.50
  b A cup of coffee costs £2
3 Square value is 28; triangle value is −13

More complex equations: exam practice

1 −6
2 3 and 7
3 a $x(x - 3)$
  b $(x + 5)(x - 5)$
  c $(y - 2)(y - 8)$
  d $y = 2$ or $y = 8$
4 a 11 **b** $2 \times 11 + 1 = 23$
5 £1.30
6 $x = 5, y = -2$
7 $x = 3, y = 5$

Block 15 Sequences

Chapter 15.1

Are you ready?

1 a 31, 37 **b** 25, 20 **c** 53, 60 **d** −30, −40
2 a 1, 4, 9
  b

Practice

1 a 60 **b** 40 **c** 1250 **d** 2
2 a

  b Add 2 to the previous term
3 a Ed needs to state the first term.
  b i The term-to-term rule is add 4 to the previous term and the first term is 5; 21
    ii The term-to-term rule is add 50 to the previous term and the first term is 40; 240
    iii The term-to-term rule is subtract 8 from the previous term and the first term is 38; 6
    iv The term-to-term rule is double the previous term and the first term is 10; 160
4 a i 2, 17, 32, 47, 62 **ii** 35, 50, 65, 80, 95
  b i 25, 70, 205, 610, 1825 **ii** 1, −2, −11, −38, −119
5 a i 7, 9, 11, 13, 15 **ii** 3, 8, 13, 18, 23
  iii 6, 12, 18, 24, 30 **iv** 7, 13, 19, 25, 31

b i The term-to-term rule is add 2 to the previous term and the first term is 7
  ii The term-to-term rule is add 5 to the previous term and the first term is 3
  iii The term-to-term rule is add 6 to the previous term and the first term is 6
  iv The term-to-term rule is add 6 to the previous term and the first term is 7
c i 205   **ii** 498   **iii** 600   **iv** 601
6 a 17, 14, 11, 8, 5   **b** Linear
  c i −130   **ii** −880
7 80 is a term in the sequences given by the rules $5n + 25$ and $100 − 2n$
8 a 356   **b** −2

Consolidate

1 a 37, 44, 51, 58   **b** 24, 18, 12, 6
  c 60, 120, 240, 480   **d** 75, 120, 165, 210
2 a 4, 10, 40, 190, 940   **b** 1, −5, −35, −185, −935
3 a 21, 30   **b** 48, 192
4 a i 8, 11, 14, 17, 20   **ii** −3, 2, 7, 12, 17
  iii 8, 6, 4, 2, 0
  b i The term-to-term rule is add 3 to the previous term and the first term is 8
  ii The term-to-term rule is add 5 to the previous term and the first term is −3
  iii The term-to-term rule is subtract 2 from the previous term and the first term is 8
5 a 45th position   **b** 803
  c

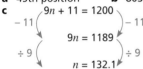

$9n + 11 = 1200$
$- 11$ $- 11$
$9n = 1189$
$\div 9$ $\div 9$
$n = 132.1$

n is not an integer so 1200 is not a term in the sequence.

Stretch

1 64, 38, 25, 18.5, 15.25
2 a 4, 12, 20, 28, 36   **b** 54, 37, 20, 3, −14
  c 0.2, 2, 20, 200, 2000   **d** −3.75, 1, 20, 96, 400

Chapter 15.2

Are you ready?

1 a 25   **b** The terms increase by 6   **c** 43
2

3 Example answer: Both sequences increase by 2 each time. The first sequence is odd numbers and the second sequence is even numbers.

Practice

1 a $4n + 1$ **b** $5n + 3$ **c** $5n$
  d $4n − 1$ **e** $7n − 3$ **f** $n + 10$
2 $4n$ and $4n − 4$
3 10th term $= 4 \times 10 + 3 = 43$
5th term $= 4 \times 5 + 3 = 23$
$23 \times 2 = 46$
$46 \neq 43$ so the 10th term of the sequence is not double the 5th term of the sequence.
4 a 418
  b

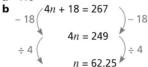

$4n + 18 = 267$
$- 18$ $- 18$
$4n = 249$
$\div 4$ $\div 4$
$n = 62.25$

n is not an integer so 267 is not a term in the sequence.

5 a Same: both sequences have a difference of 2
  Different: sequence A is ascending and sequence B is descending.
  b $14 − 2n$
6 a $13 − 3n$ **b** $70 − 10n$ **c** $29 − 5n$
7 a 901   **b** −2

Consolidate

1 a $3n + 3$ **b** $5n + 4$ **c** $8n$
  d $5n − 3$ **e** $n + 5$ **f** $11n − 2$
2 $3n − 1$: 2, 5, 8, 11
$3n + 5$: 8, 11, 14, 17
$3n − 7$: −4, −1, 2, 5
They all have a common difference of 3 between consecutive terms.
3 a $7n − 3$ **b** 3497
  c

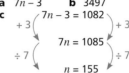

$7n − 3 = 1082$
$+ 3$ $+ 3$
$7n = 1085$
$\div 7$ $\div 7$
$n = 155$

n is an integer so 1082 is the 155th term of the sequence.
4 a $17 − 2n$ **b** $80 − 20n$ **c** $98 − 9n$
5 a Position 15   **b** 1001
  c

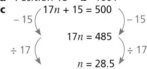

$17n + 15 = 500$
$- 15$ $- 15$
$17n = 485$
$\div 17$ $\div 17$
$n = 28.5$

n is not an integer so 500 is not a term of the sequence.
6 19th term

Stretch

1 a

b $n + 3$
c The number of squares can be calculated by taking the position number and adding 3
2 a $8n + 1$ **b** $41 − 11n$ **c** $15n − 35$ **d** $19 − 7n$

Chapter 15.3

Are you ready?

1 a $3n + 5$ **b** 305
  c

$3n + 5 = 90$
$- 5$ $- 5$
$3n = 85$
$\div 3$ $\div 3$
$n = 28.3$

n is not an integer so 90 is not a term in the sequence.
2 a Linear   **b** Not linear   **c** Not linear   **d** Linear
3 a 16   **b** −4   **c** 20   **d** $\frac{1}{4}$

Practice

1 a 13, 21, 34   **b** 34, 56, 90
  c 3, 4, 7   **d** −2, −4, −6
2 a 45, 135, 405   **b** 20, 35, 55
3 a i Multiply the previous term by 3   **ii** 1620
  b i Divide the previous term by 4   **ii** 1
  c i Multiply the previous term by 5   **ii** 125 000
  d i Divide the previous term by 10   **ii** 0.006

4 100
5 a i 0, 3, 8, 15, 24   **ii** 6, 14, 24, 36, 50
    iii 17, 8, −7, −28, −55
   b The differences between terms change so the
     sequences are not linear.
     The multipliers between terms change so the
     sequences aren't geometric either.
6 a 95   **b** 1000   **c** 169
   d 160   **e** 50   **f** 110
7 a 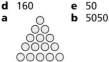   **b** 5050

Consolidate

1 a 16, 25, 41   **b** 62, 103, 165
   c 4, 5, 9   **d** −3, −6, −9
2 a i Multiply the previous term by 4   **ii** 5120
   b i Divide the previous term by 3   **ii** 2
   c i Multiply the previous term by 4   **ii** 76 800
   d i Divide the previous term by 10   **ii** 0.009
3 a 200   **b** 150
4 a 8, 11, 16, 23, 32   **b** 0, 2, 6, 12, 20
   c 0, 4, 8, 12, 16   **d** 4, 10, 18, 28, 40
5 a

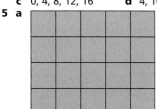

   b 9900

Stretch

1 a $1, \frac{1}{2}, \frac{1}{3}, \frac{1}{4}, \frac{1}{5}$ **b** $\frac{1}{2}, \frac{2}{3}, \frac{3}{4}, \frac{4}{5}, \frac{5}{6}$
2 $4x, 7x, 11x$
3 Table completed from left: 6, 0, −10
4 a 2498, 2502   **b** 863, 865, 1728

Sequences: exam practice

1 a

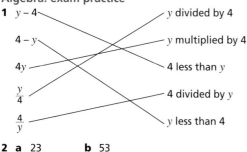

   b 16
2 a 3, 6, 9, 12, 15   **b** The 3 times table
3 5, 8
4 a 51   **b** $7n − 5$
5 3, 5, 7

Algebra: exam practice

1

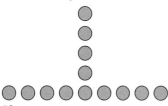

   $y − 4$ y divided by 4
   $4 − y$ y multiplied by 4
   $4y$ 4 less than y
   $\frac{y}{4}$ 4 divided by y
   $\frac{4}{y}$ y less than 4

2 a 23   **b** 53
   c 25, 12.5   **d** Divide by 2

3 −11
4 a $m = 39$ **b** $t = 2$ **c** $x = 20$
5 $w^2 + 12w$
6 a $y < 3$
   b

7 $x = 8$
8 £128
9 $x = 11$
10 $x = −4$ or $x = −5$

Block 16 Ratio

Chapter 16.1

Are you ready? (A)

1 a $\frac{3}{7}$ **b** $\frac{4}{7}$
2 For every **2** apples, there are **3** bananas.

Practice (A)

1 a

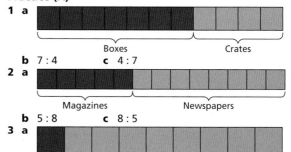

   b 7 : 4   **c** 4 : 7
2 a

   b 5 : 8   **c** 8 : 5
3 a

   b 1 : 8   **c** 8 : 1
4 a 3 : 5   **b** 5 : 3

Are you ready? (B)

1 3 : 7
2 4 : 3

Practice (B)

1 a i 6   **ii** 2   **b i** 6   **ii** 2
   c 6 : 9 and 2 : 3
2 3 : 2
3 5 : 2
4 3 : 4
5 3 : 1
6 a 3 : 2   **b** 4 : 5   **c** 4 : 3   **d** 5 : 1   **e** 1 : 3
   f 4 : 1   **g** 1 : 3   **h** 2 : 3   **i** 6 : 5   **j** 2 : 3

What do you think?

1 Example answers: 3 apples and 2 pears; 6 apples and
   4 pears; 9 apples and 6 pears (or any other multiples
   of 3 and of 2)

Consolidate

1 a

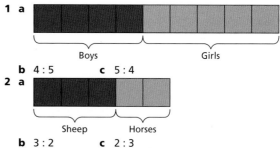

   b 4 : 5   **c** 5 : 4
2 a

   b 3 : 2   **c** 2 : 3

3 2 : 1
4 6 : 7
5 1 : 4
6 a 2 : 1   **b** 3 : 2   **c** 4 : 5
   d 1 : 5   **e** 3 : 5   **f** 3 : 1

Stretch
1 Example answers: 2 books and 7 bookmarks; 4 books and 14 bookmarks; 6 books and 21 bookmarks (or any other multiples of 2 and of 7)
2 9 blue and 15 yellow or 12 blue and 20 yellow
3 £30

Chapter 16.2
Are you ready? (A)
1 162
2 15
3 More bananas
4 1 : 50

Practice (A)
1 120 g
2 2000 ml
3 a 30   **b** 45
4 a 100   **b** 120
5 a 72   **b** 81
6 a 9   **b** 30
7 a 40 litres   **b** 56 litres
8 a 45   **b** 81
9 a 81   **b** 117
10 a 60   **b** 100
11 27
12 28
13 125 g
14 18
15 £120 and £180

Are you ready? (B)
1 36
2 6

Practice (B)
1 25
2 150 g
3 a 18   **b** 12
4 a 10   **b** 15
5 a 9   **b** 36
6 50
7 620
8 a 12   **b** 16
9 a 60   **b** 75
10 a 30   **b** 50
11 a 10   **b** 25
12 a 4   **b** 12
13 a 30 units   **b** 70 units
14 27
15 £50

What do you think?
1 20

Consolidate
1 75
2 600 g
3 5
4 126
5 15 adults and 40 children
6 a 18 apples   **b** 45 pears
7 24
8 36
9 a 9   **b** 36
10 35

Stretch
1 £640
2 a Two parts are equal so two angles are equal.
   b 80°
3 Out of the 180 cakes, 90 are vanilla and 90 are chocolate.
Out of the 420 cookies, 105 are double chocolate and 315 are chocolate chip.

Ratio: exam practice
1 a 1 : 5   **b** 5 : 1
2 a 5 : 3   **b** 2 : 3
   c 3 : 5 : 6 (or other suitable example)
3 5 : 6
4 £25 : £15
5 42 cm and 12 cm
6 50 ml
7 112.5°
8 £150
9 a 20   **b** 45

Block 17 Proportion

Chapter 17.1
Are you ready?
1 a $\frac{3}{7}$ **b** $\frac{4}{7}$ **c** 3 : 4

Practice
1 a $\frac{7}{10}$ **b** $\frac{3}{10}$ **c** 7 : 3   **d** 3 : 7
2 a 3 : 5   **b** 5 : 3   **c** $\frac{3}{8}$ **d** $\frac{5}{8}$
3 a $\frac{6}{11}$ **b** $\frac{5}{11}$ **c** 6 : 5   **d** 5 : 6
4 a 3 : 5   **b** 5 : 3   **c** $\frac{3}{8}$ **d** $\frac{5}{8}$
5 a

Blue	Blue	Green	Green	Green

   b 2 : 3
6 a

Kim	Kim	Kim	Kim	Kim	Jack	Jack	Jack	Jack	Jack	Jack	Jack	Jack

   b $\frac{5}{13}$
7 5 : 4
8 $\frac{11}{21}$
9 $\frac{1}{8}$
10 3 : 1
11 1 : 2
12 2 : 1 : 1

Consolidate
1 a $\frac{9}{13}$ **b** $\frac{4}{13}$ **c** 9 : 4   **d** 4 : 9
2 a 6 : 11   **b** 11 : 6   **c** $\frac{6}{17}$ **d** $\frac{11}{17}$
3 a $\frac{15}{23}$ **b** $\frac{8}{23}$ **c** 15 : 8   **d** 8 : 15
4 a

Lucy	Lucy	Lucy	Lucy	Ben	Ben	Ben	Ben	Ben	Ben	Ben

   b $\frac{4}{11}$
5 $\frac{7}{12}$

Stretch
1 45 : 108 or 5 : 12
2 $\frac{4}{10}$ or $\frac{2}{5}$
3 5 : 2 : 3

Chapter 17.2

Are you ready? (A)

1 20

2 $\frac{3}{10}$

3 2 : 1

Practice (A)

1 a 40   **b** 100   **c** 200
2 a 8 kg   **b** 40 kg   **c** 80 kg
3 a 32   **b** 160   **c** 5 boxes
4 a 100   **b** 225   **c** 375
5 a 120   **b** 320   **c** 480
6 a 60   **b** 180   **c** 5 packets
7 Table completed from the left: 0, 40, 200, 2000
8 Table completed from the left: 1, 2, 20, 10

Are you ready? (B)

1 £6.00
2 £1.75
3 Table completed from the left: 90, 150, 300, 450

Practice (B)

1 Shop B
2 300 ml bottle
3 Shop A
4 Box X
5 Provider A
6 Cinema B
7 X Taxis

Are you ready? (C)

1 £19.00
2 The 9-pack is better value

Practice (C)

1 a 500 g plain flour, 250 g butter, 4 tbsp milk
   b 125 g plain flour, 62.5 g butter, 1 tbsp milk
   c 625 g plain flour, 312.5 g butter, 5 tbsp milk
2 a 180 g butter, 225 g sugar, 1.5 eggs, 1.5 tsp vanilla
     extract, 270 g plain flour, 225 g chocolate chips
   b No, Tom only has enough sugar to make
     20 cookies. 150 g sugar makes 10 cookies,
     so 300 g makes twice this amount (20 cookies).
3 A
4 a ₺450
   b €2
   c ₺1500
   d $1 = ₺27

Consolidate

1 a 320 bags   **b** 560 bags   **c** 960 bags
2 a 250 points   **b** 750 points   **c** 4 levels
3 Plan A
4 Store Y
5 a 400 g spaghetti, 1000 ml pasta sauce
   b 600 g spaghetti, 1500 ml pasta sauce
6 a 6 apples, 4 bananas, 2 cups of grapes
   b 9 apples, 6 bananas, 3 cups of grapes

Stretch

1 a 50 000 tools   **b** 15 hours
   c Assumes constant production rate
2 Buying 4 packs of 9 tins and 2 packs of 4 tins
   (total cost £35.60) is the cheapest way.
3 Jack can make 4 more servings of banana bread
   than Kim.
4 2 large boxes and 8 small boxes

Chapter 17.3

Are you ready?

1 Fewer

Practice

1 a 4 hours   **b** They work at the same rate.
2 a 3 days   **b** You are told the rate is the same.
3 4.5 hours
4 36 days
5 2 days
6 6 minutes
7 1 hour

Consolidate

1 a 1 hour   **b** They work at the same rate.
2 a 5 hours   **b** They work at the same rate.
3 a 2.5 days   **b** You are told the rate is the same.
4 3 hours
5 16 days
6 2.5 days
7 10 minutes
8 2 hours

Stretch

1 Table completed from the left: 120, 80, 60, 48
2 Table completed from the left: 30, 6, 3, 1.5
3 16 cm
4 Table completed from the left: 24, 8, 3, 2

Proportion: exam practice

1 a 3 : 4   **b** 4 : 3   **c** $\frac{4}{7}$

2 a $\frac{1}{6}$ **b** 5 : 1   **c** 25

3 45
4 a 360 g   **b** 300 g
   c Because you would need half an egg
5 8.75 g
6 1 day

Block 18 Rates

Chapter 18.1

Are you ready? (A)

1 180 minutes
2 2 hours 30 minutes
3 330 minutes
4 20 tests

Practice (A)

1 a 60 miles   **b** 150 miles
   c 15 miles   **d** 105 miles
2 a 3 hours   **b** 10 hours
   c $\frac{1}{2}$ hour or 30 minutes
   d $\frac{1}{4}$ hour or 15 minutes
3 a 20 mph   **b** 40 mph
   c 25 mph   **d** 2 mph
4 20 seconds
5 150 kilometres
6 6 hours
7 6 mph
8 Jack ran faster.

Are you ready? (B)

1 20 mph
2 800 metres
3 400 miles

Practice (B)

1 a 50 mph   **b** 20 m/s
   c 50 km/h   **d** 20 mph
2 a 90 miles   **b** 270 metres
   c 300 kilometres   **d** 1200 metres or 1.2 kilometres
3 a 5 hours   **b** 3 hours
   c $3\frac{1}{2}$ seconds   **d** $\frac{1}{2}$ hour or 30 minutes

4 315 km

5 156.67 mph

6 20 km/h

7 $3\frac{1}{2}$ hours

8 The driver is going at 32 mph so is exceeding the speed limit.

Consolidate

1 a 80 miles   **b** 120 miles
   c 400 miles   **d** 20 miles

2 a 2 hours   **b** 4 hours
   c $\frac{1}{2}$ hour or 30 minutes

   d $\frac{1}{3}$ hour or 20 minutes

3 a 8.33 mph   **b** 20 mph
   c 10 mph   **d** 1 mph

4 20 seconds

5 210 kilometres

6 a 100 mph   **b** 10 m/s
   c 25 km/h   **d** 12 mph

7 a 420 miles   **b** 630 metres
   c 91 kilometres   **d** 10 kilometres

8 a 5 hours   **b** 5 hours
   c 7 seconds   **d** $\frac{1}{2}$ hour or 30 minutes

Stretch

1 Athlete B

2 47 500 000 seconds

3 $3\frac{1}{2}$ hours

4 1.8 km/h

Chapter 18.2

Are you ready?

1 g   kg

2 cm³   m³

3 60 miles

4 100 cm³

Practice

1 a 9 g/mm³   **b** 8 g/cm³
   c 250 mg/m³   **d** 8 kg/m³

2 a 130 g   **b** 200 mg
   c 2 kg   **d** 6.8 g

3 a 5 cm³   **b** 4 m³
   c 30 mm³   **d** 12 cm³

4 a 5 g/cm³   **b** 11 g/cm³
   c 6000 kg/m³   **d** 12.5 g/cm³

5 a 221 g   **b** 144.5 g   **c** 4.25 g

6 31 010 kg

7 a 20 cm³   **b** 8 cm³
   c 320 cm³   **d** 200 cm³

8 2.5 m³

Consolidate

1 a 8 g/cm³   **b** 20 mg/cm³
   c 0.5 g/cm³   **d** 0.8 g/cm³

2 a 36 g   **b** 54 kg
   c 20 g   **d** 108 mg

3 a 5 cm³   **b** 7 m³
   c 12 mm³   **d** 7 cm³

Stretch

1 19 440 g

2 9 kg/cm³

3 1 mm

4 Silver

Rates: exam practice

1 50 mph

2 4.5 miles

3 a 8 am   **b** 1.5 hours
   c It shows that it took longer for the same distance.

4 0.75 g/cm³

5 7.2 hours or 7 hours 12 minutes

Ratio, proportion and rates of change: exam practice

1 200 g

2 4 : 5 : 8

3 10 miles

4 £45 : £18 : £27

5 *Shake Shack* at £2.40 each

6 The pressure decreases.

7 468 euros

8 £9.50

9 17

10 69 mph

11 132

12 21

This page has deliberately been left blank